J. Paulo Davim (Ed.)
Additive and Subtractive Manufacturing

De Gruyter Series in Advanced Mechanical Engineering

Series Editor
J. Paulo Davim

Volume 4

Additive and Subtractive Manufacturing

Emergent Technologies

Edited by
J. Paulo Davim

DE GRUYTER
OLDENBOURG

Editor
Prof. Dr. J. Paulo Davim
University of Aveiro
Dept. of Mechanical Engineering
Campus Santiago
3810-193 Aveiro
Portugal
pdavim@ua.pt

ISBN 978-3-11-077677-5
e-ISBN (PDF) 978-3-11-054977-5
e-ISBN (EPUB) 978-3-11-054833-4
ISSN 2367-3796

Library of Congress Control Number: 2019945022

Bibliographic information published by the Deutsche Nationalbibliothek
The Deutsche Nationalbibliothek lists this publication in the Deutsche Nationalbibliografie;
detailed bibliographic data are available on the Internet at http://dnb.dnb.de.

© 2021 Walter de Gruyter GmbH, Berlin/Boston
This volume is text- and page-identical with the hardback published in 2019.
Cover image: Sky_blue/iStock/thinkstock
Typesetting: Integra Software Services Pvt. Ltd.
Printing and binding: CPI books GmbH, Leck

www.degruyter.com

Preface

Nowadays, additive manufacturing (AM) and subtractive manufacturing (SM) offer numerous advantages in the production of single and multiple components, offering incomparable design independence with the facility to manufacture components from a wide range of materials, polymers, metals, composites, and so on. AM and SM are used to fabricate products in several industries: aeronautic, automotive, biomedical, and others. Therefore, the progress in AM and SM is very important for the modern industry.

This volume aims at providing recent information on progress on advances in additive and subtractive technologies in nine chapters. Chapter 1 of this book provides information on emerging trends in AM and SM. Chapter 2 is dedicated to the fused deposition modeling using polylactic acid (PLA), thereby improving the performance (state of the art). Chapter 3 describes the development of the basic drill design for cored holes in AM and SM. Chapter 4 contains information on AM of magnesium alloys. Chapter 5 is dedicated to AM for patient-specific medical use. Chapter 6 describes stereolithography and its applications. Chapter 7 contains information on ultrasonic-assisted deep-hole drilling. Finally, Chapter 8 contains information on information and computational modeling for sustainability evaluation and improvement of manufacturing processes.

This volume can be used as a research book for final undergraduate engineering course or as a topic on AM and SM at the postgraduate level. Also, this volume can serve as a useful reference for academics, researchers, mechanical, materials, production and industrial engineers, and professionals in AM and SM and related industries. The scientific interest in this book is evident for many important centers of the research, laboratories, and universities as well as for industry.

I acknowledge De Gruyter for this opportunity and professional support. Finally, I would like to thank all the authors who contributed for this book.

J. Paulo Davim
Aveiro, Portugal
November 2019

https://doi.org/10.1515/9783110549775-202

About the Editor

J. Paulo Davim received his Ph.D. in mechanical engineering in 1997, M.Sc. in mechanical engineering (materials and manufacturing processes) in 1991, Mechanical Engineering degree (5 years) in 1986 from the University of Porto (FEUP), the Aggregate title (Full Habilitation) from the University of Coimbra in 2005, and D.Sc. from London Metropolitan University in 2013. He is senior chartered engineer by the Portuguese Institution of Engineers with an MBA and Specialist title in engineering and industrial management. He is also Eur Ing by FEANI-Brussels and Fellow (FIET) by IET-London. At present, he is professor at the Department of Mechanical Engineering of the University of Aveiro, Portugal. He has more than 30 years of teaching and research experience in manufacturing, materials, mechanical and industrial engineering, with special emphasis in machining and tribology. He has also interest in management, engineering education, and higher education for sustainability. He has guided large numbers of postdoc, Ph.D., and master's students as well as coordinated and participated in several financed research projects. He has received several scientific awards. He has worked as evaluator of projects for ERC-European Research Council and other international research agencies as well as examiner of Ph.D. thesis for many universities in different countries. He is the editor in chief of several international journals, guest editor of journals, books editor, book series editor, and scientific advisory for many international journals and conferences. Presently, he is an editorial board member of 30 international journals and acts as reviewer for more than 100 prestigious Web of Science journals. In addition, he has also published as editor (and coeditor) for more than 120 books and as author (and coauthor) for more than 10 books, 80 book chapters, and 400 articles in journals and conferences (more than 250 articles in journals indexed in Web of Science core collection/h-index 52+/9000+ citations, SCOPUS/h-index 57+/11000+ citations, Google Scholar/ h-index 74+/17000+).

https://doi.org/10.1515/9783110549775-203

Contents

Preface —— V

About the Editor —— VII

List of contributors —— XI

Prasanta Sahoo, Suman Kalyan Das
1   Emerging trends in additive and subtractive manufacturing —— 1

Moisés Batista, Ana Pilar Valerga, Jorge Salguero,
Severo Raul Fernandez-Vidal, Franck Girot
2   State of the art of the fused deposition modeling using PLA:
    improving the performance —— 59

Viktor P. Astakhov, Swapnil Patel
3   Development of the basic drill design for cored holes in additive
    and subtractive manufacturing —— 113

Ke Huang, Tianxing Chang, Yandong Jing, Xuewei Fang, Bingheng Lu
4   Additive manufacturing of magnesium alloys —— 149

Theodora Kontodina, Dimitrios Tzetzis, J. Paulo Davim, Panagiotis Kyratsis
5   Additive manufacturing for patient-specific medical use —— 199

Samad Nadimi Bavil Oliaei, Behzad Nasseri
6   Stereolithography and its applications —— 229

Van-Du Nguyen, Ngoc-Hung Chu
7   Ultrasonic-assisted deep-hole drilling —— 251

Karmjit Singh, Ibrahim A. Sultan
8   Information and computational modeling for sustainability evaluation
    and improvement of manufacturing processes —— 271

Index —— 289

List of contributors

Viktor P. Astakhov
General Motors Business Unit of PSMi, 1792
Elk Ln, Okemos, MI 48864, USA

Moisés Batista
Department of Mechanical Engineering and
Industrial Design, Faculty of Engineering,
University of Cadiz, Avenida de la
Universidad de Cadiz 10, Puerto Real, E-11519
Cadiz, Spain

Samad Nadimi Bavil Oliaei
Department of Mechanical Engineering,
ATILIM University, Kizilcasar Mahallesi,
06830 Incek Golbasi, Ankara, Turkey

Tianxing Chang
School of Mechanical Engineering, Xi'an
Jiaotong University, No. 28 West Xianning
Road, 710049 Xi'an, P.R. China

Ngoc-Hung Chu
Thai Nguyen University of Technology, Thai
Nguyen, Vietnam

Suman Kalyan Das
Department of Mechanical Engineering,
Jadavpur University, 188 Raja S.C. Mallick Rd,
Kolkata 700032, India

J. Paulo Davim
University of Aveiro, Department of
Mechanical Engineering, Campus Santiago,
3810–193 Aveiro, Portugal

Xuewei Fang
School of Mechanical Engineering, Xi'an
Jiaotong University, No. 28 West Xianning
Road, 710049 Xi'an, P.R. China
National Innovation Institute of Additive
Manufacturing, Buiding A, Door of
Metropolis, Jinye Road, 710065 Xi'an,
P.R. China

Severo Raul Fernandez-Vidal
Department of Mechanical Engineering and
Industrial Design, Faculty of Engineering,
University of Cadiz, Avenida de la
Universidad de Cadiz 10, Puerto Real, E-11519
Cadiz, Spain

Franck Girot
IKERBASQUE, Basque Foundation for Science,
48013 Bilbao, Spain
Faculty of Engineering, University of the
Basque Country, Alameda de Urquijo s/n,
48013 Bilbao, Spain

Ke Huang
School of Mechanical Engineering, Xi'an
Jiaotong University, No. 28 West Xianning
Road, 710049 Xi'an, P.R. China
2 State Key Laboratory for Manufacturing
Systems Engineering, Xi'an Jiaotong
University, No. 99 Yan Cheung Road,710054
Xi'an, P.R. China

Yandong Jing
School of Mechanical Engineering, Xi'an
Jiaotong University, No. 28 West Xianning
Road, 710049 Xi'an, P.R. China

Theodora Kontodina
International Hellenic University, 14km
Thessaloniki – N. Moudania, School of
Science and Technology, Greece

Panagiotis Kyratsis
Western Macedonia University of Applied
Sciences, Department of Mechanical
Engineering and Industrial Design, 50100 Kila
Kozani, Greece

Behzad Nasseri
Department of Chemical Engineering and
Applied Chemistry, ATILIM University,
Kizilcasar Mahallesi, 06830 Incek Golbasi,
Ankara, Turkey

https://doi.org/10.1515/9783110549775-205

Van-Du Nguyen
Thai Nguyen University of Technology, Thai
Nguyen, Vietnam

Swapnil Patel
General Motors Business Unit of PSMi, 1255
Beach Ct., Saline, MI 48176, USA

Prasanta Sahoo
Department of Mechanical Engineering,
Jadavpur University, 188 Raja S.C. Mallick Rd,
Kolkata 700032, India

Jorge Salguero
Department of Mechanical Engineering and
Industrial Design, Faculty of Engineering,
University of Cadiz, Avenida de la
Universidad de Cadiz 10, Puerto Real, E-11519
Cadiz, Spain

Karmjit Singh
School of Engineering and IT, Federation
University, Mt Helen Campus, VIC 3350,
Australia

Ibrahim A. Sultan
School of Engineering and IT, Federation
University, Mt Helen Campus, VIC 3350,
Australia

Dimitrios Tzetzis
International Hellenic University, 14km
Thessaloniki – N. Moudania, School of
Science and Technology, Greece

Ana Pilar Valerga
Department of Mechanical Engineering and
Industrial Design, Faculty of Engineering,
University of Cadiz, Avenida de la
Universidad de Cadiz 10, Puerto Real, E-11519
Cadiz, Spain

Prasanta Sahoo, Suman Kalyan Das

1 Emerging trends in additive and subtractive manufacturing

Abstract: The manufacturing sector has experienced a rapid advancement in the last few decades. From the popularization of computer numerical control machines as the face of traditional machining to the development of additive manufacturing (AM), it has been an open field in a series of innovations. Although AM is promising in terms of material savings and generation of complex parts rapidly, it cannot be a replacement for the conventional machining process, which seems better in terms of mass production. Thus, a hybridization of both the technologies seems to offer an optimal solution with respect to increased productivity and efficiency. Besides, improvement in information and communication technologies, Internet of things (IoT), robotics, and so on makes the manufacturing process self-sufficient and fully integrated. Besides, collaborative manufacturing systems are now a reality that responds in real time to meet changing demands and conditions in the factory, in the supply network, and in customer needs. All these have paved the roadmap for the implementation of Industry 4.0 (4th industrial revolution).

Keywords: Additive manufacturing, subtractive manufacturing, 3D printing, stereolithography, fused deposition modelling

1.1 Introduction

In the past few decades, few industries have been impacted by rapid advancement in technologies quite like manufacturing has. The ever-increasing demand for advanced products along with the pressure of cost reduction has led to a competitive attitude among the industrialists, which may be cited as one of the primary reasons for infusion of innovation in the manufacturing sector. Manufacturers have been faced with an "evolve-or-die" ultimatum as customers expect faster rates of innovation. Hence, apart from the traditional machining, things such as additive manufacturing (AM) techniques, smart manufacturing, and agile manufacturing, have been embraced by the manufacturing community, and continuous effort has been put to increase the effectiveness of these techniques. AM is the fast upcoming technology that has already proved its mettle and is responsible for transforming how products are designed and produced. It has gained popularity through other names such as 3-D printing, rapid prototyping (RP), additive layer manufacturing (ALM), and solid freeform fabrication (SFF) and has begun to be incorporated even in our daily lives. Subtractive manufacturing (SM) belongs to more of the conventional manufacturing techniques

https://doi.org/10.1515/9783110549775-001

based on machining of materials. Computer numerical control (CNC) is the modern machine for carrying out SM. As the advancement in the field of AM has been phenomenal in the past couple of decades, this chapter emphasizes more on its current trends and techniques. Figure 1.1 illustrates the basic philosophy of SM and AM.

Subtractive manufacturing

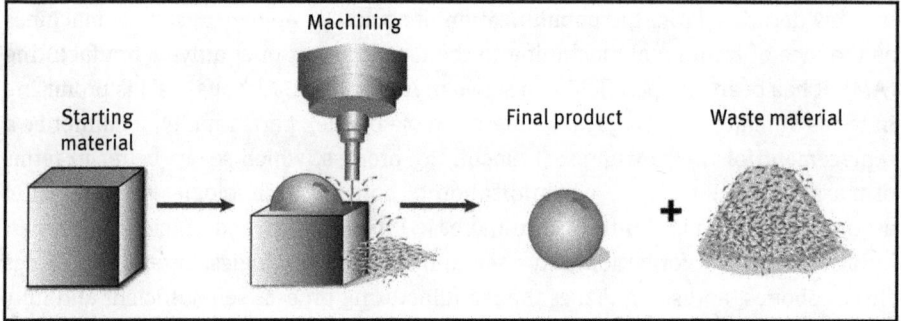

Additive manufacturing

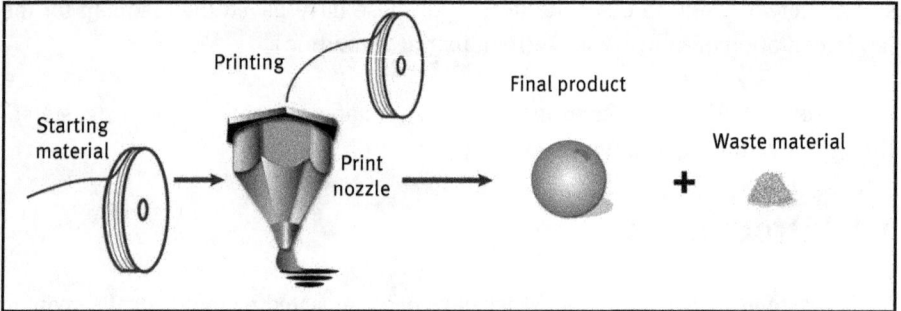

Figure 1.1: Basic principles of subtractive and additive manufacturing (source: GAO analysis).

The basic idea of AM is to first create the CAD (computer-aided design) model of the object to be build. The CAD model is sliced into individual layers. The object is then built by AM progressively layer by layer. Hence, it is obvious that costly jig and fixture, other equipment and processing which are normally required for any conventional production processes are eliminated in this case. According to the joint ISO/ASTM standard [1], AM is defined as the "process joining materials to make objects from 3D model data, usually layer upon layer, as opposed to subtractive manufacturing methodologies," which cuts, drills, and grinds away the undesired excess from a solid piece of material, often metal. AM processes, on the other hand, have some following commonalities, namely a pc to store and process the geometric information as well as to drive the fabricator, deposition of feedstock which is administered as points, lines, or areas to create a part [2]. Some of the advantages of AM include:

i. Complex parts can be fabricated without the need of costly tooling.
ii. There is no issue of tolerances such as circularity, linearity, and perpendicularity, which requires a major share of attention in case of conventional machining.
iii. Components can be produced as per requirement reducing the load on inventory as well as the lead time for critical or outdated parts that need replacement.
iv. Material wastage is very less compared to SM.
v. Easy building of prototypes and optimizing the design.

Due to these host of advantages, AM has found wide acceptance across a variety of domains, such as aerospace and aviation industry, medical and surgical devices, electrical and electronics industry, defense and military applications, automotive applications [3]. Fuel injector nozzles are of complex shapes and require multicomponent assembly, which is now preferably produced through AM resulting in significant cost savings [3]. It is also reported [3] that burner tips for mixing and swirling result in energy savings as well as their service life extends if the same are made of high-temperature materials. Now, with AM it is possible to give complex shapes to high-temperature materials. Automobile sector has been one of the major adopters of 3D printing. Apart from various spare parts of brakes, clutches, and other subsystems of a vehicle, AM has had two major points of influence on automotive applications: as a source of product innovation and as a driver of supply chain transformation. AM has also penetrated medical and dental sector, where various organs, namely, liver, kidney, heart, ear, and nose have been produced by AM. Biocompatibility of the produced organs have been significantly improved. Besides, AM has the advantage of producing individually matched organs that can be derived from the patient's own medical imaging.

The inception of AM was based on production of nonmetallic parts, especially made of plastics. This posed a hindrance in its wide acceptability as most of the practical engineering components are metallic. However, substantial development in AM metal processing techniques over the past couple of decades has changed this scenario. Production of low-cost lasers, cheap high-performance computing devices, and development in feedstock technology (for producing metal powder) have made AM a front-runner production method [3] in today's world. This is also evident from the rapid movement of the commercial AM machines out of the shelves. Although AM technique is found to be a champion among all the production methodologies, some optimization and fine-tuning with respect to material of feedstock, produce tailor-made materials, AM processes, surface finish, properties of the produced part in order to be able to yield defect-free, sound, and reliable AM components. Figure 1.2 shows some actual photographs of parts being manufactured by machining and 3D printing.

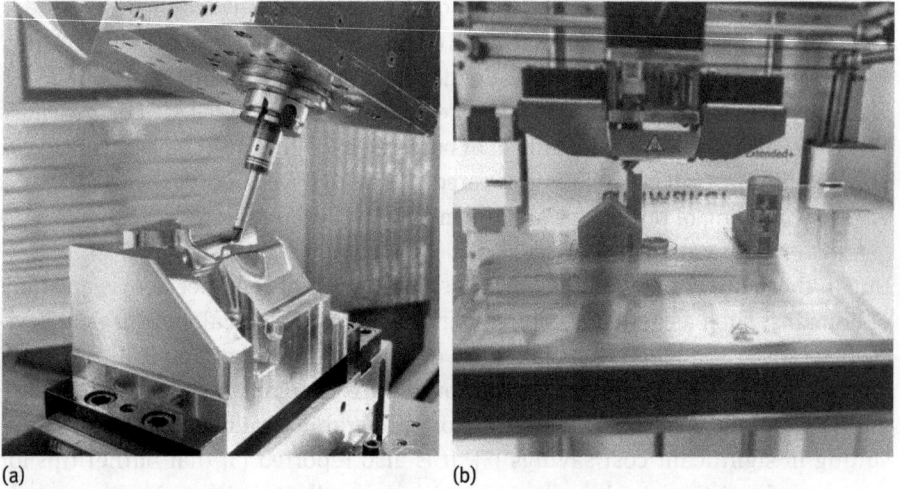

(a) (b)

Figure 1.2: Parts being manufactured by (a) machining and (b) 3D printing (creative commons).

1.2 Evolution of additive manufacturing

1.2.1 Development of 3D printing

Modern AM technology or 3D printing, as it is known, was first realized around 1975 [2]. 3D Systems which was founded in 1986 is credited with selling the first commercial AM machine SLA-I. SLA-1 as the name suggests was based on stereolithography (SLA) technique of 3D printing. Progress in laser technology and material research paved the way for the first successful demonstration of this process. According to Bandyopadhyay and Bose [4], "SLA is a system where an ultraviolet (UV) light source is focused down into an UV photo-curable liquid polymer bath where upon contact, the polymer hardens. Patterns can be drawn using the UV source to semi-cure the polymer layer. Uncured polymer stays in the bath and provides support to the part being built. After a layer of printing is done, the hardened polymer layer moves down on a build plate in the liquid medium and the next layer of polymer is available on top for the following layer. This process continues until the part is finished based on the CAD design and is removed from the liquid medium. In most cases, further curing is needed before the part can be touched." The first patent of an RP system (using SLA) was obtained in 1986 [4].

1.2.2 Advancement in other rapid prototyping techniques

Apart from development of SLA-based system, other techniques of 3D printing were also developed parallelly. Advancement in material science and associated technologies led to the development of innovative AM machines that employed materials and methods. Apart from SLS, some of the popular RP systems included selective laser sintering (SLS), fused deposition modeling (FDM), ink jet printing, and laminated object manufacturing.

SLS was developed in the University of Texas at Austin [4]. The material is spread in powder form on the substrate where a laser selectively cures the powder according to the CAD geometry. The cured layer is then lowered for spreading of next layer of powder and the process is repeated until the desired part is produced. The first SLS machine was realized in 1986 [4]. Around the same time, FDM technique was invented in which a thermoplastic is extruded through a heated nozzle turning it to semimolten state, which is then deposited on to the bed by the movement of the nozzle (layer by layer) to produce the part.

In 1994, the first 3D printer using an inkjet approach was released [4]. This technique has lot of resemblance to SLA. Here, the thermoplastic material (often photopolymer) is jetted on the build tray in thin layers. If required, the layers may be exposed to UV light for curing. High-quality wax model could be produced though this technique similar to investment casting.

The preceding text discusses a few of the founding RP systems that were developed around the 1980s. However, many other people around the world realized the significance of these processes and started research in this field. Several enterprises have developed their individual 3D printing equipment based on new technologies. These developments of AM have ushered in a new era in the field of manufacturing.

1.2.3 Transformation from rapid prototyping to additive manufacturing

At the beginning, the capability of RP machines was limited to only producing parts made of plastics and other polymers, and they were unable to handle metals and ceramics. Now, the philosophy of AM is different from that of RP as the parts produced by AM is supposed to be of actual practical use [4]. Some more iterations happened in the field of RP technology before components made of metals or ceramics could be produced. Realizing the importance of metal processing, many companies started their own ventures of developing AM machines that could produce metallic parts. An enterprise of the name of EOS made the initial metal AM machine in around 1994 [4]. The company did experimentation on SLS technique and came up with a direct metal laser sintering machine (DMLS) machine, which had the ability to sinter metal powders. The machine could produce parts made of common metals, such as stainless steel, nickel, aluminum, cobalt, and even titanium alloys. The company continued its research in this direction and went on to become one of the most successful AM companies in the world.

Contemporary to EOS, another AM technology was tested in New Mexico, which was capable of producing metal components. The technology was known as laser-engineered net shaping (LENS®) [4], which employed high-intensity laser for melting and re-solidifying the powder spread on a tray. Both the tray and laser head being a mobile metal can be deposited on the desired locations layer by layer until the component is produced.

Electron beam melting (EBM) is another technology, which was also being developed at around the same time [4]. The process is a modification of SLS, where an electron beam instead of a laser melts the powder selectively on the loading tray. Once the layer is processed, fresh powder is spread to generate the next layer and the process is continued till the object is built. Many orthopedic implants are made out of this technique. EBM has also been employed for the fabrication of complex components for aerospace applications.

1.3 Evolution of subtractive manufacturing

SM is basically machining in which the raw material in the form of a rod or block (near to the overall dimension of the desired object) is chosen, and selective and controlled material removal is done to produce the final object. Due to the philosophy of material removal, these manufacturing techniques are collectively termed as SM. Various cutting machines such as lathe, milling machine, drilling machine, and shaper that are employed for the material removal process are elaborated as follows:

– **Lathe:** A block of material (raw material) is rotated against a harder and sharper object also known as the cutting tool. The raw material is usually moved in the lateral direction.
– **Milling machine:** The cutting tool moves in a rotary manner removing material from the stock.
– **Drilling machine:** The drill bit makes contact with the material while spinning creating hole in the manner.
– **Shaper:** The cutting tool removes material from the raw material while reciprocating against it linearly.

Single or multiple modes of machining may be required based on the complexity (in geometry) of the part. Besides, this philosophy of producing part by machining is applicable to any solid material such as wood, plastic, ceramic, composites, and metal. Early machining were done manually. However, with time the process has developed and currently machining process has been highly automatized by the introduction of computer numerically controlled (CNC) machines. CNC machines use computers to control the movement and operation of the machine tools.

Wide impetus to the machining process was given by the Industrial Revolution in the late 1700s. Since then machining has evolved into a mature and advanced technology. Earlier machining was a very rudimentary with people using their hands to carve out the desired object from the raw material with frequent needs of manual forging and manual filling of metal. During the machine age (late nineteenth to early twentieth century), the traditional machining processes, such as turning in lathe, milling in miller machines, and drilling in drill presses, were developed with manufacturing of the respective machine tools. These processes are very popular and still practiced in many parts of the world even today. Figure 1.3 illustrates some of the traditional forms of machine tools.

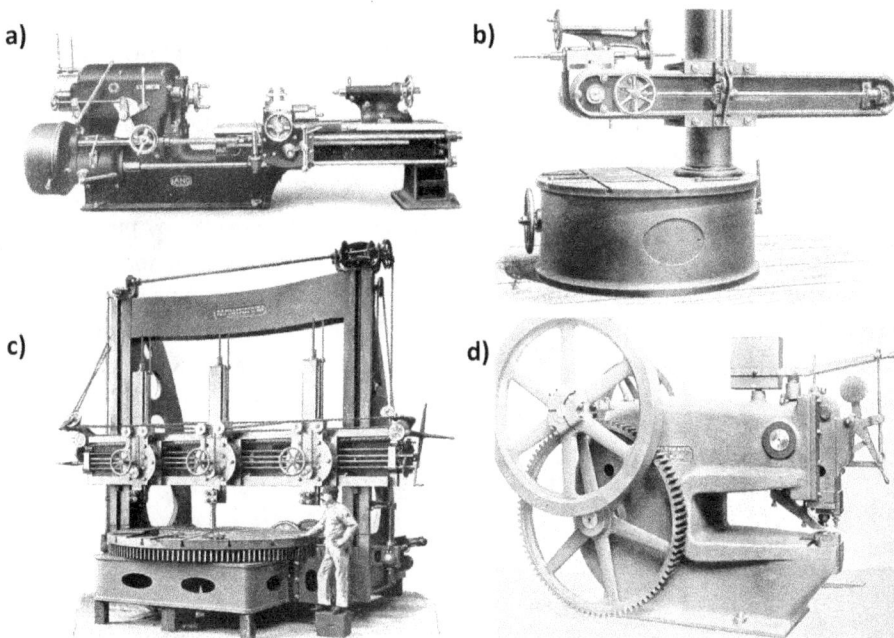

Figure 1.3: Traditional machine tools: (a) lathe (image source: Machine Tools catalogue), (b) drilling machine (source: staticflickr.com), (c) milling machine (source: staticflickr.com), and (d) punching machine (source: Flickr's The Commons).

After World War II, new machining technologies such as electrical discharge machining, electron beam machining, electrochemical machining, photochemical machining, and ultrasonic machining were developed in the arena of SM. The earlier machining processes were differentiated from the newly developed techniques by coining of terms such as conventional and nonconventional machining, respectively. The non-conventional machining was found to be useful particularly when dealing with novel and improved materials which are otherwise difficult to machine through traditional processes.

With the advent of new technological developments, demand for accurate and precision machining began to gather. Besides, the need for faster and bulk production leads to the understanding that some form of automation has to be introduced into the machining process. Post World War II, a numerical controlled (NC) machine was developed by John T. Parsons in association with MIT (Massachusetts Institute of Technology) [5]. The project is required to fabricate aircraft components that had complex geometries and also to find a more cost-effective way of doing it. Gradually with time, NC was infused into the manufacturing sector and it was not long before it established itself as the industry standard.

During the late 1970s, as computer began to be introduced for solving practical problems of mankind in various domains, the idea that computer can also control the machining process was floated. Soon the philosophy of CAD and computer-aided manufacturing (CAM) was developed, which catalyzed the progress of CNC machining to become a reality. The first commercial CNC machine was produced in 1976 and at around 1990 these machines had become the standard for manufacturing industry. The kinds of CNC machines included CNC lathe (illustration provided in Figure 1.4), CNC mills, CNC routers, and CNC grinders which are capable of manufacturing products made of plastic to steel. The initial NC machines were driven by punch codes that had a set of codes, which were known as G-codes. G-codes stand for "Geometric Code" and tell the machine when and how to move besides controlling its machining operation. The early NC models were hardwired and hence it was very difficult to modify the pre-set instructions. However, with the introduction of CNC machines, the

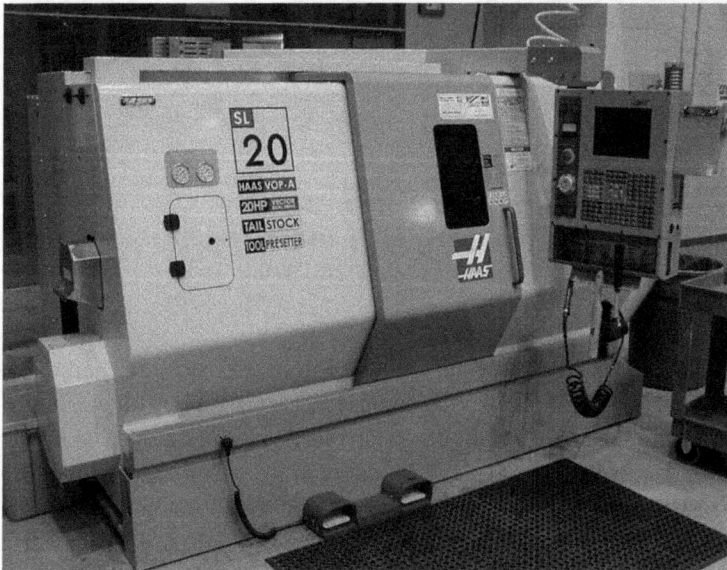

Figure 1.4: Photograph of CNC turning machine (Wikimedia Commons).

computer itself could participate in the operation and control of the machine. With time, G-codes have evolved to become very flexible allowing the operator to do real-time adjustments if needed.

The aim of both CNC and the conventional machine is one and the same, that is, to produce the desired part from a block of raw material. However, the primary advantage of CNC over the conventional machine is the degree of automation involved with the former. The CNC machine can be kept running without any human intervention in case of complex jobs that require long time. Besides, high productivity, speed, and accuracy are the key advantages of using CNC-based machines. For example, in case of conventional lathe, a skilled operator is required for a machine. However, with CNC, a skilled operator can program and operate multiple machines. With advancement in the networking and communication technologies and evolution of Internet, CNC machines can also be operated remotely if needed.

Now, people may argue that SM may seem to hold limited applicability in the era when AM machines are becoming increasingly cheaper and technologically advanced. As SM, although with the latest development is only CNC machining, one that produces part is by removing the material from a larger sized raw material. However, SM allows one to design and manufacture using the end-use materials. SM can handle any materials ranging from polymers and resins, composites, ceramics, wood, metals, and others. Based on the application, a particular material can be chosen and can be worked with to produce the part. SM seems to be suitable to manufacture objects for small and large volume production runs. Moreover, SM can give the desired surface finish to the object and is the preferred manufacturing process when certain mechanical properties are desired.

1.4 Advantages and limitations of SM

One of the advantages of SM compared to AM is that it does not have any layering in the object produced. The layering of the AM process may sometimes affect the microstructure of the final object, thereby also affecting its strength which may be a concern in case of critical components of a system. One can obtain various surface finishes in case of SM, which is possible by choosing an optimal set of machining parameters. The surface finish dictates friction and wear and can be particularly important in case the object is part of a tribological pair when integrated in the system. Besides, SM employs surface grinding as a final step before completing a particular job. Based on the coarseness of the grinder, surface roughness of the object can be varied and the desired roughness obtained. The "stepped surface" observed in case of many AM processes is eliminated in case of SM. However, the philosophy of material removal sometimes puts SM under some limitations, namely undercuts during milling operation are difficult to avoid. Moreover, machining consumes a lot of energy compared to additive process.

One of the disadvantages of operating NC machines is that the operator must have proficiency in G-code programming so that he/she can convert the geometric information of the designed model into machine-executable codes. With the introduction of advanced CNC machines, the tool path can be generated automatically and hence SM prototyping is quite comparable to that made by AM [6].

Subtractive RP can produce parts with high surface finish. Their tolerances regarding this is comparable to components made by injection molding. Thus, the components can be directly subjected to functional testing. Again the development in tool path generation software enables subtractive RP to handle complex geometries previously unimagined. Thus, it can be summarized that SM can handle a wide variety of raw materials. It yields good dimensional accuracy and surface finish with repeatability, which is suitable for mass production. However, some material wastage is also involved in machining.

1.4.1 Rapid injection molding

Rapid injection molding (RIM) is considered to be a part of subtractive prototyping. It is a rapid manufacturing technique to ensure that production is fast and quality is maintained. In RIM, the CAD model of the object is processed based on which the mold is manufactured by machining (milled). The mold is generally made of aluminum which gives an advantage regarding reducing the time for making the mold as well as tooling costs as would have been incurred in case of traditional steel molds. Once the mold is finished, thermoplastic resins are injected in it, which upon solidification produces the prototype [6]. These prototypes can be subjected to full functional tests and can also be tested for proper fittings and assembly. Besides, the molded parts have good surface finish and hence don't require any post-processing steps. The features of RIM can be visualized as follows:
- Production of prototypes in production-grade resin in 3–5 weeks.
- Any engineering-grade resin of any color can be employed.
- The molded parts are strong and be tested functionally.
- Implement bridge tooling – molds can be used for 1,000s of parts.
- If desired molds can have textures.

1.5 Current state of AM and its impact on industry

AM has proven itself and has become matured enough to be accepted by the industry. Its advantages compared to SM have grabbed the eyeball of many industries. In summary [4], AM process initiates with the production of the CAD model of the component. The CAD model is transformed into STL (STereoLithography or

Standard Tessellation Language) file, which is in general the file type recognized by most AM machines. After that the system splits the model into multiple layers in an orientation, which is easiest to build and at the same time the part remains stable when its manufacturing is underway. Then the desired material in various forms are deposited and binded layer by layer to produce the component. By this methodology intricate shapes can be produced with relative ease and using a wide variety of materials. Because of its ability of producing complex geometries, the potential for the application of AM is huge. Some of the beneficial aspects of AM is presented as follows [7]:

I.   Parts can be produced directly from the design without minimal intermediary steps
II.   Custom-made parts are possible to be manufactured with no extra tooling or cost
III.   The produced part can be put in actual function
IV.   Post processing after manufacturing is minimal
V.   Manufacturing of flexible and lightweight components with hollow or lattice structure is convenient
VI.   Manufacturing waste is minimal; hence, maximum material is utilized
VII.   The overall product development and realization time are very less
VIII.   Manufacturing can be carried out on demand, thus reducing inventory costs
IX.   Excellent scalability

1.5.1 Advantages of AM

1.5.1.1 Limitless design

AM has had significant influence on the manufacturing process of many products across various industries. AM is able to manufacture parts that were impossible to be made through SM. As AM can manage to produce the final object without undergoing much machining and processing, intricate shapes that are otherwise difficult to produce without the traditional methods can be made by AM [4]. This gives the flexibility to the designer, who can focus on the functional aspect without paying much heed to the manufacturing aspects. The designer has only to monitor the overall size of the object and whether it can be fit in the AM machine. Apart from the machine, no major tooling is required for AM. However, for some objects post production machining may be required to obtain the desired surface finish. As AM does not generate much material waste, the problem of disposing the waste material is also not a problem. Statistically it is reported that AM can reduce the material usage by about 75% which in turn can reduce the manufacturing time and cost by around 50% [4]. The enormous savings possible by AM have made the process attractive to manufacture around the world.

1.5.1.2 Versatility in manufacturing

Versatility in building parts is another aspect that makes AM process so lucrative. Reworking on parts is very easy in case of correction of flaws and changing the design for optimal use. This can be very difficult in case of manufacturing methods. Thus, AM can make working prototype very quickly, which can also be tested to see whether it can serve the intended function. Thus, as previously mentioned, components can be manufactured as per requirement. The designer can also modify the design or try a design optimization. Moreover, if a requirement for a custom-made part comes up, it can be delivered very quickly without disturbing the normal manufacturing process [4].

1.5.1.3 Obtaining enhanced performance by alternating material

Nowadays parts made of various materials such as plastics, metals, composites, and ceramics can be made with AM. However, plastics remain the most popular material as they have been the most studied and researched AM material. Work is also underway to investigate whether AM can be used as a material processing technique, so that materials with desired combination of properties can be developed. AM has been employed to bond metals and ceramics to create composited with enhanced wear resistance properties [4]. Through AM it is possible to produce a layer of coating on a particular material to increase its thermal and wear resistance, which may otherwise be very poor. Moreover, in case of a component that has broken or some of its material removed, AM can be used to bond the parts together or it can add material just at the desired location to restore the functionality of the part. Thus, the maintenance cost can significantly brought down by adopting AM as the parts needed can be discarded [4].

1.5.2 Incorporation of AM in modern manufacturing

The versatility of AM has led the modern-day industries to embrace them with open hands. Many companies have already created facilities dedicated to AM. GE Aviation has a facility where around 60 EBM machines and direct laser sintering machines can be installed [4]. AM has found huge application in aerospace industries, which deal with complex part and also require high reliability. AM allows for optimization of components, reducing their weight, lowering material loss, and thus increasing the overall profit. A lot of other aerospace manufacturers have also begun the induction of AM and integration of the same in their production process. Thus, AM has started to get its due attention; however, as of now mainly big enterprises have installed the large AM fabricators, which can give functional parts (mainly metallic) ready for use. Besides, there are issues such as surface finish of the parts as well as the material properties that are being fine tunes. But overall AM technology has matured enough to be able to produce components ready for use.

1.5.3 Growth of CAD: impact on manufacturing

The philosophy of CAD is to make use of computer systems for the creation, alteration, and analysis of a design. Ivan Sutherland is credited with the development of the first CAD software in 1960s, whereas PRONTO is the first commercial CAD software. The present day has seen the growth and commercialization of various CAD packages. Some of the popular CAD software are SolidWorks, Autodesk Inventor, Catia, Creo, and others. In the contest of AM, CAD has found its importance sky rocketed. One of the revolutionary aspects of AM is that it is capable of producing ready-to-use objects from CAD model. Development in CAD now makes possible to virtually design anything. Any theoretical model can be converted into digital form. The CAD file is converted to STL file and the part is produced by the AM method. Due to this people can now design more complex and efficient components. Even at the development phase of CAD, designers still had to keep in mind the manufacturing aspects of the design while designing components. However, that constraint has now been lifted.

1.5.4 Current manufacturing challenges

Although lot of optimism is prevalent about AM, there are major hurdles that have to be crossed, so that the technology receives extensive acceptance. Some of the prominent challenges are mentioned as follows [8]:

1.5.4.1 Deficiency in AM-compatible design knowledge

Although AM is able to convert any product design to reality, a certain amount of rethinking is required to apprehend the full potential of the technology. In case of manufacturing by AM, first of all the model should be oriented in its most stable position. Moreover, in case of overhanging parts, supporting may be required. Thus, the design should have the necessary compatibility to be manufactured by AM methodology. Besides, the design can be made so that unnecessary materials can be eliminated to the best extent possible. Reinforcements if required can also be planned in the design itself as AM doesn't suffer from the limitations of SM or traditional machining.

1.5.4.2 Higher manufacturing cost

One of the major obstacles for the extensive use of AM is its higher manufacturing costs. The cost structure of AM and SM is significantly different from each other. In case of AM the tooling costs like that experienced in case of SM are not experienced. However, these costs (in SM) can be managed and the process can be made cost-effective if the volume of production is increased. In case of AM, an initial cost estimate is based on cubic centimeters of the product manufactured. The cost factors involved in AM can be listed as follows:
– cost of AM machine
– material cost
– energy cost
– post-processing cost

It has been studied that with polymers and plastics, the volume threshold where AM is profitable is increasing. But in case of metals, AM incurs more production cost compared to conventional machining.

1.5.4.3 Restricted industrial scaling of AM

Even after the adoption of AM worldwide, large-scale production with AM is almost nil or observed in a very few cases. One of the reasons for this is that AM machines are more suited for prototyping. Production in series or large-scale production is still not profitable with AM. Many components are too big to be produced with AM. A handful of instruments can be found around the world, which can handle large components but they too are stand-alone systems working with a single material at a time. GE Aviation is one of the few enterprises that has employed AM for large-scale production of fuel nozzles for turbofan engines. However, adoption among others is very slow. Thus, there is a need to develop the AM machines suitable for large-scale production. For these to happen, the production costs need to be lowered. Besides, the AM system should have features such as process stability, quick changeovers, in-process quality control, and higher reliability. The system should also be easy to maintain and repair in case of breakdown.

1.5.4.4 Threats on cyber security and IP rights

AM systems relying largely on digital and computer-based systems are vulnerable to some risks of which the SM are almost free of. There are risks of the design being stolen as the actual parts are produced from a digital CAD file. There are chances of the design being copied even from the actual model (if it is replicable) after it is produced and sold. Besides, as the data is stored digitally in the computer or sometimes in cloud to have better communication between the supplier and the customer, there looms a danger from cyber-attacks. Detailed discussion regarding this is presented in one of the following sections.

Some of the other general challenges faced in AM [7] lies in the development of a robust, autonomous, user-friendly, safe, and integrated system so that parts can be manufactured quickly and with the necessary accuracy (dimensional). Besides, the surface finish of the finished product should be acceptable and the overall process should not consume too much amount of energy. Variations in the quality of the product from machine to machine as well as from batch to batch are again a challenge and result mainly from the lack of a proper understanding between the AM process parameters and the product quality. In case of metallic objects, prevention of surface oxidation is another factor that should be kept in mind, otherwise the product quality may suffer. Providing inert gas atmosphere is one of the solutions currently in use against oxidation. Apart from this, metals require higher laser power in comparison to polymers, which add to the overall cost of the AM process. Table 1.1 lists some of the technological limitations of AM.

Table 1.1: Technological limitations of AM [8].

Design and engineering	Manufacturing	Service
Lack of design knowledge (e.g., long-term performance of materials and design for 3D printing)	High production costs (e.g., material costs and limits on production speed)	Lack of industry-specific testing procedures (e.g., for production processes)
High risk of design pirating through users	Limitations on size (for specific AM technologies)	Lack of structural regulations in supplier networks
	Limitations on product quality (e.g., in range and combination of materials, resilience, and surface finish)	Risk of supply chain disruption
	Dependence on small number of machine suppliers	

One of the promising aspects for AM is that companies that manufacture AM systems are actively engrossed in improvement of their machines. Many technical institutes, R & D laboratories, and government organizations are also showing interest in AM technology and trying to address the limitations of the process in their own way. This way the workforce is getting trained, and the skill gap for AM compatible design is also getting gradually bridged. Next-generation machines are expected to reduce the production cost of AM due to the fact of patent expiration of many related technologies. Moreover, due to improvement in the accuracy of the parts, post-processing needs are also getting reduced.

Manufacturers of AM machines are improving in-process control of the machines as well as incorporating advanced quality diagnostics into the system. These are some progressive steps toward making AM suitable for large-scale production. According to some manufacturers of large AM machines, the material properties of the products obtained from AM are quite comparable to that obtained through SM techniques.

One of the vital needs of AM is the availability of good quality raw materials with desired properties. Low-quality materials result in bad parts unsuitable for the application for which they are made. Polymers frequently used in AM such as nylon, acrylonitrile butadiene styrene (ABS), and polyether ether ketone (PEEK) have been improved with them becoming more heat resistant and displaying more compatibility to the AM process. Parts made with these materials also show improved strength and reliability. Metals such as steel, aluminum, titanium, as well as precious metals such as gold and silver have also been developed to be suitable for manufacturing through AM pathway.

1.5.5 Other issues in manufacturing

1.5.5.1 Projection-based manufacturing issues in large-scale production

Through large volume production, the standard distribution of goods and products can be carried out [4]. Large volume productions require a strong inventory and can yield cheaper products as well as higher production rates. However, this system has some limitations. If a projection-based manufacturing philosophy is adopted, it can result in wastage of goods in case projection is not accurate as demand does not remain constant and are vulnerable to change because of various reasons. This can further lead to loss of jobs. The above problem can be easily solved if parts were produced on demand.

1.5.5.2 Issues related to centralized manufacturing

If a product is mass produced in a particular country, then rest of the world is dependent on that country for supply of that product. Now, if a country gets affected by some unforeseen issues such as natural calamity or war, the supply of the product gets influenced or at worst gets stopped. This has already been experienced in automotive industry when Japan was struck by the deadly tsunami in 2011 [4]. This situation can be avoided if the production process is distributed across various regions and across the world. Now, distribution of the production process is related to the concern about the degradation of the product quality. With AM, however, this issue can be handled as if the CAD model is shared and equipment available, then quality products can be produced.

1.5.5.3 Generalized designs preferred for business

The existing philosophy of production is to manufacture goods that are of interest to majority of the people in the society. Initially a market survey is conducted to note the choices and preferences of the people. After that based on the general choice, the specification of the product is decided. Finally, the product is manufactured and made available in the market. Thus, it may so happen that in many cases the available product does not exactly fit the needs of a particular customer [4]. The manufacturers also try to fit their designed product within the specification given to them and hence end up making products that are not optimal as the information given to them may not be always reliable. Moreover, when dealing with large-scale industrial manufacturing system, it may be really challenging to produce custom-made products as the tooling and other related accessories are fixed and recognized. Hence, through this system of manufacturing, the customer settles for a product which is just adequate but not perfect. Through the advent of CAD and AM, this problem can be solved; as customized products can be produced quite easily, people getting their perfectly suited product is only a matter of time.

1.6 Various AM techniques

The various techniques of AM have been briefly described in this section. Utilizing the techniques wherever suitable, AM has had a great advancement which is continuing with the advent of newer techniques.

1.6.1 VAT photopolymerization

When a photopolymeric resin is exposed to UV light, it gets cured and thus hardens. This technique has been widely implemented to manufacture products in an additive manner [9]. To simply put, a vat of photopolymeric resin is cured by UV light according to the sliced geometry of the CAD model. Subsequently, the next layer of resin is introduced and cured and process is continued until the total model is realized.

1.6.1.1 Stereolithography

In SLA, a vat of resin (photopolymer) is exposed to UV laser due to which the resin gets photochemically solidified based on the layer of the CAD model. Once solidified, the platform with the vat is lowered to the thickness of one layer (around 0.05–0.15 mm) and new layer of resin is spread on top of the previously processed one. Likewise, the process is repeated till the solid object of the CAD model is obtained. The solid object is again cured by UV light so that their mechanical properties are enhanced. Finally, the completed part is washed with a cleaning agent to remove the wet resin from the surface of the object.

1.6.1.2 Direct light processing

Direct light processing (DLP) technique for producing parts is almost similar to SLA method except that the former employs a more conventional source of light such as an arc lamp with a liquid crystal display panel. The panel is fitted such that it covers the entire surface of the vat with resin. During building, the 3D layered image of the object is displayed on the resin layer with the help of the projector. The exposed part of the liquid resin gets hardened. After that, second layered image is cast on the fresh layer of resin. The process is repeated until the 3D model is complete and the vat is drained of liquid, revealing the solidified model. DLP AM is faster and can print objects with a higher resolution.

1.6.2 Powder bed fusion

Powder bed fusion (PBF) employs the principle of fusion (sintering or melting) of material in powder form using a thermal source of energy (usually laser or electron beam). Layer-by-layer sintering is done sequentially to finally give rise to the object. Spreading of powder in thin smooth layers is done by a mechanism using either a roller or blade that is integral to PBF system. Once the part is finished, it is covered in loose powder from which it is taken out and cleaned. Through PBF, plastic as well as metallic parts can be produced.

1.6.2.1 Selective laser sintering

SLS is a PBF process that employs a laser power source to heat and fuse the powdered material (typically nylon or polyamide). Arrangement is made so that the build platform is lowered to a single layer thickness so that the next layer of powder could be spread and processed. Upon completion of the job, the part is cleaned and subjected to post processing if required.

1.6.2.2 Selective laser melting and direct metal laser sintering

Both SLM and DMLS are used to produce metallic parts. SLM is nearly of the same concept as SLS but only the powder is fully melted in case of the former. It is to noted that the metallic powder must be prepared nicely so that the granules are perfectly shaped in order to even out while spreading. As the powder is melted and re-solidified, it allows the formation of new crystal structure in the material. Porosity can also be redefined in the solidified material. Thus, further costly and time-consuming post-treatment steps are eliminated. Also through SLM, fully dense parts can be produced in a direct manner. In case of DMLS, the metallic powder is heated until they reach near melt temperature but not melted. Under this condition the powders fuse together chemically. DMLS can work with alloys, however, the common material used in DMLS is aluminum or titanium powder. Ultra-thin layers of fused powders are produced one on top of each other. Support structures may sometimes be required while fabricating parts by SLM or DMLS based on the geometry of the object.

1.6.2.3 Electron beam melting

In EBM, metal powder is completely melted and fused using a concentrated beam of high-energy electrons. Through local meting and re-solidification, the sliced geometry of the object is obtained. A vacuum atmosphere is preferable so that electrons don't collide with the gas particles. Besides, oxidation can be eliminated if the process is carried out in vacuum. The advantage in using this process is that it generates lesser amount of residual stresses resulting in lesser distortion and warping. Thus, requirement of support structures is less. EBM is also an energy-efficient process and can produce parts faster than SLM and DMLS.

1.6.2.4 Multijet fusion

Multijet fusion is a more cost-effective way to 3D print complex parts very quickly. At first, a thin layer of powder is spread on the platform. Then droplets of the fusing agent along with the detailing agent are deposited from a nozzle based on the required geometry of the part. The detailing agent inhibits sintering and is spread near the edge of the part. After that thermal energy is applied to fuse the powders where the fusing agent was spread. This is repeated layer by layer to give rise to the geometry.

1.6.3 Extrusion-based AM

Extrusion-based AM has gained tremendous popularity due to cheaper setup and hardware. In this process, the material feedstock is extruded through a nozzle onto the build tray. The nozzle defines the voxel size for the system. Many times the feedstock is heated and converted to a semimolten state before extrusion. The nozzle moves according to the G-code of the CAD model and builds the part layer by layer. Sometimes the nozzle deposits the feedstock at room temperature followed by curing.

1.6.3.1 Fused deposition modeling

Most of the 3D printing machines that we see around employ FDM technique to build the parts. Here, the feedstock (thermoplastic material) is supplied in the form of filaments. The filament is melted by heating at the nozzle and forced out of it. The nozzle is guided by the G-code generated from the CAD model of the object. As the nozzle continues to deposit material onto the build tray, the tray is also lowered so that the part is built layer by layer. The material solidifies upon cooling and the part is taken out of the tray. In order to get adhesion between interlayers, the temperature of the extruded material should be maintained just below its solidification point. For some materials, the heating of the build tray (bed) may sometimes be required in order to improve the print quality. Heating of the bed keeps the plastic warm, thus preventing warping. Warping commonly happens at the edges of the part due to uneven cooling compared to the plastic inside the part. In case of parts with features such as internal holes, blind holes, undercuts, and hollow parts, support structure may be required during building which are removed later on. Thus, proper planning and strategy have to be decided regarding the position of the part and the side from which the manufacturing should start. Sometimes, removing the support structures also may be difficult based on the intricacy of the part geometry. At present, two types of support structures are employed: one which can just be washed away and the other which has to be removed by breaking from the part.

FDM can be employed to build parts made of composite materials. In those machine dual-nozzle heads are present through which feedstocks of two different materials are extruded [10]. One of the nozzles builds the physical part, whereas the other one is used to provide reinforcement which bonds with the primary material of the physical model. Interestingly, the FDM technique has also been used to develop functionally graded materials using the dual nozzle setup.

1.6.4 Material jetting processes

This method is quite similar to 2D ink jet printers. Here a print head dispenses droplets of photosensitive polymer on the build tray, which are then cured by high temperature or UV light. After that the tray moves one layer down and the next layer is built. This way gradually the whole part is realized. Materials such as photopolymer and metals that can be cured or hardened with high temperature or when exposed to UV light can be used as a material in this process.

1.6.4.1 Jetting technique

In case of material jetting processes, the photosensitive polymer is sprayed onto the build plate through a print head having multiple tiny nozzles. Due to this, jetting processes can deposit uniform layer of polymer quite rapidly compared to processes where the material is deposited through a single nozzle. After deposition, the polymer droplets are cured by exposure to UV light. Support is normally essential while building through jetting techniques. The support is printed simultaneously with the part but of dissolvable material that can be conveniently washed away once the job is printed.

1.6.4.2 Jetting of nanoparticles

Nanoparticle jetting is comparatively a newer AM method developed by a company named XJet. This method uses a unique liquid dispersion methodology, where liquid droplets containing suspensions of the selected nanoparticles of the build material as well as the support nanoparticles are jetted onto the build plate to form layers corresponding to the sliced layer of the CAD model. The liquid suspension plays the role of the base material which is evaporated by increasing the temperature of the build envelope. Once the part manufacturing is finished, the support material (made of special material) disintegrates easily and is removed. Finally, the part is subjected to short overnight sintering. Through this process, metal and ceramic parts can be manufactured with high-quality and dimensional precision.

1.6.4.3 Drop on demand

This process gain uses a setup with dual nozzle tips. The build material is deposited through one nozzle, whereas through the other support material is introduced. The support material is normally dissolvable and can be easily removed. The nozzle head follows the instruction from computer and produces one layer.

Once a layer is produced, a fly-cutter that is integrated to the setup skims the surface of the layer to ensure that it is perfectly flat and ready to receive the next layer of the material.

1.6.5 Binder jetting processes

Binder jetting process typically uses a print head, which sprays a liquid binder on a thin layer of powder (build material) according to the path defined by the computer. The powders which receive the binder are bonder together [2]. After that the bed is lowered and the next layer of the powder is sprayed. In this manner, gradually the total part is realized. After completion of the job, the part remains encapsulated in powder and in this condition it is subjected to a curing process through which it gains strength. After that the excess powder is removed by cleaning with air jet. Depending on the material a post-processing curing step may be required. Metal-binded parts may require heat treatment or many times they are infiltrated with low melting temperature metal like bronze. In case of ceramics, the infiltrant is usually cyanoacrylate adhesive. The main advantage of this process is that as the process occurs at room temperature and thermal related distortions are avoided. One of the limitation of the process is that the parts produced may not have the desired mechanical strength and hence their applicability is compromised.

1.6.6 Direct energy deposition

Direct energy deposition (DED) covers a range of processes such as directed light fabrication, laser-engineered net shaping, direct metal deposition, and 3D laser cladding. The basic methodology here is to heat, melt, and bind the build material (powder or wire form). The thermal energy is supplied by a focused laser or electron beam. Control of grain structure of the build material is possible, which can lead to its application in repair jobs. Besides, large volume parts can be made with this process. Two techniques of DED are as follows:

1.6.6.1 Laser-engineered net shape

LENS system consists of nozzle for dispensing powder, a laser head for melting the powder, and an inert gas tubing to provide inert atmosphere so that oxidation can be prevented. One of the uniqueness of this process is that the build table moves suitable to get the powder melted according to the geometry, whereas the laser head remains stationary. The inert gas also improves the interlayer adhesion.

1.6.6.2 Electron beam additive manufacture

Here the build material in the form of powder or wire is fused together with the help of electron beam. Here, a vacuum environment is required so that electrons don't hit the gas particles as preventing oxidation. The efficiency of heating by electron beam is considered to be more efficient compared to laser.

1.7 Modern machining techniques: CNC machining

CNC machine is the upgraded version of traditional machining processes. Here, the machining tools are driven by a pre-programmed computer program that guides the tools to the necessary locations from which material removal takes place. The main advantage of CNC machining is that the processes are fast and highly automatized. CNC system consists of the actual machine tool and a computer that plays the role of the controller unit of the machine. While in case of the NC machines, the program (in G-code) had to be fed into the system with the help of punch cards, in case of CNC the program can be directly entered to the integrated computer which commands and drives the machine thereafter. Basically, CNC machines allow to pre-program the speed and location of the tool and drive them through the installed software. The software can then run the machine like a robot operator without much human involvement. Due to the presence of the computer, the programs can be written and edited as per the will of the operator and can also be repeatedly used. Apart from these, human errors in machining can be avoided and parts with high precision and dimensional accuracy could be produced. Due to these host of capabilities, CNC machining has found wide acceptance worldwide and has become the industry standard.

In comparison to the NC machines, CNC machines can add more flexibility to the manufacturing process. Computational capability is one of the features that was absent in case of the previous generation NC machines. However, with the development of computation systems and sensors, the CNC machines could not only robotically drive the machine tool but can also be used to perform small calculations and analysis like interference detection and simulating the program before actually running it. Nowadays, due to advancement in networking and introduction of Internet of things (IoT), CNC machines have become smart and can even be operated remotely if desired. The normal protocol for CNC operation nowadays is to at first model the part in a CAD software. From that, the G-code for producing the part is generated by the CAM software, which is also responsible for manufacturing the part.

Traditional machine tools such as lathe, milling machines, and drilling machines are all available in CNC variety. As today's applications require more and more complex parts to be made, multiple machining processes are becoming essential for making a single object. Thus, many CNC manufacturers are producing multimode CNC machines, which combine several functions in a single system. Thus, loading and unloading time, machine setting time, and others are saved considerably. Finally it can be said that CNC machines can manufacture parts consistently, which might have been difficult to achieve using a human operator.

1.8 Manufacturing workflow process

Workflow process is a series of activities, which should be followed in order to achieve a business outcome. Generally, this process is linear and proceeds in a sequential manner following the existing rules of business. Many times the work flow process mapping is illustrated by graphical means, that is, using flow diagrams.

In case of manufacturing, the process workflow denotes the steps that start from receiving of the raw materials and end with the making of the final product [11]. Enterprises around the world are continuously on the look out to optimize their process workflow so that quality product is manufactured smoothly with the least amount of time. In order to improve the product quality, tools such as total quality management and six-sigma technique are implemented. Just-in-time manufacturing is adopted as a management strategy to control the inventory.

1.8.1 Fundamental steps in manufacturing workflow

Various products are manufactured in diverse ways and require different sets of activities. Manufacturing of food product will definitely not follow the same route as required to manufacture automobile components. But the general manufacturing workflow followed by most of the industries around the world is as follows:

1.8.1.1 Concept development and production planning

Upon product finalization, preliminary concept about its manufacturing, generating ideas, brainstorming, and so on are part of this phase. Besides, decision making about the procurement of the material, its transportation and logistics, and also its proper stocking should be made. Material management should be properly done to ensure smooth supply of the material in desired form and quantity.

1.8.1.2 Decide manufacturing process and engineering

In this step, various manufacturing steps and routes are decided. Also the associated facilities and systems required for manufacturing are developed.

1.8.1.3 Manufacturing and assembly

This phase is about making the manufacturing setup ready for production. The materials are tested for their quality, and production runs are carried out to manufacture some initial models of the designed components. In case of complex system, trial assembly is carried out to detect any defect that may have been ignored. Once finalized, full-fledged manufacturing can be done.

1.8.1.4 Quality assessment and reliability

It is a responsibility on part of the manufacturer to give quality assurance to its customer about its product. Thus, quality checks and inspection are carried out on the finished products. Besides, quality check also ensures that the manufacturing processes are followed nominally and no deviation is encountered. A checklist about the various aspects to be observed as per the quality assessment should also be maintained and updated time to time.

1.8.1.5 Facility maintenance and repair

Once the production run is smooth it is also necessary to maintain the production unit so that safety of the overall facility is not hampered. Besides, maintenance results in lower breakdown of machines, which can in worst case can halt the production run. Certain protocols should be developed in order to ensure that the facility is running optimally.

1.8.2 Additive manufacturing workflow

There are a variety of AM processes available, which have successfully built various parts. As already discussed, these processes employ various techniques to produce the parts. Most processes, however, use a build material which is cured by some form of thermal energy. Now, in some processes, the build material is in the form of powder, whereas others employ material in the form of filament or wire. Besides, laser is used to melt the material in some processes, whereas in others electron beam is used. Some processes involve spraying binder onto the powder (build material) using an inkjet-type printing head. Broadly the AM workflow is summarized in Figure 1.5.

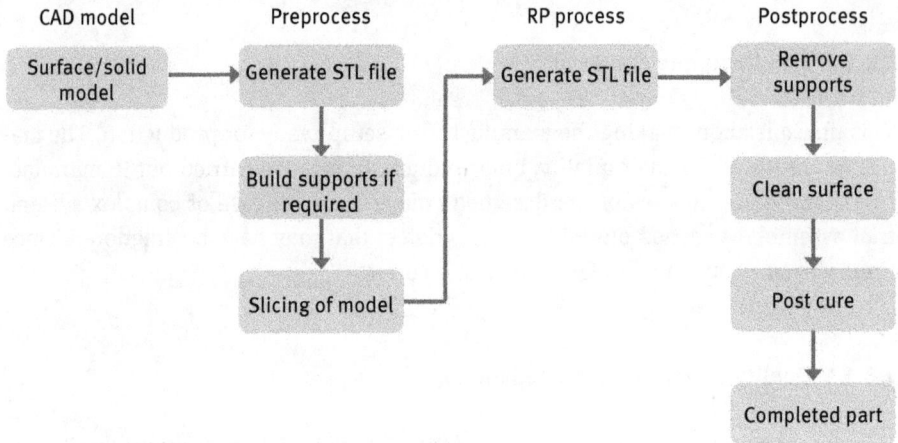

Figure 1.5: AM process workflow [12].

1.9 Emergent trends in additive manufacturing

1.9.1 Penetration in biomedical sector

AM has found uses in medical sector for some years now through different applications. With advancement in surgical techniques, organ transplant has become quite nowadays. However, the main problem that the organ transplantation sector faces is the availability of suitable donors when needed. It is the result of extensive research in tissue printing by AM that organ printing is gradually becoming a reality. External organs such as ear and nose have already been 3D printed and even successfully implemented in many patients. However, research is currently ongoing to print major internal organs such as heart, kidney, and liver. In biomedical sector, AM has had major success in orthopedics and dentistry. AM is particularly useful for bone ingrowth capability [13]. Moreover, AM is able to make complex parts out of metals making it ideal to fabricate various implants and prosthetics. In medical sector, application of AM can be majorly categorized as follows:
– Creating medical models for training
– Production of surgical implants and guides
– Manufacturing of external aids for persons with disabilities
– Printing of biological organs and devices

One of the challenges faced in medical sector is the individuality of the organs or patient-specific organs and devices. In AM, customized patient data that are available from various diagnostic tests such as magnetic resonance imaging, computer tomography scans, X-rays, 3D scanners, and ultrasonography are used to build full anatomy of the patient [13]. After that AM machine converts this 3D data to realize the part or organ. Hence, the printing occurs in three major steps:

- *Image acquisition*
  Through various diagnostics instruments as already mentioned.
- *Post-processing of the image*
  The acquired images are converted to 3D data. The CAD software converts the 3D data to the STL file, which is then fed into the printing machine.
- *3D printing and model realization*
  AM technologies then realize the model from the model.

1.9.1.1 Application of AM in medical sector

The various areas of biomedical sector where AM has contributed till date are as follows [14]:

- Printing of the near equal models of the human organs, which helps in training of doctors before performing the actual procedure
- Printing of bones that can replace defective bones
- Printing of dental and other implants of higher strength and quality
- Reducing the weight of implants
- AM has been used to reconstruct fractured skull and nose
- Produce patient-specific organs and devices with suitable material for better matching and reducing chances of rejection and other postoperative infections
- Surface finish of various implants are found to be excellent
- For aesthetic surgery

1.9.1.2 Various AM techniques and materials employed for bioprinting

Jet-based techniques based on 3D tissue engineering have been observed as a viable option for printing of the structure of an organ layer by layer to form a cell scaffold. After that cell seeding (introducing cells of interest) is done directly into the scaffold structure. In some cases, researchers have printed organs by layerwise deposition of cell (or collection of cells) into a 3D gel with sequential maturation of the printed construct into perfused and vascularized living tissue or organ [15]. For complex organs such as heart, components, namely, valves are individually attempted to be developed and is a subject of active research. Some printed organs such as bladder and urine tubes are advancing toward clinical trials.

An important aspect of organ transplantation as well as bioimplants is its compatibility so that the same is not rejected by the body, and the postoperative complications can be avoided. The potential material candidates for biomaterials include synthetic polymers (curable), synthetic gels, and naturally derived hydrogels [16]. Besides, for 3D printing, materials such as alginate or fibrin polymers are used. The polymers are integrated with adhesion molecules that support the physical attachment of the cells. These polymers are employed to maintain structural stability as well as be open to cellular integration.

The first application of 3D printing in the biomedical sector was prosthetics and it has repeatedly seen success stories. The stability of custom-made total hip prosthesis can be enhanced through AM [17]. Moreover, the original functionalities of the joint can also be completely restored by this process. Combination of some degradable or allogeneic scaffolding with cellular bioprinting have potential to create biologic prosthetics that can serve as transplantable tissue. It has been reported that medical 3D-printing market might reach 983.2 million dollars by the year 2020 [18]. Figure 1.6 shows some of the 3D-printed human body parts and organs.

1.9.2 Application in construction industry

As AM has been developed, several of its processes seemed suitable for the construction industry to adopt and it did. It started from manufacturing of a few fundamental components of building such as aesthetically decorated parts. Gradually, constructor started using AM to make abstract structures that were very difficult or time consuming to produce through conventional means. When AM could print bigger parts, entire building has been printed through this technique. One of the advantages of this is that in case of construction in harsh environment, labors need not spend much time on site. Instead the construction is mainly undertaken offsite and just assembled onsite. AM can also shrink the supply chain when a part needs expedited delivery. Besides, customized parts can be easily obtained through AM. AM has the potentiality to bridge the skill gap necessary in construction industry. By the application of AM techniques, the job left is quite common and can be handled by an average skilled labor. Overall, AM allowed for faster, reliable, and accurate constructions especially for complex items.

The AM technologies that have been applied in construction industry mainly included [19] binder jetting, extrusion process, PBF, and DED. With the ability to handle larger varieties of materials, AM is found to be of potential to aid in the construction industry. The material that AM is required to deal with for the construction purposes are [19]:
- cementation material,
- polymer materials, and
- metallic materials.

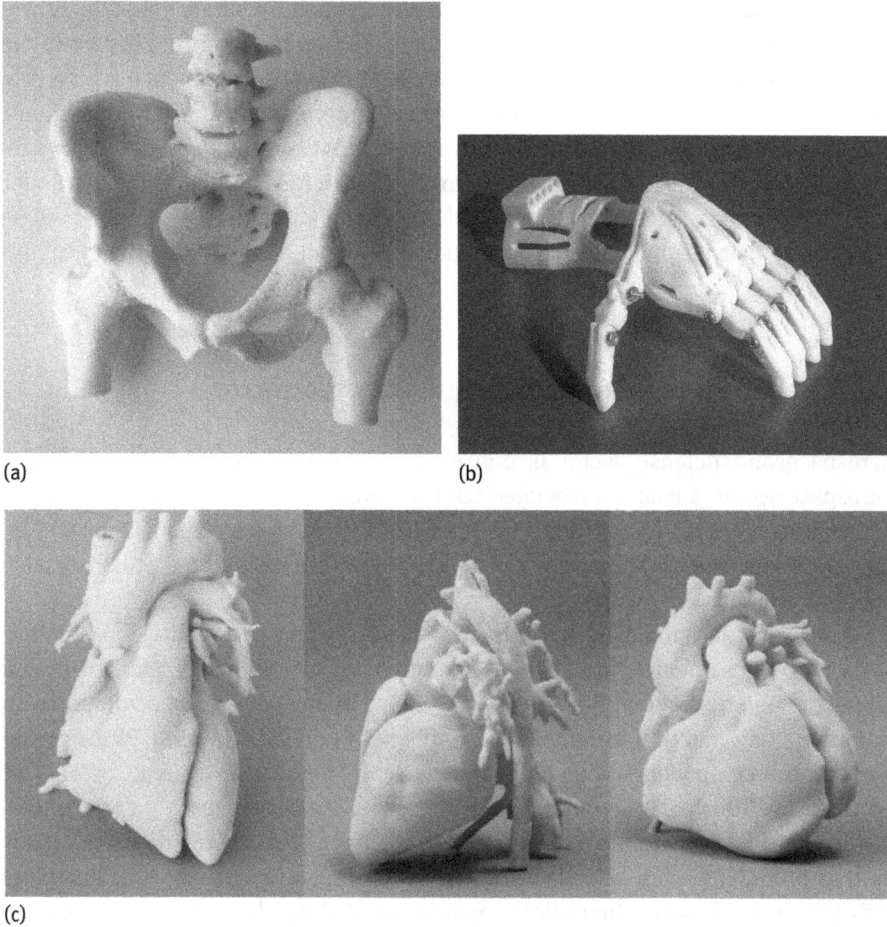

Figure 1.6: Three-dimensional-printed (a) human pelvis, (b) prosthetic hand, and (c) whole human heart models (creative commons).

AM in construction has made it possible to produce novel forms that were previously unimagined or difficult to realize. Architects can now produce more complex interiors and exteriors. Again in case of construction work, formwork accounts for around 35–60% of the total cost of concrete structure [19]. The cost is mainly associated with the realization of the formwork as well as its disassembly once the structure is strong enough to stand on its own. Now, as the complexity of the constructed structure increases, the complexity of the formwork also increases. More planning, design, and labor are required to realize the formwork, which increases the overall cost of the construction. Moreover, for specific geometries, formwork becomes specific, which implies it can't be reused in future construction and literally becomes waste. With 3D printing, the formworks can be easily manufactured and hence cost involved with concrete-based construction is surely to come down. Moreover, with AM, customized parts can be produced based on the choice of the customer increasing the customer satisfaction. Moreover, components can be produced on demand as has already been mentioned in one of the previous sections. Topology optimization is another aspect in which AM has the potential to contribute in construction sector. Through this engineers can freely optimize the design for maximum strength requiring minimal material even though the design may become complex. Details of topological optimization have been addressed in one of the future sections.

1.9.3 Penetration in aerospace and defense sector

Aerospace and defense sector is one of the early adopters of AM techniques. Aerospace sector is mainly associated with the production of complex parts and in low volume which is found to fall in line with the principles of AM. Through AM, aerospace and defense industries can now produce stronger parts, which are also lighter in weight.

Aerospace design generally begins with development of a model of an aircraft or its part. These models are also often subjected to aerodynamic testing in wind tunnels to obtain the preliminary data that helps in developing the actual aircraft. Material jetting and SLA are the AM technologies often used in these industries to produce the components with high dimensional accuracy, with extra detail and good surface finish. These qualities are necessary in an aerospace component so that it can fulfill its required objective and that too reliably. AM can be used for jobs such as producing a full-sized landing gear, nozzles, and engine and turbine parts with full detail. Besides the critical components, AM can be employed to develop aesthetic parts like interiors of the spacecraft.

With the implementation of AM in the aerospace sector, intelligent lightweight structures can be manufactured, which has resulted in weight reduction of 40–60% compared to the previous generation aircrafts. This has resulted in significantly less fuel consumption and also lesser CO_2 emission. Moreover, for longer space travel, there has been plans to carry a 3D printer, so that any critical part which requires replacement can be printed and used. Table 1.2 shows some of the recommended materials for aerospace applications along with their AM processing techniques.

In case of defense industry, various parts can be built in a cost-effective way. AM has already been used to manufacture parts of missiles and other systems. Besides, AM has been planned to be used for printing of parts required for replacement in case of damage in the battlefield. Besides, AM can be employed for producing regenerative medicine in case of grievous injury in the battle field.

Table 1.2: Recommended materials for applications specific to the aerospace industry.

Application	Requirements	Recommended process	Recommended material
Air ducts	Flexible ducts and bellow directors	SLS	Nylon 12
Cabin accessories	Customized functional knobs	SLA	Standard resin
Casted metal parts	Metal parts casted using 3D printed patterns	SLA and material jetting	Ca stable resin or wax
Metal components	Consolidated, lightweight, functional metal parts	DMLS/SLM	Titanium or aluminum
Full size panels	Large parts with smooth surface finish	SLA	Standard resin
Engine compartment	Heat-resistant functional parts	SLS	Glass-filled nylon
Lights	Fully transparent, high-detail models	Material jetting and SLA	Transparent resin
Bezels	End-use custom screen bezels	Material jetting	Digital ABS

1.9.4 Development of an enriched exchange format

The STL has been used for transferring information of CAD model and the software of the AM machine. This language was originally developed by 3D Systems in 1987. STL file is very simple and contains information only about the surface mesh, but has no provisions for expressing color, texture, material, and so on. Moreover, there is no provision to define physical units in STL file, which sometimes creates ambiguity. Although several file formats have been proposed over time, none could gain such popularity so as to replace STL as standard language of AM. There are primarily two reasons responsible for these failures: First, till now STL has been able to serve the AM community fairly well. There has been no requirement in terms of functionality, which is beyond the capacity of STL. Second, the proposed general formats included features many of which are irrelevant to the field of AM and hence make coding unnecessarily complex.

A few potential enriched exchange formats have emerged, namely, additive manufacturing file format (AMF), which have the provision to define material, composite, metadata, and STL 2.0 [18]. The development of AMF and STL 2.0 is significant particularly for the jobs requiring combination of intricate geometries and multiple materials [20].

1.9.5 Manufacturing interoperability

Most of AM machines can produce small parts in stand-alone systems. Implementing them in production line makes the things complicated as AM technologies are still not adapted to this [18]. Some companies have experimented by combining small AM machines to create a facility capable of printing large components that are made of ceramics and metals. This way the build speed could be increased along with the decrease in lead time. Moreover, supply chain is the prevailing organizational structure in the manufacturing sector. This structure consists of global network of suppliers, logistic providers, transporters, producers, retailers, and customers. These entities should be able to share technical as well as business information uninterruptedly among themselves. Although this information was previously communicated through paper mode, nowadays the same is communicated electronically and in a correct manner.

1.9.6 Topological optimization

Topological optimization is a mathematical tool that optimizes the material distribution in a design satisfying the strength requirements. This is required in order to fit the part in a given space or in general to save the cost. The philosophy of topological optimization is based on idea of removing material from a solid block in order to minimize or maximize say mass while satisfying a set of constraints such as maximum stress or displacement.

Topological optimization has been prevalent in aerospace automotive and construction industries. Although topological optimization results in significant cost savings, they were difficult and time consuming to implement using the traditional techniques of manufacturing engineering. With the introduction of AM that can directly generate the parts from the CAD model, implementation of this optimization has been relatively easier. It is reported that GE has employed AM to manufacture fuel nozzles as single components. This has resulted in 25% reduction in the weight of the nozzle as well as five times increase in strength [19]. It is further reported that AM-made nozzle exhibited higher combustion efficiency compared to the same manufactured by SM. In case of construction industry, topological optimization often results in increased complexity of the structure, for example, different sized beams to be made based on different load demands [19]. Before AM the designs were often optimized based on simplicity to manufacture which was favorable from the standpoint of lower time and cost, as well as minimized the chances of error. With AM, making customized parts was not of a problem anymore and hence constructors emphasized on topological optimization for construction. It has also been reported that some companies have carried out optimization on their constructions and the void space saved out of it was utilized for thermal installation and sound proofing [19].

1.9.7 Materials for AM

In case of manufacturing, the raw material should be supplied in a state compatible to the process. AM involving special processes is not an exception. First of all, plastics are still the dominant material in case of 3D printing though metal printing is gradually picking up pace. In case of vat polymerization, the build material should be liquid thermoplastic polymer which will get cured under the application of UV laser. Again in case of extrusion-based processes, the feedstock should be supplied in filament form so that the same could be extruded and melted to deposit on the build tray. Further for some materials only nozzle heating is sufficient, whereas for some material such as ABS, the bed also must be heated. Once the part is built, it is customary to subject the part to test condition to assess its performance before implementing it in service. Sometimes, AM-made parts are post processed to enhance microstructure and reduce porosity.

1.9.7.1 Polymers (thermoplastics)

Thermoplastic polymers are the material of choice for AM processes such as material extrusion and PBF. In both the processes, the polymers are melted and the bond strength between the layers is achieved due to adhesion between thermal layers, which is caused by the successful interdiffusion and re-entanglement of the polymer across the layer interface [21]. However, in case of extrusion processes, material supplied in filament form is extruded through the nozzle and gets deposited on the build tray. In case of PBF, the material should be given in powder form. Moreover, amorphous thermoplastics are preferred in case of material extrusion, whereas semicrystalline polymers are the choice for PBF [2]. The choice of amorphous polymers in extrusion process is due to their low shrinkage during solidification [22]. Polymers such as ABS and PLA (polylactide) are generally used in case of extrusion process. These materials are capable of softening over a wide range of temperature (till glazing temperature) and forms a high viscous semisolid that is ideal for extrusion through a nozzle of tip diameter ranging between 0.20 and 0.50 mm [2].

In case of PBF, UV laser is employed to melt and fuse the semicrystalline powdered feedstock. Nylon is one of the ideal feedstock materials for this process. Through optimal setting of the machine temperature, the material that is melted by the laser remains in a molten state as well as in thermal equilibrium with the rest unmelted powder. Material gets solidified after the build, which minimizes the chances of residual stresses. Now, in case of FDM, strength in x- and y-directions is greater than that in the z-direction. To make the process efficient, researchers are searching for a means to develop strength in the z-direction. This is vital in products such as load bearing prosthetics. A few of the popular thermoplastics are discussed below.

1.9.7.1.1 Acrylonitrile butadiene styrene

ABS is a thermoplastic polymer commonly used in 3D printers that employ FDM or fused filament and fabrication techniques. It is very popular as it has great plastic properties. Moreover, it is very light weight and has good impact strength besides being abrasion resistant. ABS can survive the attack by many chemicals. ABS has a relatively high melting temperature (above 200 °C) [23], which makes it resistant to warping and cracking. Thus, objects like casings are made by ABS. ABS is a typical material for making low-cost prototypes and models.

1.9.7.1.2 Polylactide

PLA filament is the popular material used in 3D printing. It is nontoxic (derived from natural sources such as sugarcane or corn starch) and is available in various colors and blends. It is also warp resistant but weaker than ABS. Thus, objects printed with PLA can be reworked to introduce additional features such as holes. Besides, PLA can give very good surface finish especially if grinded and polished. PLA is used to produce plastic films, bottles, and medical devices that need to be biodegradable. PLA contracts when heated and hence is also suitable to use as a shrink wrap material.

1.9.7.1.3 Nylon

Nylon is a polyamide and is strong and flexible. It can also resist chemical attacks and display splendid material memory. Nylon is, however, hygroscopic and the printing is preferably done in vacuum or at elevated temperatures. The melting temperature of nylon is quite high and the surface exhibits a very low coefficient of friction. This is the reason for it to be the material of choice for manufacturing of gears. Besides, nylon has found applications ranging from consumer electronic to adventure sports.

1.9.7.1.4 Polycarbonate

Polycarbonate (PC) is a thermoplastic polymer with carbonate group in its chemical structure. It is light weight, dense, and has splendid tensile strength. The impact resistance of PC is about 10 times more compared to certain acrylics. That's why PC is often the material of choice for making bulletproof jackets. Besides, they are scratch resistant and are used to manufacture data storage devices such as CDs and DVDs. Being heat resistant, this material is also used in parts that are subjected to elevated temperature during service conditions such as intake manifold. PC is also used in electrical components as they are known to insulate electricity. In case of PC, the temperature of the nozzle is required to be maintained around 260–300 °C [23].

1.9.7.1.5 Polyvinyl alcohol

Polyvinyl alcohol (PVA) is mainly used for making the support structures in case of overhangs or features such as internal holes/cavities. The reason behind this is that

PVA is water soluble and can be easily removed once the part is printed generally by dissolving in warm water. PVA is used in dual nozzle printers with some other materials such as PLA as the primary material. It is to be noted the adhesion of PVA with other materials such as PLA, ABS, or nylon is very good.

1.9.7.2 Metals

AM deals with thermoplastic largely to manufacture product. However, there was always a demand for metals to be used as a printing material so that working parts can be manufactured. With progress in research, commercial AM machines are available, which can handle metals, ceramics, glass, composites, concretes, and so on [23]. AM machines have also been developed, which can deal with biolinks mainly to create tissues and artificial organs.

Depending on the actual process used, metal feedstocks are in the form of powder or wire. The AM processes through which metal parts can be fabricated are SLM, DMLS, EBM, and so on, which are different types of PBF technology. Metals that are available in powder form are steel, titanium, alloys, aluminum, copper, super alloys, and others. Precious metals such as silver, gold, and platinum can also be found in powdered form. Availability of metals in wire feedstock option is even more wide-ranging. It is obvious that to be available in wire form the metal must have good ductility. Steel, stainless steel, and pure metals such as titanium, niobium, molybdenum, and aluminum are available in the form of wire feedstock.

In case of AM with metals, the same is sintered or fully melted to produce parts. When sintered, metallic powders partially melt and fuse with each other. In case of EBM, however, the powder is fully melted. This helps in reducing porosity which is helpful in case of performance in elevated stress and temperature conditions as in aerospace applications. Besides, elimination of porosity enhances the corrosive resistance of the metal. Figure 1.7 illustrates some metallic parts and components that have been made by AM. With metal AM, the prospect is huge as functional engineering parts can be produced very fast. A few well-known metals in AM are discussed in the following text.

1.9.7.2.1 Titanium

Titanium is one of the strongest and hardest materials available. High strength-to-weight ratio is one of the important features of titanium, which makes it suitable for aviation-based applications. Besides, titanium has the added advantage of resisting corrosion and is durable which also makes it the material of choice for implants in human bodies. Titanium is an apt material for AM-based manufacturing as the same is costly, and minimization of its waste is a necessity. EBM method melts the titanium

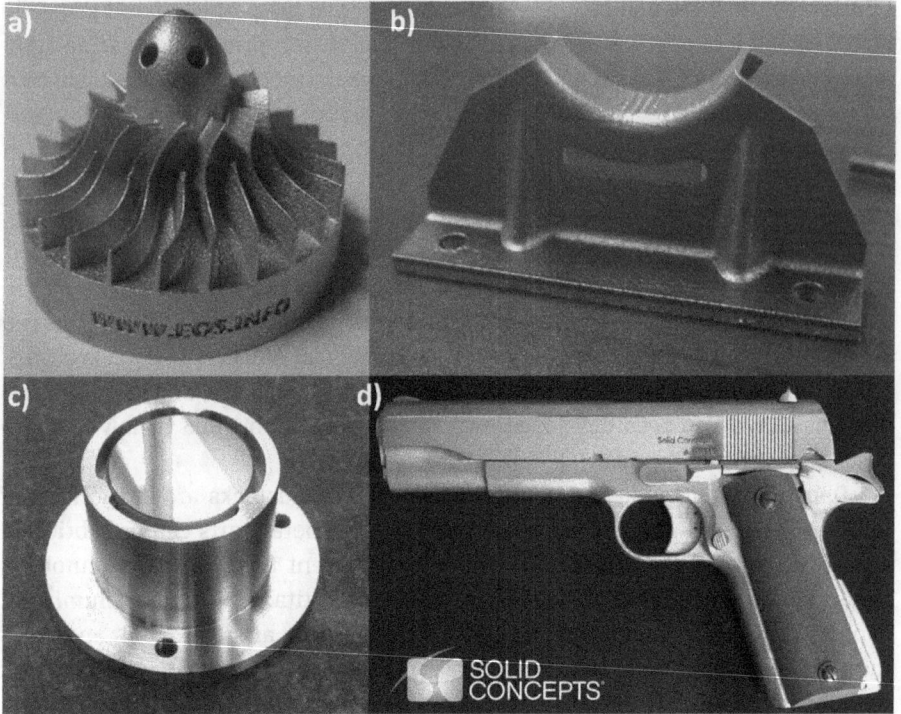

Figure 1.7: Various metallic objects made by AM: (a) turbine (Wikimedia Commons), (b) prototype of pump bracket for helicopter (image source: coastguard.dodlive.mil), (c) prototype of mirror mount (image credit: NASA/JPL-Caltech), and (d) 1911 pistol (image credit: Solid Concepts Inc.).

(supplied as wire) and deposits it on the build tray to manufacture parts of jet engines and turbines. According to Reuters [24], GE can save $3 million per 787 Dreamliner aircraft if it could apply 3D-printed titanium (and alloys) parts. GE has successfully realized a functioning miniature model of 3D-printed jet engine through 3D printing employing titanium components. NASA has demonstrated the elevated and low temperature-withstanding capabilities of titanium-based engine.

1.9.7.2.2 Stainless steel

Stainless steel is one the commonly used engineering materials. It possesses the strength of steel and at the same time is durable and corrosion resistant. Stainless steel is also ductile and biocompatible. EBM processes stainless steel powder to make components for jet engines and rockets. Besides, the material is also used to fabricate the parts of the nuclear facilities. 316L is low carbon steel, which has been found suitable for producing nuclear pressure vessels as it is weldable, corrosion resistant, and has high strength.

1.9.7.2.3 High-performance alloys

Superalloys are generally nickel-based alloys having a combination of high tensile strength, creep strength, and rupture strength. Besides, they have splendid fatigue corrosion and thermal fatigue resistance. Inconel 625 and 718 are two popular varieties of superalloys. Inconel 625 is good for load-bearing applications, where the temperature can reach above 800 °C for a short duration. Inconel 718 is suitable for aircraft turbine engines and ground-based turbines (components such as blades, casings, and fasteners). Besides, Inconel is also corrosion resistant and find use in naval-based applications.

1.9.7.2.4 Aluminum

Aluminum is a ductile, light, and corrosion-resistant metal. Being relatively soft and having a low melting temperature, aluminum is one of the early metals to have been processed through AM. Aluminum can be fabricated into parts by mainly DMLS where it is sintered and SLM where it is melted and fused. As aluminum can be made into sheets with thickness as low as 50 µm, it is a suitable material to be built by layer-wise technique which is the main philosophy of AM. Fine detailing is also possible when aluminum is used as the built material. Three-dimensional-printed aluminum parts have found use in automotive and racing cars.

1.9.7.2.5 Precious metals

Precious metals such as gold, silver, and platinum can also be processed though AM to turn them into jewelry. Partial melting and fusion of fine metal powder are carried out in DMLS machine according to the shape of the intended jewelry which is fed into the system through a CAD model. Once the process is complete the actual jewelry is taken out from the powder in which it remains encapsulated. Table 1.3 presents the metals used in AM according to their applications.

1.9.7.3 Ceramics

Ceramics are popular due to their superior mechanical strength particularly at elevated temperature conditions. Ceramics have been used in engineering applications such as aerospace, automobile, and energy especially in gas turbines, jet and rocket engines, heat exchangers, and battery. Now, many of these components are complex in shape and giving such shape to ceramics is challenging as ceramics are brittle, have low toughness, high crack sensitivity, and also have high hardness. Ceramics are also used for decorative purposes such as pottery and showpiece. Dentistry is another field where ceramics have found application.

Among AM techniques, PBF was one of the first technologies to have processed ceramics. Apart from that SLA, binder jetting, and sheet lamination are also used to

Table 1.3: Popular AM alloys and applications [3].

Alloys ⇒ / Applications ⇓	Aluminum	Maraging steel	Stainless steel	Titanium	Cobalt chrome	Nickel superalloys	Precious metals
Aerospace	X		X	X	X	X	
Medical			X	X	X		X
Energy, oil, and gas			X				
Automotive	X		X	X			
Marine			X	X		X	
Machinability and weldability	X		X	X		X	
Corrosion resistance			X	X	X	X	
High temperature			X	X		X	
Tools and molds		X	X				
Consumer products	X		X				X

make ceramic parts. Very intricately and finely detailed porcelain objects are preferably created through SLA. This is because SLA is known to produce high-resolution parts and ceramics are vulnerable during post-processing steps (in case modification is required). For this purpose, ceramic paste of ceramic powder added with a suitable photopolymer is employed. Curing and glazing processes take place as usual. The properties of the final product depends on the ration of ceramic particles and the photopolymer in the paste. Besides, there should not be large aggregates in the paste in order to produce defect-free ceramic parts.

The available ceramic materials have already displayed good compatibility toward the AM processes. Harder ceramic materials such as boron carbide (B_4C) and titanium boride (TiB_2), which have found applications in high value-added armors are also attempted to be made into compatible form so as to be compatible with AM processes. Figure 1.8 shows some gold and ceramic objects created by 3D printing.

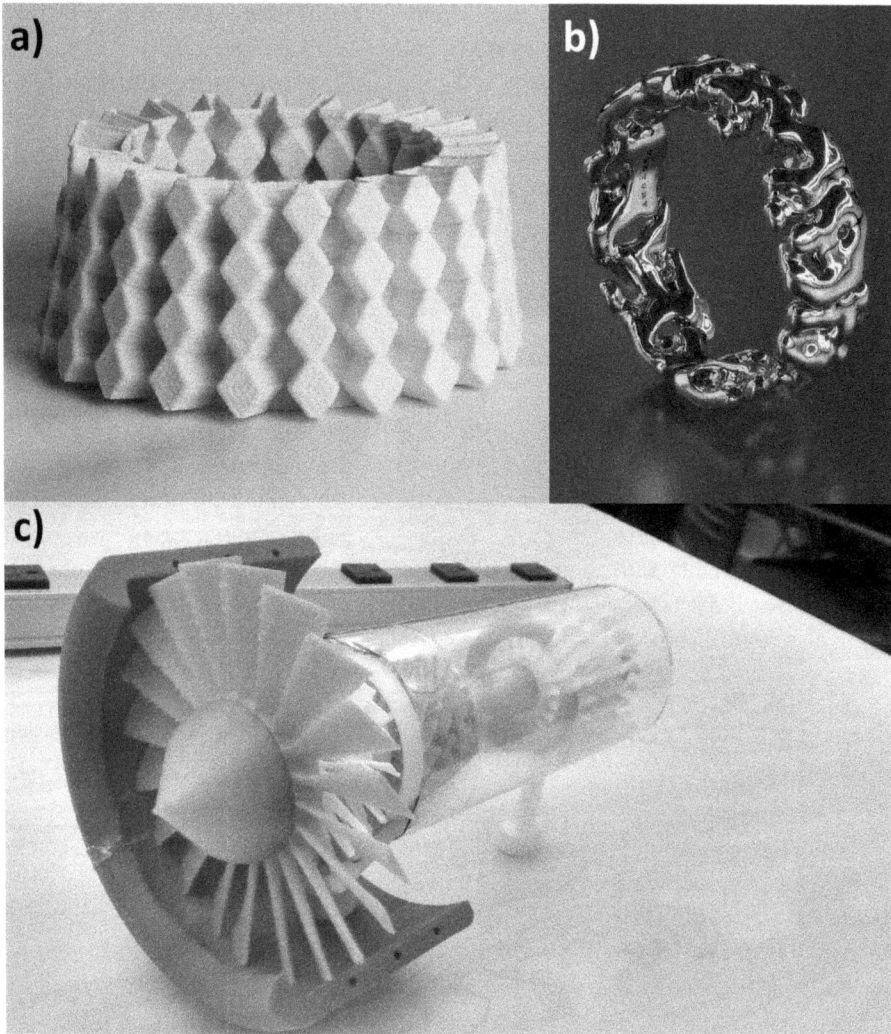

Figure 1.8: Three-dimensional-printed gold and ceramic objects: (a) ceramic showpiece (image credit: Studio Under, (b) gold jewelry (image credit: Emmanuel Touraine, edition Ventury), and (c) ceramic jet engine turbine model (image credit: CMitchell).

1.9.7.4 Glass

Glass parts are mainly made using SLS technique, where glass powders are partially melted and fused with each other. Glass has mainly found applications in optics as well as cutlery. Due to its stability under elevated temperature and high transmissivity, glass in the form of fused quartz is also tough to be a potential candidate for optical, communication, and electronic segment.

MIT' Mediated Matter Group has developed an AM method (G3DP) in which transparent glass is heated above 1,000 °C and tuned into molten stage. The molten glass is then sprinkled with the help of a nozzle aluminum zircon silica. It is to be noted that this system can make complex shapes. Hence, we can see that a lot of materials can now be processed through AM and the list is continuously expanding as the technological know-how increases.

1.10 Recent developments in subtractive manufacturing

The field of SM has not remained stagnant but progressed with the advancement in communication technologies such as Internet, cloud computing, robotics, and artificial intelligence. The availability of high-performance computing power, advanced hardware, and others have only propelled the CNC-based manufacturing forward. At present, SM can make intricate parts with relatively high precision. The manufacturing processes have become highly automated with minimum human interference, which also result in quick manufacture of components.

1.10.1 Effect of communication technology

Manufacturing is undergoing a digital revolution. All types of machinery – old and new – are being embedded with sensors, switches, and intelligent controls to generate data and send it over the Internet, all in the service of making factories smarter. Its the Industrial IoT (IIoT). The advent of optical fibers has also led to a jump in the data transfer speed and helped in the cause.

Attempts to make factories more intelligent are nothing new. Supervisory control and data acquisition (SCADA) systems have been on factory floors for years. Through networked data communications and graphical user interfaces, they gather data on the processes, send it to computers, and issue commands to connected devices. But SCADA systems don't "talk" to other systems, like logistics or production, nor is the data that SCADAs generate analyzed in any meaningful way.

A huge amount of useful data is trapped in factory floor machines. If captured, the data could be applied to improve operations, reduce costs, and make for a safer workplace. The ability to predict when a machine needs servicing instead of waiting until it breaks down, for example, could reduce overall maintenance costs by about 30%, and also could lead to nearly 70% fewer breakdowns, according to an IIoT report from Accenture, a global management consulting company. The idea behind IIoT is to connect independent things – machines, robots, and humans – and use that intelligence to get much more value from them together than you can individually.

1.10.2 Use of automation and robotics in manufacturing

Automation crept into the manufacturing sector along with NC machines. Gradually with the progress in technology, development in communication and networking, and advancement in artificial intelligence, the automation in manufacturing has undergone a sea change, which has also experienced huge improvement in factory production. The initial philosophy behind having automation in manufacturing is as follows:

– Increase of labor productivity and reduce labor cost
– Elimination of routine, clerical, and mundane jobs
– Mitigation of the shortage of skilled labor
– Improvement of worker safety by decreasing hazardous jobs
– Improving product quality by maintaining the production processes with a tighter tolerance
– Reduction of the lead time in manufacturing
– Accomplish jobs beyond the capacity of humans

Many of these estimations have been fulfilled with the introduction of automation in manufacturing. Over the last few decades, automation has transformed the industry floors, the economics of manufacturing, and has also affected the nature of employment made in manufacturing sectors. However, increasing demand for more efficient products led manufacturers go for more refinements in the automation techniques and also to introduce robots in manufacturing. At present, manufacturing sector is on the point of a new era in automation primarily based on progress in electronics and robotics, artificial intelligence, machine learning, big data, and IoT. This has enabled the manufacturing sector to perform with efficiencies like never before. In many cases, with the present automation, machines are able to outperform humans. These also include some of the tasks based on cognitive abilities. However, mindless automation may not prove to be cost effective in the long run. Hence, an optimal balance should be thought and automation should be provided in the rightful areas and with the required amount.

1.10.2.1 Introduction of robotics in manufacturing

Introduction of robots have completely metamorphosed manufacturing. Robot can move heavy objects and materials alone, which previously required a combination of humans and cranes. Besides, robots are able to perform various preprogrammed tasks in manufacturing in a repeated manner without slowing down due to fatigue. They are also engaged in dangerous jobs such as handling of toxic chemicals or hot parts that can cause injury to the worker. Moreover, robots can do jobs with high precision and accuracy, which leads to improved product quality. Such reliability is possible only through the application of robotics. Robots can also be regularly upgraded for better reliability and to suit newer jobs/methods. Robots also help in increasing the safety of the workplace. The primary limitation of integrating robots into business is their high cost of investment. Besides, the maintenance cost can also increase the overall cost of production. However, the long-term return on investment makes robots the perfect investment in manufacturing. The following are some trends that are driving the industry worldwide:

- *Material handling robots*
  The primary task assigned to the modern-day industrial robots is material handling. Material handling comprises movement of raw materials as well as transfer of parts, packing, palletizing, loading, and unloading. A new type of robots known as cobots (collaborative robots) are being introduced in the manufacturing sector. These robots can supplement human labors and work alongside them, thus increasing efficiency and reducing chances of hazards. Their potential to revolutionize production lines is growing day by day.

- *Welder robots*
  Welding is another popular job given to robots. The robots working here are mainly involved in spot welding and arc welding. Jobs: The advantages of welding robots are that they carry out their jobs with unmatched precision and repeatability, thus increasing the output as well as its quality. The weld robots are broadly of two categories: one uses a pre-fed set of instructions to carry out the job, and the second uses a machine vision or both to do their jobs. As robotic weld setups are becoming cheaper, it is becoming easier to install them and even relatively small manufacturers are introducing them in their production line. Welding robots are fast, efficient, and are also equipped with smart systems such as collision avoidance [25].

- *Robots for assembly*
  Assembly is one of the important jobs in any industry. Robots involved in assembly line are expected to do jobs such as fitting, fixing, inserting, tightening, and disassembling. The advancement in various sensor technologies such as force sensor, torque sensor, and tactile sensors has helped in making the assembly robots more efficient and have popularized their applications. In many cases, assembly robots are quicker and better in assembling parts than humans. Again the use of machine vision system makes the robot self-sufficient with respect to detection of correct component. Based on the intricacy of assembly, the robots are equipped with various sensing technologies so that the robot is cost-efficient. Assembly robots are nowadays very popular in automotive industries producing various automobiles such as car, truck, and bus. Figure 1.9 illustrates robotic assembly line for a car manufacturing unit.

- *Robots for dispensing jobs*
  Jobs like painting, spraying, and applying adhesive are done by dispensing robots. Compared to humans, these robots have better control over the fluid flow and can also dispense them in the exact location and in the required amount.

Figure 1.9: KUKA robots in car assembly line (creative commons).

Dispensing robots decrease manufacturing time but help in improving product quality. Robots can also be used to remove paints/layers from objects. Besides, these objects can also be used for surface preparation, grinding, and smoothening from metallic surfaces. The material finish of such components is quite high. Moreover, robots are fast and help in increasing production rates, which ultimately lead to higher profit. Workers are also not exposed to harmful dust and fumes when removing material.

– *Robots for inspection*
Robots have also been employed for inspection and quality checks such as detection of flaw, error, and fault. The basis of the inspection robots is also mainly machine vision, which is becoming increasingly powerful. Inspection robots can detect the correct part and can measure to see whether it matches the dimension specified. Within a reasonable tolerance, these robots are very effective.

With the associate accessories becoming cheaper, the affordability of manufacturing robots is increasing. Even small-scale farms are also installing these robots to increase their productivity. Thus, there is a high demand for these robots that are now produced in masses. These robots use standard parts that are straightforward to install. Communication between the robots are also very good making them perfect for assembly line installations. The robots are also durable and rugged, which attract minimal maintenance cost. Hence, it won't be exaggerated to say that robots are the future of manufacturing.

1.11 Additive and subtractive manufacturing: applications

Since the discovery of AM a few decades back, AM has grown into a mature technology today and its market has also grown significantly. AM that started as a means to build porotypes rapidly can now produce functioning models. The application of AM is found in variety of industries ranging from space to toy, which epitomizes billion dollar industry [7]. The decreasing cost of AM machines is one of the reasons for its penetration in the manufacturing sector. The machine that can be bought in 500 dollars today costs around 1 lakh dollar three decades back [7]. In future, AM will be more focused in producing real-life functional products for which the research and development of the AM processes and associated materials is underway. A few significant applications of AM are given in Table 1.4. SM has been with mankind since the early days of civilization, and its application touches the life of each and every individual in this planet. Hence, its application has not been listed separately in this chapter.

At present, AM processes are more suited to high value products but produced in low scale. AM disregards unit labor cost and traditional economies of scale. However, AM introduces flexibility in the manufacturing system with possibilities of mass customizations in the design within a very short period of time. Again the initial cost of investment for installing AM system is quite high even today and it requires a highly skilled person to operate. Moreover, there is still a dilemma whether the entire decision making should be handed over to machines. Thus, it can be said that the objective of AM is not to replace SM as beating the latter in terms of cost-effective mass production is still hard. Rather AM should be used in a justifiable manner and to make parts that are either difficult or not cost effective to be manufactured by SM. This type of balanced application of AM and SM is already showing potential through the application of them by global enterprises such as Airbus, Boeing, GE, Ford, and Siemens [7].

1.12 Environmental impact

According to the Brundtland report, sustainable development "meets the needs of the present without compromising the ability of future generations to meet their own needs" (United Nations 1987). Sustainable development is based on three pillars such as economic development, environmental development, and social development. These pillars are also popularly known as people, planet, and profits. There is interaction between these entities and they also affect each other as in real-world application. Sustainable manufacturing in case of industries imply decreasing load on the environment. Environment awareness is also spreading among customers, which

Table 1.4: Applications of AM [7].

Sector	Applications
Car industry	– Integration of many parts in a unified composite part – Construction of production means – Production of spare parts and accessories – Fast standardization
Aerospace/aeronautics	– Production of accessories of complex geometry – Control of density, mechanical properties – Production of lighter accessories
Construction industry	– AM of concretes for conventional building – Novel design of functional concretes such as self-cleaning concrete, high-performance concrete – Building construction using materials available in the locality – Cement-free building – Low-cost, low-energy building – Very fast building
Medicine/ pharmaceutical industry	– Planning of surgical operation with the use of accurate anatomic models that are based on computed tomography or the magnetic resonance imaging – Development of adjustable orthopedic implants and prosthetics – Use of printed simulated corpse for medical training in anatomy – Printing of biodegradable living tissues for tests during the development phase of the medicinal product
Sports industry	– Production of accessories of complex geometry – Creation of adjusted protective equipment for better application and use – Creation of prototypes of multiple colors and composite materials for product testing

also influences their decision regarding buying. There is even a pressure on the industrialists to manufacture sustainable products. Moreover, the environmental legislations are also becoming stringent day by day. Thus, in order to satisfy all, manufacturers are forced to make their manufacturing processes aligned toward sustainability and at the same time stay competitive in the market.

Any company wishes to reduce the costs of their product in order to avoid higher prices. It is wise to keep low prices at early stages of product development as the same decides the future cost of the product. Now, raw materials and energy costs account for nearly half of the total cost of the product [26]. Thus, the resources must be optimized judiciously for a company to be sustainable as well as stay competitive simultaneously. To identify the environmental impact of different stages in the life of a product, life cycle assessment (LCA) is used. Through LCA, systematically one can reduce the transfer the effect form one stage to other stage of the product's lifecycle. This in turn helps in understanding the benefits and hazards of ALM (advanced laser material). In a nutshell, LCA approach quantifies all the basic stages of a product's life cycle from "cradle-to-grave," in terms of in-flows and outflows of materials and energy. This implies the material and energy that the product/process consumed and emitted.

1.12.1 Impact of conventional machining

Machining consumes a lot of energy. However, the energy consumed can be difficult to measure and estimate for various components manufactured through different production processes and environments. It is reported that CNC milling machines consume 85% of the energy for running the various parts of the equipment and hence electrical requirement per component is inversely proportional to the material removal rate [27]. It is further reported that for cutting steel, CNC milling machines required only 7% more energy than running in air [28]. This is due to the energy consumed by the auxiliary equipment, namely, coolant pumps and compressor. Thus, percentage of utilization of machine tool is very significant with respect to environmental impact. It does not imply that if the machine tool processes a part faster, it is environmental friendly. But it has to be ensured that the machine rather than remaining idle processes parts as is feasible for it. Other factors such as usage of water and release of toxins also have to be considered to truly estimate the effect of the tool on the environment. Release of cutting fluid is one of the environmental concerns for machine tools. Cutting fluid provides necessary lubrication and cooling effect at the part tool interface. It facilitates proper chip removal and corrosion protection. However, after machining, the disposed cutting fluid which is many times toxic causes multiple environmental impacts. Apart from this worker injury due to accident is another concern. Even though much of the processes have been automatized, but human involvement has not been completely eliminated. Thus, operators are vulnerable to occupational hazard.

1.12.2 Impact of additive manufacturing

As demand for manufacturing processes that are environment friendly is growing stronger by the day, AM processes are also scrutinized and compared with more traditional forms of manufacturing, such as machining and injection molding. Now, some inherent advantages of AM include reduced material wastage, higher material efficiency, greater part flexibility, as well as manufacturing flexibility [29]. However, AM and its associated process have some impact on the environment and many of which have not yet been studied. As AM is an energy-driven process, some of the works in the literature have considered this aspect. It has been studied that electrical energy consumption is directly dependent on the duration of the job in 3D printing

[30]. Moreover, the most significant parameters is the production time and minimizing the same for each AM system was recommended.

A few studies have been conducted, which deal with the material toxicity associated with AM process. Faludi et al. [27] have reported the toxicological and environmental hazards as a result of handling, using, and disposing of the materials employed in AM. It has been observed by Drizo and Pegna [31] that most materials used in RP processes can't be reclaimed. They specifically tell that in SLA, the whole vat of resin becomes unrecoverable after exposure to UV light. Again in case of SLS, the ratio of fresh material to recycled powder is around 20–50%. Thus, one can get an idea about the generated waste after each build.

1.13 AM versus SM: when to choose between the two?

Although there are fundamental differences between AM and SM, they are not mutually exclusive. In many cases, both are used alongside each other in various stages of manufacturing and development of the object. For example, in case of development of a prototype, initial models during the concept development phase may be easier and convenient to develop in plastic and using AM processes such as SLA or SLS. This is because plastic-based models apart from being cost-effective are also suitable to produce smaller and complex parts through AM route. Even functioning prototype can also be made to check the design. However, during the later stages of the development of the product, SM processes are preferable as larger batches of product need to be manufactured. Besides, the availability of various surface finishes and speed of manufacturing are some of the reasons for choosing SM over AM. The procedural differences have been sufficiently highlighted in the previous discussions. However, the differences that a manufacturer would look out before implementing AM in the production line are listed in Table 1.5.

1.13.1 Choosing between AM and SM

With SM in the background, AM has evolved into a somewhat mature technology today. Manufacturing sector has now gained enough experience to decide which process is the best for a particular job. And most of the time it is seen that manufacturers can't choose a single technology for once and all. Instead SM and AM processes are often found to complement each other especially in the manufacturing of jig, fixtures, brackets, mold templates, pattern, and other tooling. However, there are some aspects that should be kept in mind so that effective utilization of the two manufacturing methodologies can be achieved. There are four broad considerations that may serve as a guideline for making decision.

Table 1.5: Differences between AM and SM from manufacturer's perspective.

Heads	AM	SM
Equipment costs	Starting price of professional desktop 3D printers (plastic) – $3,500 Starting price of industrial AM machines (metals) – $400,000	Starting price of small CNC machines for workshops – $2,000 Prices of advanced industrial-grade machines depend on the number of axes, features, part size, and tooling needed for specific materials
Training	Minor training required to operate desktop printers on setup, maintenance, operation, and so on Industrial AM systems need dedicated staff and extensive training	Small CNC machines need moderate training for software, job setup, maintenance, machine operation, and finishing Larger, industrial SMs require dedicated staff and extensive training
Facility requirements	Desktop printers can be adjusted in tables available in office and benchtop systems in a workshop environment (moderate space) Industrial AM machines need a dedicated space or room with HVAC control	Small CNC machines are suitable for workshops Industrial systems require a larger, dedicated space
Ancillary equipment	Tools and (some automated) systems for cleaning, washing, post-curing, and finishing depending on the process	Various tooling. More advanced systems automate some processes like tool changing, chip clearing and handling, and coolant management

1.13.1.1 Material aspect

1.13.1.1.1 *For plastics:* AM
The initiation of AM was with manufacturing of plastic parts. AM can handle all the common plastics as well as some special variants as well. As already discussed, the list includes but not limited to:
- ABS
- PLA
- PVA
- Nylon
- Resins

In the recent past, ceramics and metals have also been included in the list. But these materials need special high-end machines for processing, which are still out of the reach of the mass and are suitable to take care of all production needs.

In case of low-volume casting methods, the silicone mold can be made by AM and then used to cast different materials.

1.13.1.1.2 *For larger variety of materials:* **SM**

SM can make products out of almost any material. It is a proven and rugged technology, which mankind has been using for ages. SM should be preferred mainly for the following materials:

- Metals and alloys such as aluminum and steel
- Wood
- Foams like polystyrene or structural foam
- Ceramics

SM offers a lot more flexibility in terms of materials compared to AM. In case of manufacturing a large part, it may be wise to go with SM than producing it with AM. Besides, similar material giving cost effectiveness in manufacturing through the other route is always preferred.

1.13.1.2 Quantity of parts to be manufactured

1.13.1.2.1 *For low-scale manufacturing:* **AM**

AM is a fast manufacturing technique but its unit production cost is small only for low-volume production runs. Thus, AM methods are suitable for making prototype and concept models and in quantities of 1–10 parts. AM, however, can produce customized parts even in the same run cost effectively, which may not be economically feasible using SM methods such as turning or milling.

1.13.1.2.2 *For large-scale manufacturing:* **SM**

For producing in large scale or for mass scale production, SM is still the preferable approach. With introduction of CNC machines and automation, the production speed of SM has increased considerably. It can now produce parts with high dimensional accuracy and in a repeated manner. Moreover, molds for injection molded components which are made of steel are produced through SM.

1.13.1.3 Geometry of part and surface finish

1.13.1.3.1 *For intricate geometry:* **AM**

When the geometry of the part is very complex, AM is the preferred choice as it produces the part directly from the CAD model unlike the need of the cutting tool (of SM) to enter into very thin internal features. AM can produce parts that are impossible to be generated by SM. However, the part accuracy, dimensional resolution is governed by factors such as minimum layer thickness and the type of material. Besides, surface finish is a bit of concern for parts manufactured by AM. To address these issues, high-end AM machines with high quality materials are required, which again increases the cost of production.

1.13.1.3.2 *For high-dimensional accuracy and repeatability:* SM

To manufacture the commonly encountered features such as holes, threads, flats, and counterbores, SM is still the preferred choice which gives high accuracy and repeatability. Besides, SM can produce very good surface finish in metallic parts.

1.13.1.4 Schedule of project and part revisions

1.13.1.4.1 For quicker lead time and prototype iterations: AM

AM processed parts have faster lead time compared to machined parts. Application of AM in the production process can decrease the production cycle as well as allow multiple revisions of the design more quickly compared to SM. In case of SM lot of time is eaten in setting up the machine such as reprograming the tool path and setting up of work piece. The best practice would be to do all the preliminary design development using AM and move toward SM for mass production only when the design is freezed.

1.14 What does the future say?

1.14.1 Combining additive and subtractive manufacturing: hybrid techniques

Each of the manufacturing processes has its own merits and demerits. Often it is found that one process complements the other and hence give an optimal situation when acting in unison. This experience has led manufacturers to combine two or more manufacturing processes giving rise to hybrid system. In case of AM and SM too hybridization is seen to give promising results. However, before elaborating the discussion of the exact philosophy of hybrid manufacturing process, the following statements are made [32]:

- In a hybrid manufacturing system, several manufacturing technologies are combined to yield the optimal output. In case the constituent processes belong to the same manufacturing technology, the combination is defined as subhybrid process.
- In case of hybrid manufacturing the constituent processes should be interactive toward each other, which means that the effect of changing one process should

influence the subsequent process and the final output. For example, in case of laser-assisted turning, initially the laser processing makes the material soft which makes subsequent turning quite easy and fast [33]. Thus, each of the processes shall have effect on the other and shall be interdependent.

- The processes constituting the hybrid manufacturing should directly act on the workpiece which is being produced.

As already discussed, one of the criteria for choosing between AM and SM is to consider the complexity of the part [34]. AM is preferable when the part is of highly complex geometry, whereas CNC machining is recommended for parts having conventional shapes and features. This is because CNC can produce parts within a very tight dimensional tolerance, which is desired criteria in many critical assemblies and systems. This philosophy, that is, choice of the manufacturing process taking into account the complexity of the job and the productivity is depicted in Figure 1.10.

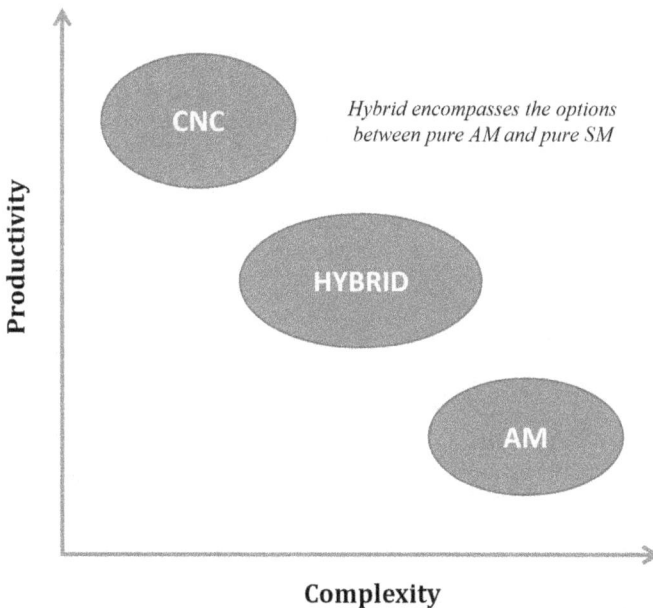

Figure 1.10: Hybrid manufacturing to be optimal in terms of productivity and complexity [34].

However, part complexity should not be the only criteria for judgment as it does not take into account all the advantages and limitations of both the processes. In case of SLS, build powder is sintered by laser layer by layer to produce the part. Now the same part can be manufactured one order of magnitude quicker by CNC machining [34]. Now, again the speed of AM can be increased by having a higher energy input, which results in thicker layer to be deposited. But with thicker layer the surface finish may not be acceptable as illustrated in Figure 1.11. Thus, there exists a trade-off between productivity and surface finish in this case. Even if the process is slowed down to get better surface finish, many times a post machining session may still be required to get the desired finish on the parts. Thus, instead of being competitive, both AM and SM are now seen as complementary techniques, which have a higher gain in terms of productivity. Thus, by hybridization, the two technologies can be merged to have the benefits of both the techniques in the production process. By suitable planning and judgment it is possible to manufacture parts in required quantity, with high resolution and quality and at the minimum time and cost. This would be the true realization of hybridizing both AM and SM.

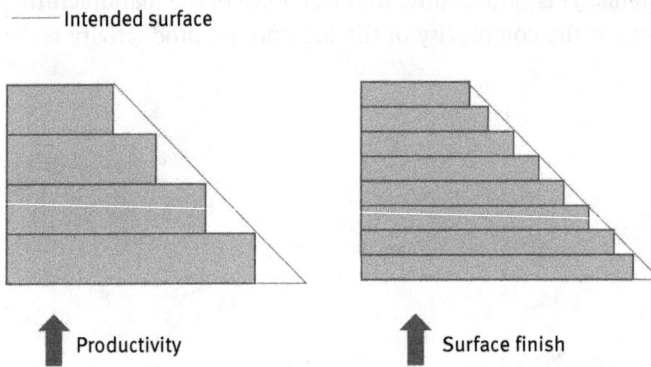

Figure 1.11: Processing speed versus surface finish: process dilemma for AM.

1.14.1.1 Applications

One of the examples cited by ASME about hybrid manufacturing is to reduce the idle time of expensive CNC machines by the incorporation of 3D printing heads into the CNC machines. This way on demand the CNC machine can transform into a state-of-the-art 3D printer and its usability can be increased. However, it is to be noted that this method is not the exact merger of the two techniques. A better refinement would be to at first print the part using AM and then decide whether to go for milling operation. This type of convertible systems is useful for small industries who need not invest in two separate machines. Some of the applications of hybrid manufacturing are described as follows:

1.14.1.1.1 Laser cladding and SM
Some researchers have attempted to retrofit traditional milling machines with a laser cladding unit. The main objective of this is to incorporate the flexibility of

laser cladding as well as to have the capability of producing higher surface finish using milling and at the same time reduce the setup time [32]. Jeng and Lin [35] fabricated metal and alloy injection molds by conducting laser cladding and milling operations in series. The incorporation of a five-axis laser cladding unit with a five-axis milling machine has been reported in the literature [36]. In this setup when it is required to deposit some feature in the horizontal direction the entire workstation can be rotated avoiding the need of supporting material (during deposition) and reducing the processing time. Five-axis CNC machine that has normal responsibilities such as drilling, milling, and grinding has been employed to machine parts made by laser cladding [37]. Mognol et al. [30] conducted topological and dimensional analysis and suggested the features that can be produced by laser cladding and the ones produced by high-speed milling.

1.14.1.1.2 Arc welding and SM

The process works on the same principle as the previous combination, and only the laser cladding is replaced by arc welding [32]. Couple of gas metal arc welding guns has been used to deposit various materials, and CNC milling has been used to make injection mold inserts [38]. Face milling has been employed to machine each layer built by metal inert gas and metal active gas welding [39]. Xiong et al. [40] studied the mechanism of plasma arc deposition and integrated the plasma torch on a milling machine.

1.14.1.1.3 Shape deposition manufacturing and SM

In shape deposition manufacturing (SDM), molten metal is deposited along with the support material. Upon cooling the metal gets solidified and support material is removed. Cooper et al. [41] reported to have employed SDM to drop wax beads to build shape of the part roughly. Then the rough shape is cured and finally milling is done to obtain the final dimensions of the part. Lanzetta and Cutkosky [42] employed the combination of SDM and milling to build smooth and sculpted 3D contours of dry adhesives, which could be used to aid human and robotic climbing.

Hybrid manufacturing also has the potential to result in significant cost savings by salvaging the scrapped parts in machining due to machine crash or breaking of the cutting tool. Instead the part can be scanned to assess the damage and if found suitable could be repaired using AM deposition techniques.

1.14.1.2 Hybrid system challenges

Although a hybrid system seems to be an optimum solution for the modern day, manufacturing needs balancing of both speed and productivity. However, some of the challenges in its implementation are as follows [34]:

- **Tethering** – In some machines the cladding head has multiple tethering to laser source, to powder feed supply and the inert gas supply. This limits the flexibility of users in changing the nozzles as well as moving between adding and removing material.
- **Laser concerns** – As lasers require clean environments to operate, adequate measures must have to be taken regarding this.
- **Alignment** – In case of some of the retrofitted machines, the nozzle axis may be located off the spindle center line creating two different tool center points. If the machine tool is not very accurate, moving from AM to SM or vice versa may introduce error.
- **Light tolerance** – Traditional CNC machines are not light tight. Thus, if a laser is fitted to it, safety of the workers may be an issue.
- **Media abrasiveness** – Presence of powdered metal from AM system may get mixed with coolant and lube oil and enter into rotational and sliding joints in the machine. In this scenario, heavy wear and tear may occur, resulting in premature failure of ball screws and guide ways.
- **Ancillary equipment** – Laser and nozzle are not the only equipment that needs to be fitted. Along with them, powder feeders, fume protection systems, and cooling circuits also need to be provided which may make the system very complex.

1.14.1.3 State-of-the-art hybrid manufacturing systems

It has been reported that with the development and application of multiaxis 3D printing machines, many of the present limitations of AM can be eliminated and the part quality could be improved [43]. These machines can give higher degrees of mobility (4–6 dof) compared to commercially available AM machines and techniques (3 dof max). The presence of these additional degrees of mobility can reduce the manufacturing time since lesser setup changes are required. Moreover, higher part quality can be expected as the parts need not be relocated and repositioned frequently eliminating the risk of stacked tolerances. Figure 1.12 shows a multiaxis arm whose head can be interchangeably made into AM head and SM head. Figure 1.13 shows both surface printing and milling operations running on the machine.

It is already discussed that through hybridization of AM and SM, material waste can be reduced, and part quality can be improved as well as productivity could be increased [43]. Future trends in hybridization include development of novel process planning methodologies toward improving the environmental sustainability, dimensional performance, and reducing the manufacturing cost. In addition, closed-loop control methodologies can also be investigated to further improve the quality of parts [43]. A hybrid machine produced by Mitsui Seiki can swap laser heads similar to swapping of cutting tools [34]. Material removal by laser ablation through a

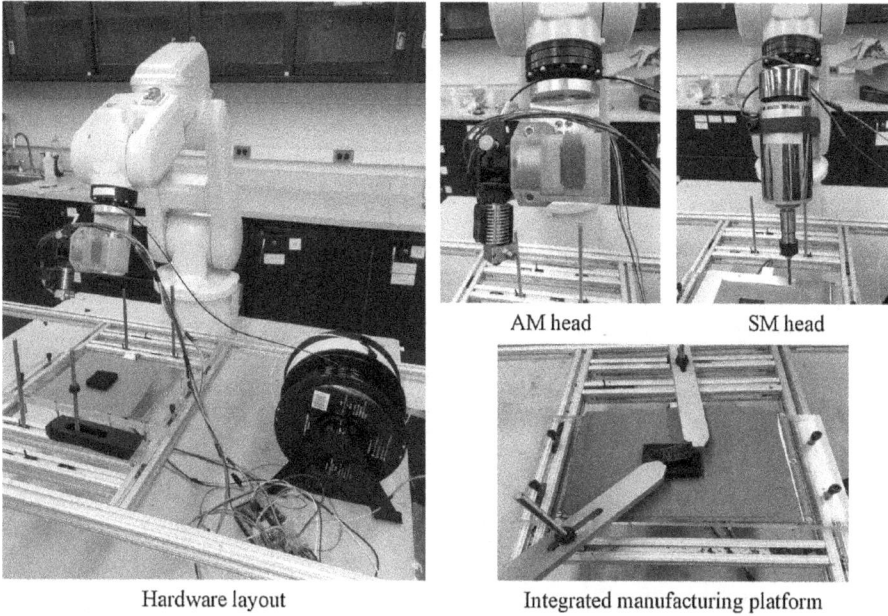

AM head   SM head

Hardware layout   Integrated manufacturing platform

Figure 1.12: Hardware layout of hybrid additive subtractive manufacturing process [43].

(a)   (b)

Figure 1.13: Freeform surface printing and milling: (a) freeform printing and (b) freeform milling [43].

micromilling process requires 100 W of power. A higher capacity lase of say 2 kW is gain required for laser deposition for AM. Now, scaling down of 2 kW laser to 100 W is difficult, which severely affects the flexibility of the system. In future developments, attempts may be made to develop systems that can switch between high-power and low-power laser systems as per the requirements. Future systems could do laser ablation to micromill components, perform additive deposition, and traditional and laser-based metal removal.

1.15 Closure

SM has been the basis of progress and development of mankind so far. However, since the inception of AM around 50 years back, the technology has been constantly expanding, growing, and advancing with much enthusiasm. The manufacturing sector has seen the abilities of AM and has went all out to develop it and too quickly. Results are that many new types of RP and AM methods have been created and printing of metallic parts has become a reality. But it has been realized by the manufacturing sector that without a revolution in the technology, AM can't become the sole manufacturing process around. Currently the best utilization of AM is seen as to complement the conventional machining process. Thus, hybridization of both the techniques have come up, which is found to give optimal solutions with respect to productivity. In reality, the future of additive lies in a hybrid manufacturing system, one that combines additive and subtractive techniques for ultimate optimization. Moreover, due to the advancement in robotics, communication, and associated technologies, smart manufacturing is now a reality. All these have led society to see the fourth industrial revolution, that is, Industry 4.0.

References

[1]   ISO/ASTM, S., Standard Terminology for Additive Manufacturing – General Principles – Part 1: Terminology. 2015, ASTM: West Conshohocken, PA.
[2]   Bourell, D., et al., Materials for additive manufacturing. CIRP Annals, 2017, 66(2), 659–681.
[3]   DebRoy, T., et al., Additive manufacturing of metallic components – Process, structure and properties. Progress in Materials Science, 2018, 92, 112–224.
[4]   Bandyopadhyay, A.E. and Bose, S.E., Additive Manufacturing. 2016, Boca Raton: CRC Press.
[5]   The Evolution of CNC Machining. December 23, 2018; Available from: https://www.baronma chine.com/news/the-evolution-of-cnc-machining/.
[6]   Langnau, L. Subtractive Manufacturing: What You Need to Know. December 20, 2018; Available from: https://www.makepartsfast.com/2011-make-parts-fast-handbook-subtractive-prototyping/.
[7]   Tofail, S.A.M., et al., Additive manufacturing: scientific and technological challenges, market uptake and opportunities. Materials Today, 2018, 21(1), 22–37.
[8]   Jörg, B. and Richard, K., Additive manufacturing: A long-term game changer for manufacturers. December 15, 2018; Available from: https://www.mckinsey.com/business-functions/operations/our-insights/additive-manufacturing-a-long-term-game-changer-for-manufacturers.
[9]   Ben, R. Additive Manufacturing Technologies: An Overview. December 10, 2018]; Available from: https://www.3dhubs.com/knowledge-base/additive-manufacturing-technologies-overview#/vat-photopolymerization.
[10]   Muthu, S.S.E. and Savalani, M.M.E., Handbook of Sustainability in Additive Manufacturing: Vol. 1. 2016, Singapore: Springer Nature.
[11]   Opsdog. An Introduction to Manufacturing Process Flow Charts and Workflows. December 12, 2018]; Available from: https://opsdog.com/categories/workflows/production.

[12] Prakash, K.S., Nancharaih, T., and Rao, V.V.S., Additive Manufacturing Techniques in Manufacturing -An Overview. Materials Today: Proceedings, 2018, 5 (2, Part 1): 3873–3882.

[13] Javaid, M. and Haleem, A., Additive manufacturing applications in orthopaedics: A review. Journal of Clinical Orthopaedics and Trauma, 2018, 9(3), 202–206.

[14] Javaid, M. and Haleem, A., Additive manufacturing applications in medical cases: A literature based review. Alexandria Journal of Medicine, 2018, 54(4), 411–422.

[15] Mironov, V., et al., Organ printing: computer-aided jet-based 3D tissue engineering. Trends Biotechnology, 2003, 21(4), 157–161.

[16] Skardal, A. and Atala, A., Biomaterials for integration with 3-D bioprinting. Annals of Biomedical Engineering, 2015, 43(3), 730–746.

[17] Rahmati, S., Abbaszadeh, F., and Farahmand, F., An improved methodology for design of custom-made hip prostheses to be fabricated using additive manufacturing technologies. Rapid Prototyping Journal, 2012, 18(5), 389–400.

[18] Badiru, B.A., Valencia, V.V., and Liu, D., Additive manufacturing handbook: product development for the defense industry. Systems Innovation Series, ed. B.A. Badiru. 2017, Boca Raton: CRC Press Taylor & Francis.

[19] Delgado Camacho, D., et al., Applications of additive manufacturing in the construction industry – A forward-looking review. Automation in Construction, 2018, 89, 110–119.

[20] Paterson, A.M., Bibb, R.J., and Cambell, R.I., Evaluation of a digitised splinting approach with multiple-material functionality using Additive Manufacturing technologies. in Annual International Solid Freeform Fabrication Symposium – An Additive Manufacturing Conference, 2012.

[21] McIlroy, C. and Olmsted, P.D., Deformation of an amorphous polymer during the fused-filament-fabrication method for additive manufacturing. Journal of Rheology, 2017, 61(2), 379–397.

[22] Drummer, D., Cifuentes-Cuéllar, S., and Rietzel, D., Suitability of PLA/TCP for fused deposition modeling. Rapid Prototyping Journal, 2012, 18(6), 500–507.

[23] What materials are used in Additive Manufacturing? December 5, 2018]; Available from: https://www.ge.com/additive/additive-manufacturing/information/additive-manufacturing-materials.

[24] Alwyn, S. Printed titanium parts expected to save millions in Boeing Dreamliner costs. December 10, 2018]; Available from: https://www.reuters.com/article/us-norsk-boeing/printed-titanium-parts-expected-to-save-millions-in-boeing-dreamliner-costs-idUSKBN17C264.

[25] Len, C. Robots in Manufacturing Applications. December 14, 2018]; Available from: https://www.manufacturingtomorrow.com/article/2016/07/robots-in-manufacturing-applications/8333.

[26] Nörmann, N. and Maier-Speredelozzi, V., Cost and Environmental Impacts in Manufacturing: A Case Study Approach. Procedia Manufacturing, 2016, 5: 58–74.

[27] Faludi, J., et al., Comparing environmental impacts of additive manufacturing vs traditional machining via life-cycle assessment. Rapid Prototyping Journal, 2015, 21(1), 14–33.

[28] Diaz, N., Redelsheimer, E., and Dornfeld, D., Energy Consumption Characterization and Reduction Strategies for Milling Machine Tool Use, 2011, Berlin, Heidelberg: Springer Berlin Heidelberg.

[29] Huang, S.H., et al., Additive manufacturing and its societal impact: a literature review. The International Journal of Advanced Manufacturing Technology, 2013, 67(5), 1191–1203.

[30] Mognol, P., Lepicart, D., and Perry, N., Rapid prototyping: energy and environment in the spotlight. Rapid Prototyping Journal, 2006, 12(1), 26–34.

[31] Drizo, A. and Pegna, J., Environmental impacts of rapid prototyping: an overview of research to date. Rapid Prototyping Journal, 2006, 12(2), 64–71.

[32] Zhu, Z., et al., A review of hybrid manufacturing processes – state of the art and future perspectives. International Journal of Computer Integrated Manufacturing, 2013, 26(7), 596–615.

[33] Sun, S., Brandt, M., and Dargusch, M.S., Thermally enhanced machining of hard-to-machine materials – A review. International Journal of Machine Tools and Manufacture, 2010, 50(8), 663–680.

[34] Robb, H. Hybrid system combines additive subtractive manufacturing. December 15, 2018]; Available from: https://www.aerospacemanufacturinganddesign.com/article/hybrid-system-combines-additive-subtractive-manufacturing/.

[35] Jeng, J.-Y. and Lin, M.-C., Mold fabrication and modification using hybrid processes of selective laser cladding and milling. Journal of Materials Processing Technology, 2001, 110(1), 98–103.

[36] Zhang, J. and Liou, F., Adaptive Slicing for a Multi-Axis Laser Aided Manufacturing Process. Journal of Mechanical Design, 2004, 126(2), 254–261.

[37] Hur, J., et al., Hybrid rapid prototyping system using machining and deposition. Computer-Aided Design, 2002, 34(10), 741–754.

[38] Song, Y.-A. and Park, S., Experimental investigations into rapid prototyping of composites by novel hybrid deposition process. Journal of Materials Processing Technology, 2006, 171(1), 35–40.

[39] Karunakaran, K.P., Sreenathbabu,A., and Pushpa, V., Hybrid layered manufacturing: Direct rapid metal tool-making process. Proceedings of the Institution of Mechanical Engineers, Part B: Journal of Engineering Manufacture, 2004, 218(12), 1657–1665.

[40] Xinhong, X., et al., Hybrid plasma deposition and milling for an aeroengine double helix integral impeller made of superalloy. Robotics and Computer-Integrated Manufacturing, 2010, 26(4), 291–295.

[41] Cooper, A.G., et al., Automated fabrication of complex molded parts using Mold Shape Deposition Manufacturing. Materials & Design, 1999, 20(2), 83–89.

[42] Lanzetta, M. and Cutkosky, M.R., Shape deposition manufacturing of biologically inspired hierarchical microstructures. CIRP Annals, 2008, 57(1), 231–234.

[43] Li, L., Haghighi, A., and Yang, Y., A novel 6-axis hybrid additive-subtractive manufacturing process: Design and case studies. Journal of Manufacturing Processes, 2018, 33, 150–160.

Moisés Batista, Ana Pilar Valerga, Jorge Salguero,
Severo Raul Fernandez-Vidal, Franck Girot

2 State of the art of the fused deposition modeling using PLA: improving the performance

Abstract: One of the main processes within the additive manufacturing is fused deposition modeling, due to its versatility, its high capacities, and its low cost. However, this process has certain restrictions that enhance its expansion in the industrial sector such as porosity in the structure and reduced surface quality, in addition to the anisotropy generated by the trajectories and deposition of the layers. Also the current trend in the world is to reduce the consumption of petroleum-based polymers, so alternatives must be found in biodegradable materials to move the process toward an ecosustainable industry.

Therefore, this chapter compiles an introduction to the state of the art of this technology of such potential with the use of a biodegradable polymer, polylactic acid. It considers the characteristic defectology of this process, the main parameters, as well as the possible applications in which it is implanted and in which it could be implanted. Finally, different possibilities are considered for improving the surface quality of the parts generated with this technology by means of different types of post-processing.

Keywords: Fused Filament Fabrication, Polylactic Acid, Green manufacturing, Biodegradable polymer, Sustainable industry, Manufacturing parameters

2.1 Additive manufacturing

Production is understood as the production of objects through the transformation of a starting material into a product through the application of energy [1, 2]. With the arrival of the industrial revolution at the end of the eighteenth century, the way of seeing manufacturing was completely changed. Craftsmanship was left to be able to use manufacturing processes and automation in series, thanks to the application of the steam engine in production. It is considered that this is the first industrial revolution, based on the acquisition of energy through coal. A second revolution is distinguished in the late nineteenth century, with the use of oil and electricity, and even a third when talking about the use of nuclear energy combined with any of the previous (twentieth) century. The rest of innovations could be considered as a technological revolution (robotization, renewable energies, etc.).

https://doi.org/10.1515/9783110549775-002

It could be considered a technological revolution, the invention of a new form or manufacturing process, as is the case of additive manufacturing (AM). Although it should not be understood as industrial revolution by itself, since it does not vary the sources of energy subtraction, but it offers the possibility of making parts of high complexity regardless of any tooling [3].

ASTM has defined AM as a "process of adding or joining materials, usually layer by layer, to create objects from 3D CAD models, unlike their opposite the machining, in which material is cut or removed." Some synonyms of AM are 3D printing (3DP), rapid manufacturing, additive processes, layered manufacturing, or freeform manufacturing [4]. This definition is applicable to all classes of materials, including metallic, ceramic, polymeric, composite, and biological systems [5].

AM is understood as the group of manufacturing processes that, from a three-dimensional virtual model (VM), add material sequentially and selectively layer by layer at those points where it is needed. This means that from any computerized digital model, its physical representation can be obtained through the use of these technologies. This manufacturing could be understood as a technology that reproduces any object that is desired as long as it can be designed using CAD (Computer Aided Design) software [1, 6].

On the other hand, AM is the formal term acquired by the technique that is used to be called rapid prototyping (RP) and what has been colloquially called 3D printing, the latter being somewhat incorrect, since the term 3D printing comes from 3DP that corresponds to a specific family of processes and not to the entire branch. The term rapid prototyping is used in a variety of industries to describe a process quickly and to once create the representation of a product before the final release or commercialization. In other words, the emphasis is on the creation of something quickly and that the output is a prototype or base model from which other models are derived and, ultimately, the final product. AM, however, is a much broader term that encompasses many more applications [2, 7, 8]. This means that the technology of AM can be present at any stage of the life cycle of a product, from the prototyping or reproduction of a product, to mass production or repair of parts [3, 9].

Currently, AM equipment is commonly known as 3D printers, which are commercially available for less than $500, and also allows the manufacture of 3D objects even for domestic use. In the same way, the development of digital 2D printing together with desktop publishing has revolutionized communication and information technology, and the development of AM technologies together with the "Internet of things" has the potential to revolutionize the computer-aided manufacture of complex objects and multifunctional material systems. While conventional manufacturing is governed by processing constraints related to industrial mass production, AM is intrinsically agile, allowing a more rapid change in the design and manufacture of custom objects designed to meet the demands of people and specific applications [10, 11]. In addition to this agility, the AM is a set of processes more ecological, since it generates less waste in both the manufacture itself and for the generation of tooling (Figure 2.1).

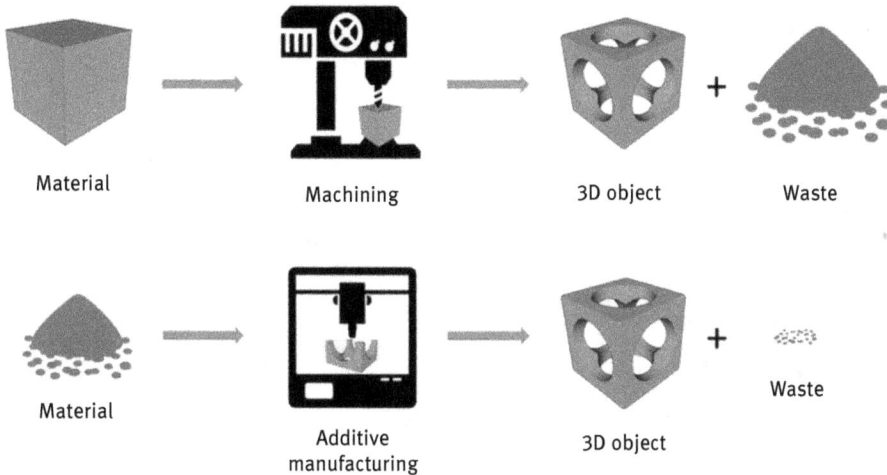

Figure 2.1: Additive manufacturing process versus conventional machining.

2.1.1 Evolution of additive manufacturing

While this technology emerged more than four decades ago, it has recently been appreciated as the important commercial manufacturing technology that it is. For this reason, numerous processes, methods, and materials within this broad group that includes the term AM are currently booming and developing. This is mainly due to the expiry of patents [12]. From 2009 to 2015, most of the key patents of the mostly used processes have expired, such as FDM (fused deposition modeling), SLS/DMLS (selective laser sintering/direct metal laser sintering), and SLA (stereolithography). This has caused, and will continue to do so, a greater diffusion of these technologies for any person or entity that wants to introduce it in the market.

All this means that, although a multitude of processes of AM have been developed and improved since the end of the 1970s, the first publication referring to the use of these technologies did not appear until the 1980s [13], and there is a very pronounced growth in the number of investigations and publications based on said lapse of patents (Figure 2.2).

AM is at the center of the strategic discussions of the European Research Association in the Future Factories of the Future Research Association due to its ability to customize parts, its high flexibility, and its efficiency in the use of resources [14]. For this reason, the development of this type of technologies has experienced a great increase since they were introduced in the market, leading a set of families of manufacturing processes that are alternatives to conventional manufacturing. This has generated a paradigm shift in the way of conceiving manufacturing. As mentioned, this set of technologies in its beginnings was called rapid prototyping.

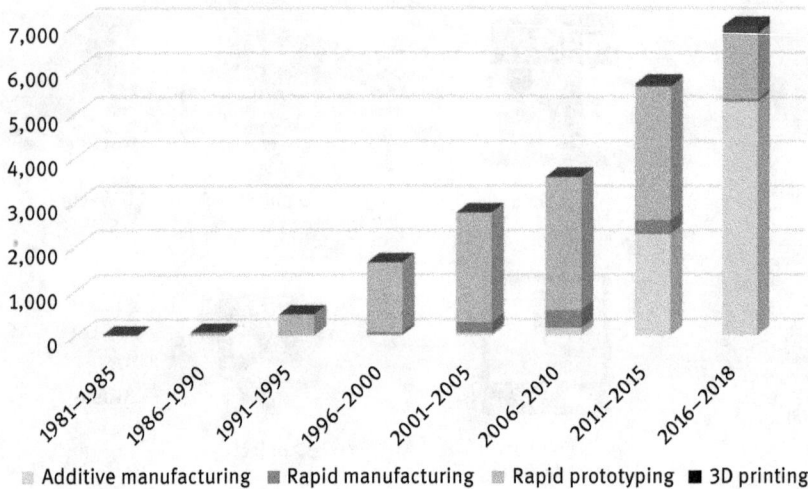

Figure 2.2: Increase of research in the field of additive manufacturing according to its different terminologies included in the *Journal Citation Reports*.

However, today there are different nomenclatures for these processes, some mentioned, and that do not mean the same [15–18].

As shown in the graph, if there is a small number of articles, a boom begins in the twenty-first century, which is strongly increased with the expiration of several patents. In turn, it is distinguished that the increase in studies mention the AM, being less and less the percentage of investigations that make reference to the term rapid prototyping.

Among the most studied aspects, as in many other types of processes, is the increase in quality, as well as the reduction of time and costs. AM still does not reach the quality of other processes. However, there is already the integration of AM processes and machining, although they are expensive and scarce solutions in the market yet. In addition, it has begun to create free software, portals where designs are shared, forums on AM, and so on. Today it is possible to find drawings of machines made with Lego®, aluminum profiles, wood, and others, or even buy kits to make your own machine.

All the above causes an increase in the capital invested by the company in this area. This great expansion in the market and the increase in the existing competition have caused a marked decrease in the costs of machines and consumables, which in turn favors the appearance of studies, greater investment, and in general, a greater interest of society, thus increasing the capital invested in both products and services [19] (Figure 2.3).

The Wohlers 2016 report states that the industry of so-called 3D printers exceeded 5.1 billion dollars, with more than 278,000 desktop 3D printers (less than 500 dollars) that were sold worldwide in 2015. In 2018, it is estimated that these

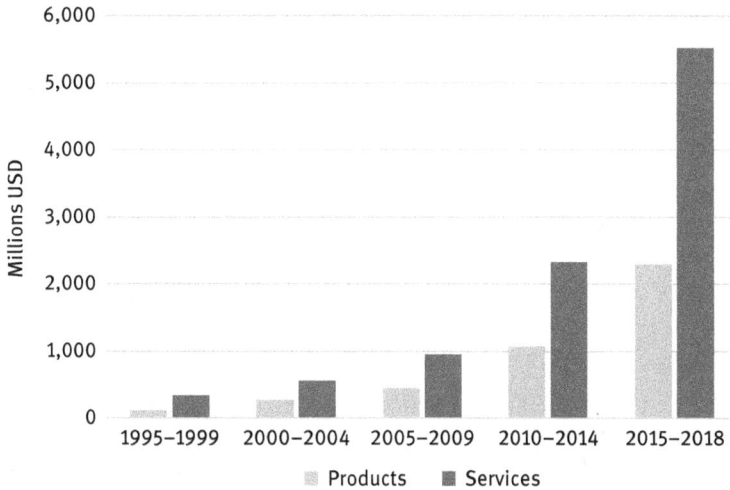

Figure 2.3: Global income of AM products and services between 1995 and 2018.

data have almost doubled, and may be even higher, since despite extensive research and conducting surveys it is difficult to control, for example, the purchase of kits and spare parts [19]. This means that new users can quickly make contact with this family of processes, causing industrial environments to focus their attention on these processes.

The AM industry increased to 17.4% in world revenues in 2016, compared to 25.9% in the previous year. Much of the recession came from the falls of the two main systems manufacturers in the business. Together, they represent 1.31 billion dollars (21.7%) of the industry of 6.063 billion dollars invested in AM. If these two companies were excluded from the analysis, the industry would have grown by 24.9% [20].

On the other hand, numerous processes and variants of them have now been developed within the large group of AM. The one that has been considered first of these techniques was SLA, although it was not distributed by a large company, 3D Systems®, until 1987. Another of the first techniques invented was SLS. This was patented in 1979 and started to be marketed and distributed by EOS GmbH® in 1990.

The most common AM technology was and remains that of FDM. It was invented and patented in 1989 by S. Scott Crump and marketed in 1991 by the company it co-founded, Stratasys®. This merged with a leading market company, Objet®, and became the largest AM equipment and materials factory that exists today [20]. Its high expansion is due to the low cost and easy use and installation.

From the 1990s, AM has undergone innumerable changes, generating many different techniques and variants of them, allowing more and more types of materials and better qualities, although it is still a field with much capacity for improvement in tolerances and qualities, as well as in manufacturing times and costs [21].

This number of processes and variants causes a complex classification. In this way, there are different classifications. One of the first ways to classify the processes of AM takes into account the purpose of the parts (rapid casting, rapid prototyping, rapid tooling, rapid manufacturing). On the other hand, these processes are classified based on the nature of the materials (polymers, waxes, metals, etc.). At the same time, other researchers order the processes of AM taking into account the material addition (powder or liquid bed, injection, lamination of solid material, etc.).

In 2009, ASTM International Committee F42 on Additive Manufacturing Technology defined a series of terms to distinguish these technologies from their analogues for manufacturing [22] (conformation and subtraction of material) and classify the different processes within the AM: material extrusion, material jetting, binder jetting, sheet lamination, Vat photopolymerization, and powder bed fusion.

If the different classifications are grouped, a classification could be established that encompasses all aspects. For this reason, a classification based on current but more complete standards is proposed, which takes into account the nature and contribution of the material, as well as the forming technology. Figure 2.4 shows a diagram of the main processes of AM classified according to the mentioned criteria.

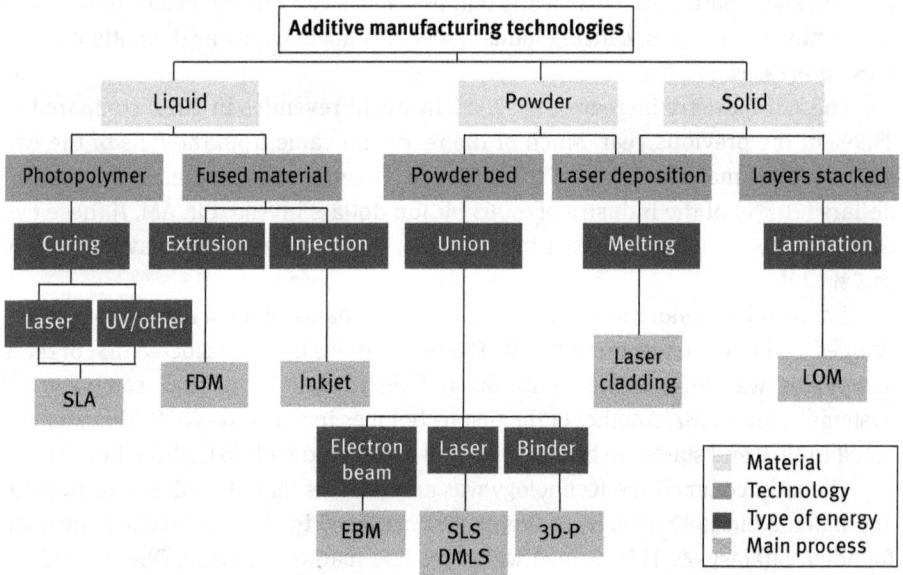

Figure 2.4: Classification of the main processes of additive manufacturing.

The following section provides a brief description of the main processes of AM destined to the conformed plastic or polymeric materials, since they are the most extended and developed at present.

2.1.2 Manufacturing of parts

Regardless of the process of AM chosen, there is a series of common steps in the creation of a part [2] (Figure 2.5).

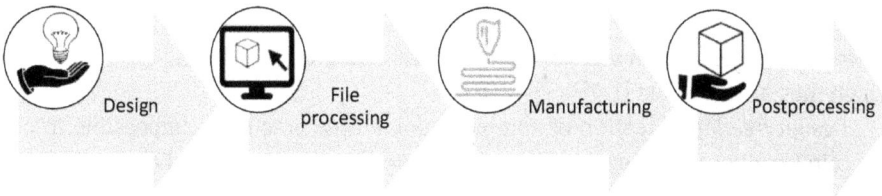

Figure 2.5: Generic flow diagram in the process of additive manufacturing.

In all cases, it is based on a VM, which can be designed or obtained by means of three-dimensional digitalization techniques. In a design for AM, two general aspects must be taken into account as in any manufacturing process: functional design and design for manufacturing. In the case of AM, this consideration is very important since it allows to optimize the functional performance of the parts. This causes a series of advantages such as saving weights, times, or costs, but also the appearance of a series of aspects to be taken into account such as the use of supports, the anisotropy, or the internal structure.

The model must be transformed into an exchange file format that is understandable by the treatment software. The most common is the use of STL (STereoLithography or Standard Tessellation Language) files. The said format consists of a closed surface composed of a mesh of triangles that defines the geometry of an object. The smaller the size of triangles, the higher the resolution obtained in the final file. Currently, there are more specific formats for AM, such as the AMF (additive manufacturing file) or 3MF (3D manufacturing format), which offer greater benefits (less space, data storage, textures, etc.). However, the use of the standard format against these others still prevails.

Before and after this transformation, it is possible to repair the meshing, especially when starting a scan from a 3D scan. If the mesh is not correct, the processing of the file cannot be carried out. Once the final file of the model has been obtained, it must be sectioned in layers (slice) in CAM software. The result will be a G-Code file that includes the manufacturing trajectories of each layer and the parameters that define the manufacturing conditions. The height of that layer as well as the main manufacturing parameters vary according to the technology used.

One or several postprocessing treatments are required to the manufactured part, depending on the technology used. These treatments can be tasks for surface finishing requirements or the extraction of the supports. In some processes such as

3DP it is also necessary to infiltrate products. Also in other processes such as metal laser sintering, thermal operations are carried out to relieve stresses and improve the properties of the part, and even its machining [23].

2.1.3 Advantages and disadvantages

There are numerous advantages in these technologies, depending on the use to which they are intended [1, 7]:
- *Design freedom*. Creation of complex geometries, practically impossible to obtain by other manufacturing processes (very thin walls, internal channels, interior angles, etc.).
- *Flexibility*. Changes in the original model can be automatically implemented to production, without the need to adapt molds or tools.
- *Integration of components*. Assembly operations, welds, joining elements, and others can be eliminated. This represents a reduction of lean time, cost, and weight.
- *Stock reduction*. At any time a spare part can be manufactured without having to carry out orders and waiting times, and there is no need for spare parts.
- *Environmental sustainability*. No toxic chemicals are used, the amount of material waste is reduced, and they also allow the introduction of biodegradable materials.
- *Variety of materials*. Depending on the requirements of the application and the type of additive technology that is used, there is great diversity in metallic, polymeric, ceramic, and even bioorganic.
- *Weight reductions*, thanks to the design of lattice-type internal structures, impossible to manufacture by other technologies.

Despite the great number of advantages that these processes offer compared to other materials forming technologies, there are certain drawbacks derived from these processes. There are some specifications to each technology, but some of the common and most important ones are briefly explained as follows [1, 7]:
- *Properties*. There are technical limitations to achieve certain properties or qualities in the final product without the need for post-processing operations.
- *Certification*. Being relatively new technologies, there is still a deficit in the development of standards and in the certification of materials, processes, and final products.
- *Training*. Currently there is a lack of knowledge of these technologies in all phases of the life cycle of a product.
- *Costs*. The cost of raw materials, machinery, software, and personnel remains high due to the relative novelty of these technologies.

- *Lean time.* They are slow manufacturing technologies compared to other processes. This does not apply to other times such as preparation times.
- *Limitation in manufacturing volume.* There are limits of maximum and minimum sizes of manufacture. In this type of processes, the manufacturing time is triggered for excessively large volumes.
- *Design problems.* Most current CAD/CAM programs are not yet designed for AM. The true innovation and distinction that can be obtained with these processes is not adapted in the great majority of these software. In many cases, these software do not have advanced functions, such as controlling the porosity or gradual density of a solid. In the same way, the current simulation and analysis programs carry out their activity under the hypothesis of homogeneous materials, although this is not the case for products manufactured by layers, since they present different behaviors according to the construction axis (anisotropy).
- *Post-processing.* Depending on the technology they are more or less expensive, but most of them are necessary.

2.1.4 Application sectors

There is a wide range of applications for these technologies, depending on the process, material, and parameters chosen. It can be used to cover objects, to repair areas of damaged parts, to add material with special characteristics in an area, to create prototypes, or directly to create pieces from scratch. Above all, these technologies are used in sectors where the value added to the product is important, personalization, products that are difficult (or impossible) to obtain through other processes, or in the initiation of a novel element in the market. The main areas that use these technologies are reflected in Figure 2.6.

The medical industry covers approximately 20% of all the use of these technologies. It is therefore the sector that takes advantage of these technologies. Medical implants of titanium or custom polymers and intelligent ceramics, artificial organs, orthopedic products, bioactive bone, structures to favor the growth of skin tissues, as well as surgical tools are examples of elements currently manufactured with these technologies [24] (Figure 2.7).

The group of consumers is referred to people with their own machine that makes countless elements, even personalized products, and mainly involves low-cost machines. Although it is currently an important sector with a great expansion, the industrial inclusion of these processes will relegate it to other minor positions.

On the other hand, the sector of molds and dies within industrial machines/tools, also known as rapid tooling, uses AM for its ease of making internal cooling channels and with free geometries, hybrid molds, and so on. Sometimes the technology is not used to manufacture the mold in its entirety, but it is applied to some

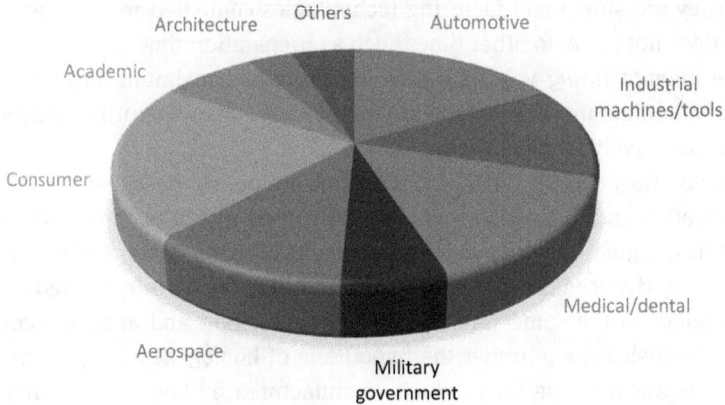

Figure 2.6: Main additive manufacturing sectors in the last decade.

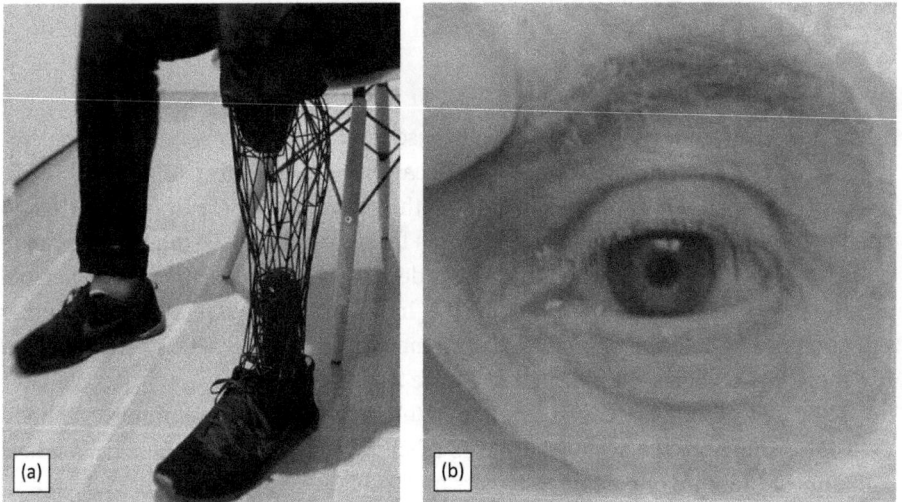

Figure 2.7: Additive manufacturing applied to medicine [25]: (a) exo-prosthetic leg and (b) eye prosthesis.

areas, inserts, since otherwise its manufacture would be very complex or the term would not be competitive.

In the case of the jewelry, textile, furniture, or art industry in its most general way, it mainly takes advantage of additive manufacturing in terms of freedom in design and the low infrastructure required to pass from being designer to be at the same time manufacturer and marketer of the design (Figure 2.8). For example, today there are machines developed especially for the jewelry sector, which allow to obtain masterpieces that are then cast to obtain the final piece in the desired material [26].

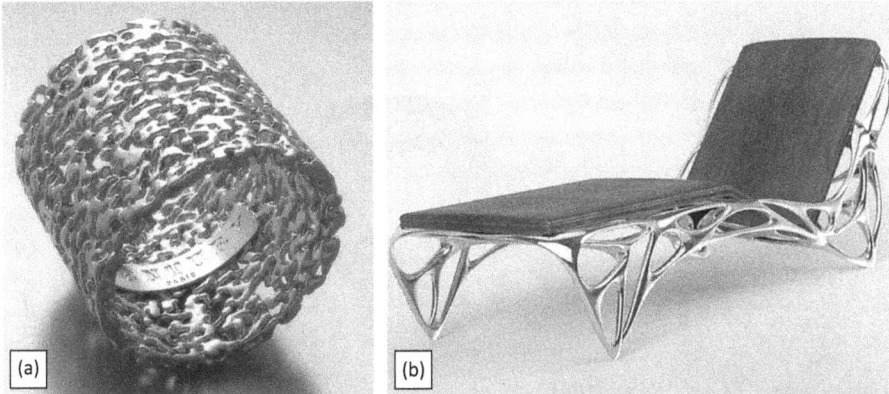

Figure 2.8: Examples of design parts: (a) jewelry part [27] and (b) furniture [28] by Emmanuel Touraine.

On the other hand, education is a sector that is increasingly making use of these technologies, since the use of prototypes is more explanatory than a simple photograph or video as support in teaching (Figure 2.9).

Another sector in which there is a need for an optimal compromise between mechanical strength and weight is the aerospace. In addition, the customization and the need to use complex geometries make AM an interesting process in the production of certain aeronautical components (Figure 2.10). Some examples are lightened structures, with internal channels, spare parts in space, all low-volume products, custom pilot equipment, and so on.

All this means that, in sectors where it is necessary to manufacture small batches or customized pieces, or very complex pieces [29], AM already has a clear and expandable market niche. In the same way, it is faster in terms of design, preparation, setup of the machine, and programming than other conventional processes; therefore, in sectors where flexibility or agility is required, it is a real alternative despite being slower than other methods in manufacturing. For this reason, it is currently used to make prototypes, pieces in small batches, or products that cannot be obtained through other technologies (Figure 2.11).

In addition, the possibility of repair or manufacture of interchangeable components in places of low availability of storage added to the advantages already mentioned causes its expansion in any sector. An example of this happened years ago in the United States Navy, where the first 3D printer was installed aboard the USS Essex. Another example is carried out by NASA, where the ability to print expeditionary structures designed to measure on the ground and using locally available materials even in space is exploited (Figure 2.12).

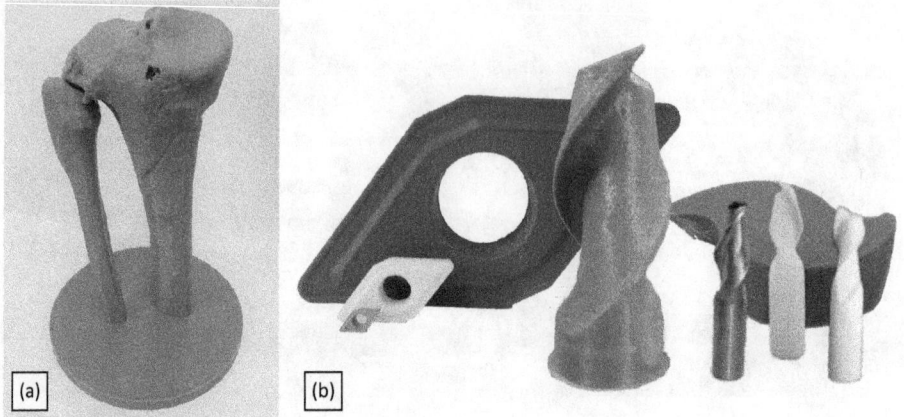

Figure 2.9: Example of pieces for the teaching sector of different subjects: (a) medicine and (b) manufacturing.

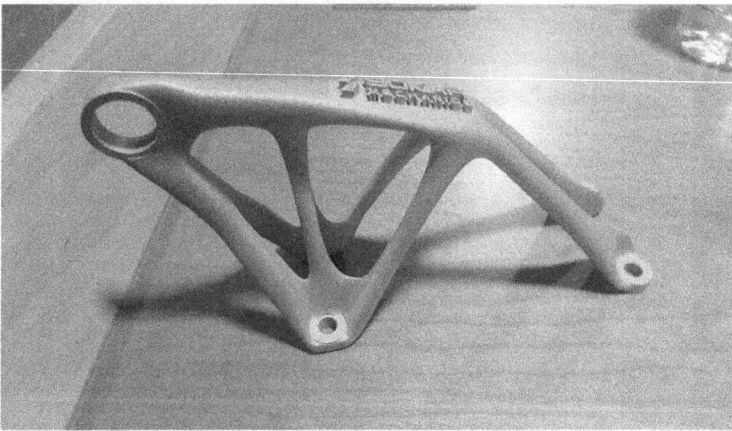

Figure 2.10: Aerospace structure created by additive manufacturing and topologically optimized.

AM has a great potential in many industrial sectors, replacing or complementing traditional manufacturing methods. Despite the opportunity that AM represents and while the consumer market is growing rapidly, there is resistance in the industry for its capture. In this sense, it is important to take measures that can create awareness of these technologies in Europe and in the world. In order to create the demand for these technologies, it is necessary to identify where the deficiencies and limitations are that prevent the absorption of AM by the industry. Hence, the first hybrid equipment for the supply of material with start-up or finishing processes is already being commercialized.

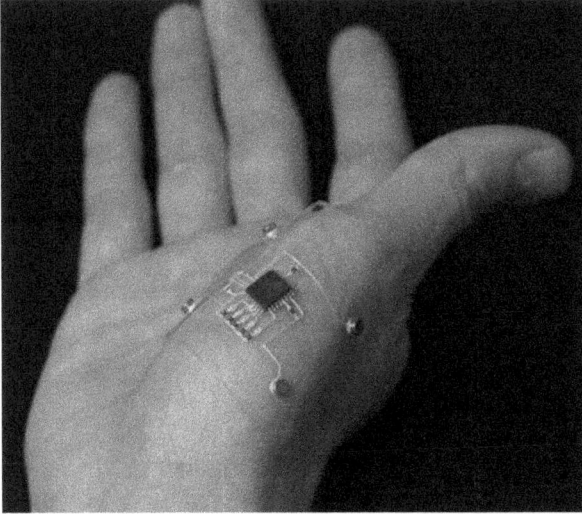

Figure 2.11: Examples of products made with 3D multiprocess printing: hybrid 3D printing for cheap wearables [30].

Figure 2.12: NASA project: Additive Construction with Mobile Emplacement (ACME) [31].

2.2 Polymeric additive manufacturing technologies

Compared to traditional polymer forming processes such as injection molding and subtraction techniques such as CNC machining, AM is slower. However, it offers special features that make it especially attractive for these materials, since it allows

CAD-guided manufacturing of multifunctional material systems with complex shapes and functionalities, including biosystems [32].

Despite the significant progress that has been made in recent years, in the case of polymers there are still a number of challenges that must be addressed to establish that AM is a large-scale manufacturing tool. Many of these challenges are related to the insufficient properties of the material (anisotropy, porosity, stability, corrosion, etc.). However, in polymeric materials, there is a large number of processes that increasingly reach requirements for final pieces.

Table 2.1 lists the main AM technologies that work with polymers, with some of their most relevant characteristics [11]. The most typical values are shown, being possible to increase the ranges and materials in most cases.

On the other hand, 51% of service providers surveyed by Wohlers Associates [10] provide parts in polymeric materials, while 19.8% provide metal parts. The remaining 29.2% offer construction services for metal or polymer parts (Figure 2.13). Given the current demand and use of plastic materials in the AM, this chapter is focused on this group of technologies, specifically in the FDM.

2.2.1 Fused deposition modeling

The additive process where the molten material is deposited from a polymeric filament is known as FDM known by its acronym FDM, FFF (fused filament fabrication), or FLM (fused layer manufacturing). This is one of the most important and most widespread processes within the large group of additive technologies. Its low cost and the wide variety of materials make it a highly accessible process [34, 35].

This process was first commercialized in 1991 by Stratasys®, which initially fed its machine with acrylonitrile butadiene styrene (ABS) and named it as FDM [36]. Stratasys® established a trademark with that name; therefore, in 2006 the RepRap Project members coined the term FFF as an alternative to provide an expression that was not legally restricted in its use [37]. However, it was not until 2009 when the main patent related to this process expired [38] and until 2011 when the free term began to be used by the scientific community, currently maintaining a predilection for FDM acronyms even in the use of open source (Figure 2.14).

In this method a thread, generally thermoplastic, is passed through the inside of a heated injector. The material is forced to advance along the nozzle heating up to reach its fusion. The molten material is injected through the orifice of the head and then deposited on the work platform or on the previous layer [39–42] (Figure 2.15).

In FDM, the solid filament is extruded in a semimolten state, which solidifies on the surface at a temperature below the glass transition temperature of the material. The temperature drops sharply from the melting to the glass transition in 0.5 s [43, 44]. As a consequence, a volumetric contraction takes place, developing a weak bond and porosity in the structure.

Table 2.1: Main processes of polymeric additive manufacturing and their main characteristics.

Process		Diagram	Volume/volume maximum (mm)	Resolution (µm)	Typical materials
Light curing	Top		250 × 250 × 250 800 × 330 × 400	50–100	Acrylates/epoxides
	Below		150 × 80 × 300 100 × 100 × 100 300 × 300 × 300	25–100	
Multiphoton lithography			5 × 5 × 1 100 × 100 × 3	0.1–5	Acrylates
Selective laser sintering			250 × 250 × 250 1,400 × 1,400 × 500	50–100	PA12/PEEK
Material jetting and polyjet			300 × 200 × 150 1,000 × 800 × 500	25	Acrylates
Binder jetting and 3D-printing			200 × 300 × 200 200 × 250 × 200 1,000 × 600 × 500	10–100	Conductive/dielectric inks starch/PLA/ceramic

(continued)

Table 2.1 (continued)

Process	Diagram	Volume/volume maximum (mm)	Resolution (µm)	Typical materials
Laminate LOM		170 × 220 × 145	200–300	PVC/paper
Material extrusion		200 × 200 × 200 1,005 × 1,005 × 1005	50–150	Different thermoplastics

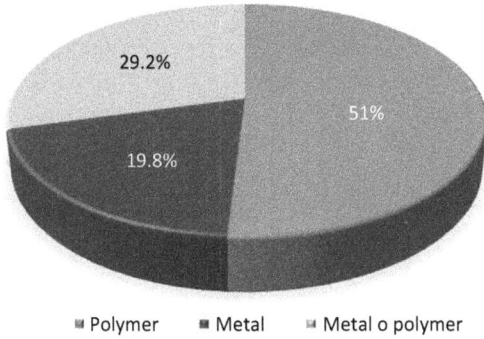

Figure 2.13: Additive manufacturing service providers distributed according to the type of material they offer [33].

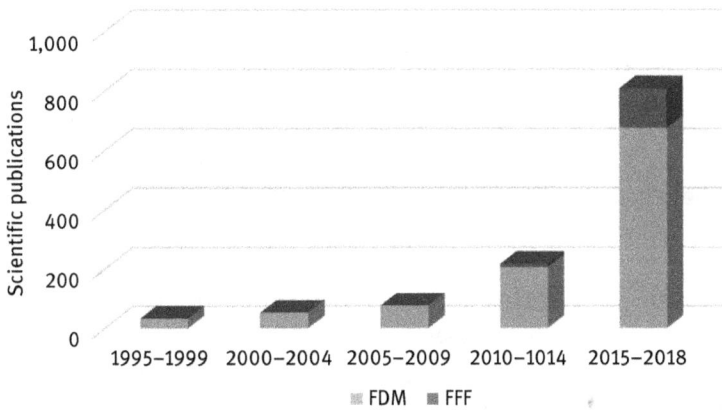

Figure 2.14: Use of the acronym FDM against FFF in the bibliography indexed in Journal Citation Reports.

The internal structure of an FDM part is not significantly different from that of a fiber reinforced composite material, as it can be interpreted as a composite structure with layers of thermoplastic polymer fibers stacked vertically in different arrangements, in addition to a small amount of air.

Another point to keep in mind of this process is the fact that it is slow compared to other manufacturing techniques. To this we must add the limited surface quality if contrasted with other processes. The surface quality of these pieces is mainly related to the height of the deposited layer. The most common machines that use this technology use a layer height (resolution) of between 0.05 and 0.1 mm. The parts manufactured therefore have a rough feel. However, current studies show that both surface quality and geometric tolerances can be improved by optimizing the process parameters or different post-processing techniques [45].

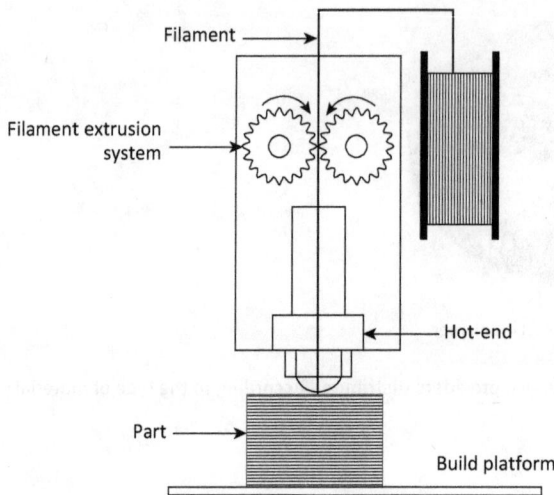

Figure 2.15: Scheme of the FDM process.

Although it has been mentioned that the process is based on the extrusion of a thermoplastic, more and more materials are processable by this technology, including polymers doped with reinforcements that can be in short fiber, or particles (metallic, ceramics, sands), food, and other bioorganic materials. So much so that the traditional definition of this process associated with the use of polymeric filament is increasingly broad. Therefore, in some cases, changing the type of material could be considered variants of the initial process. One of the examples was created by the inventor Enrico Dini (D-Shape). The work area is currently the largest available, for this reason it was the first machine that built buildings through a layer-by-layer process even from debris [46–48] (Figure 2.16).

Tissue engineering, and the final printing of whole organs or parts of the body, is an active area of research throughout the world. This technique is also known in some cases and for some applications such as 3D fiber deposition [50, 51] (Figure 2.17). Currently, there are a lot of animal organs printed, and bladders for humans that have been designed from the patient's own stem cells. One of the most studied topics with more publications to date is the ability to generate a porous and uniform structure by manipulating the parameters of the FDM process, which have made the process highly attractive to generate structures of biomaterials. Such structures are highly desirable to direct the proliferation, growth, and development of cells [52–54].

On the other hand, there are projects to develop a personalized food chain based on FDM for the elderly in nursing homes, hospitals, assisted living facilities, or even at home (served by the nursing services, Figure 2.18). This group of people requires a careful supply of certain foods due to their nutritional status and health.

Figure 2.16: 3D printed house of the company D-Shape [49].

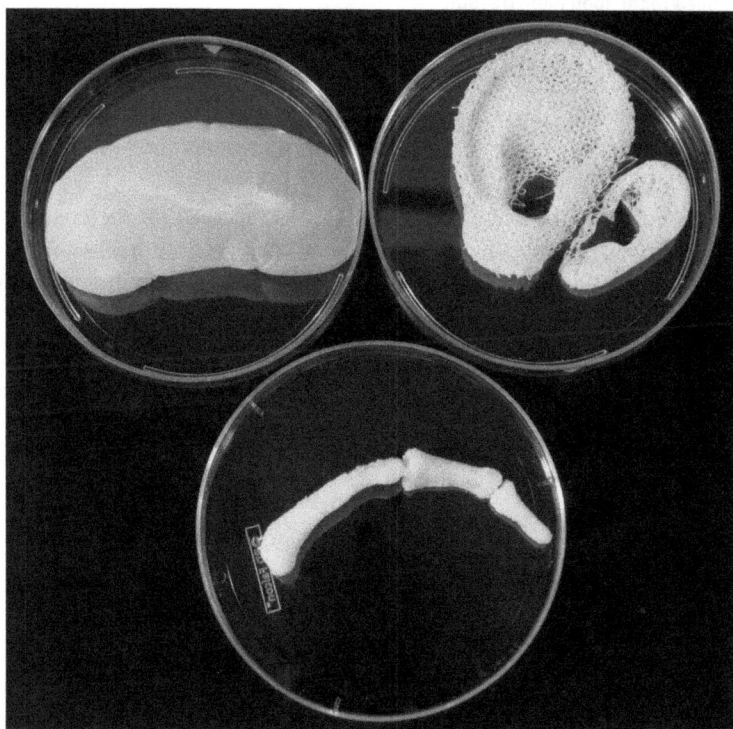

Figure 2.17: Bioorganic printing using FDM [55].

Figure 2.18: FDM applied to the food industry [60].

In addition to this, this idea of AM is used in the "new kitchen" to give distinction to the products. In these cases, what matters is not the nutrients, but the texture and appearance [56–59].

As in any AM technology, in FDM a series of stages is followed for the creation of parts that are not too far from those explained above. The general process of designing and manufacturing a part using FDM is included in Figure 2.19. It is important to emphasize that in this process most of the stages are carried out digitally.

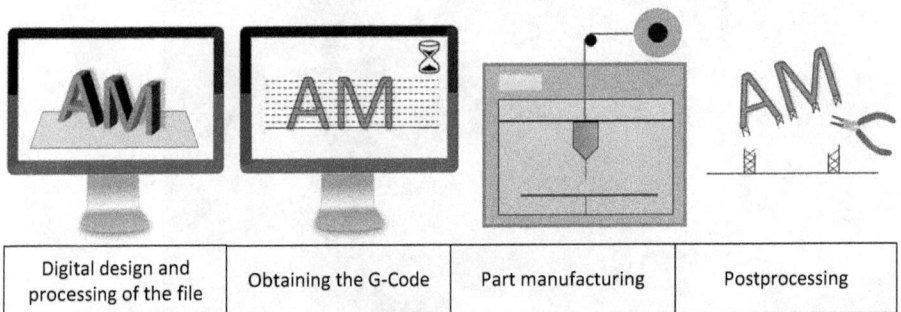

| Digital design and processing of the file | Obtaining the G-Code | Part manufacturing | Postprocessing |

Figure 2.19: Stages in the process of designing and manufacturing a piece using FDM.

2.2.2 Parameters in FDM

The great diversity of variables that exist in this process can be catalogued according to different criteria. In the bibliography, a standard classification is not carried out, likewise each software uses separation criteria of each group of different parameters. For this reason, and based on the preliminary classifications found, three groups of main variables are established that can define the result of the process: printing parameters, trajectories and support structures, and conditions of material and environment. In this section, we will briefly describe the most important variables classified in the three groups mentioned, as well as their possible repercussion in the finished parts manufactured.

2.2.2.1 Group 1: printing parameters

- *Layer thickness*. It is one of the main parameters in any of the AM technologies and one of the most studied. Its value must be selected according to the diameter of the nozzle. The smaller the thickness of the layer, the piece will have better surface finish although this value cannot be lowered infinitely for the same nozzle [61, 62]. Traditionally it is a fixed parameter although in the latest software it is possible to perform a dynamic thickness control according to the details of the piece to be manufactured.
- *Filament width*. It has a direct relation with the locality of the finishing of the parts and the details that present [63].
- *Extruder temperature*. It is the temperature at which the material is formed. This depends directly on the material used and is a variable widely analyzed in various materials [64–66].
- *Print speed*. It is the speed with which the head moves. Within this there is the possibility of modifying the speed depending on whether the filling of the part, supports, auxiliary structures, the skin, or perimeter of the part is being manufactured, or it is a vacuum displacement (without extrusion). This and the layer thickness mainly define the manufacturing time [64].
- *Feeding speed*. It is the speed of material input in the extruder. It allows to define the extrusion speed.
- *Extrusion speed*. It is the speed at which the extruded thread exits through the nozzle. It defines the flow of the material, and its control is important since it controls the lack or not of the material.
- *Flow or volumetric flow*. It is the amount of material that is extruded per unit of time. A correct use of the flow together with a good choice of temperature can reduce the porosity, dimensional deviations, and roughness [52, 67].
- *Gap/offset*. It is the space between the nozzle and the work table. If this space is too large (+) the thread will not remain homogeneously deposited, and if it is too small it can impede the correct flow of the material or the overflow of the

same, getting to damage the head or even the manufacturing platform, or inducing dimensional deviations positive.

- *Retraction*. It corresponds to the recoil of the thread inside the head to avoid an unwanted extrusion especially in trajectory changes. If a small backspace is used, residual rows may appear; if it is too large, it can prevent the correct flow of the material.
- *Overlap*. It refers to the percentage of a thread that fits over the previous one. It depends on the flow and the manufacturing trajectory. It is an important parameter to achieve densities close to 100%.

2.2.2.2 Group 2: trajectories and support structures

- *Raft*. It is one or several first layers that can be deposited on the work table. It must be used in pieces whose height-section ratio is too high to avoid a separation between the part and the platform while it is being manufactured.
- *Brim*. It is similar to the raft, with the only difference being that they begin right at the perimeter of the part. Its function is to improve adherence to the work table of the most sensitive areas.
- *Skirt*. They are auxiliary trajectories that occur around the piece with a separation to it and its function is to ensure a correct extrusion of material before starting the manufacturing process of the part.
- *Prime pillar*. It adds a small single-walled cylinder that favors the cooling of one layer for the deposition of the next. They must be used for the manufacture of small section parts.
- *Supports*. It is the material that is deposited under the overhangs. Depending on the technology, they are usually placed from a certain angle and length of the cantilever. Density and material can be varied to facilitate extraction.
- *Perimeter*. Amount of material that is deposited to form the walls or the exterior of the part.
- *Internal structure (infill)*. The part can be emptied so that it is not completely dense. The filler material confers characteristics on the element such as structural rigidity or toughness. You can modify both the geometry and the density and arrangement or orientation of the fill [42, 68, 69].
- *Orientation*. This is very important since it affects the quality of the surface, the production time, the mechanical properties, as well as the requirement of support structure, and therefore, in the economic-functional performance [70].

2.2.2.3 Group 3: conditions of material and environment

- *Build platform*. It is important that the material with which it is made, since it directly influences adhesion. In the case of not having a platform that favors good adhesion, it is necessary to resort to fixing coatings (tapes,

adhesives, etc.). In addition, there is the possibility of working with plat-
forms that may be heated or not. Likewise, it is necessary to take into account the leveling of this platform, since it favors the correct deposition of
the material [67, 71–73].
- *Environmental temperature.* There must be controlled atmospheric conditions
  depending on the coefficient of thermal expansion of the material. If the temperature difference between the extruder and the environment is too great,
  there will be defects associated with the volumetric contraction of the material
  and the stresses generated in the interlayer.
- *Humidity.* There are materials highly influenced by humidity. For this reason,
  the humidity that exists in the storage, during the manufacture, and in service
  of some parts manufactured with FDM must be controlled.
- *Ventilation.* The fans, in addition to dissipating part of the heat, help control and
  regulate the temperature around the manufacturing head, creating an appropriate environment around it. They are very necessary above all in open machines.
- *Diameter of the filament.* It is usually related to the diameter of the nozzle that
  will be used. Currently there are many manufacturers and materials available
  for FDM, so you can get coils of many diameters; however, the most common is
  the 1.75 mm. Manufacturing tolerances of the filament can influence the fluidity of the material.
- *Nozzle diameter.* It has a direct repercussion with the diameter of the filament
  without extruding and extruded. Currently they are sold in many sizes; however, there are materials that cannot be manufactured with very small nozzles,
  so the achievable quality will be lower.
- *Material.* As in any process, the material is the main variable in manufacturing.
  The modification of the rest of the parameters depends on it. In addition,
  within the same material, there may be characteristics that influence the choice
  of parameters [66].

2.2.3 Limitations and defects of FDM technology

A defect is denominated as a deviation that a part displays made by any technology, in comparison with the virtual or theoretical model of the same one. At the
same time, getting a part without flaws is not possible, since depending on the evaluation scale, there will always be some type of deviation.

The parts made using FDM technology usually have some very characteristic
defects that can be mainly related to the inadequate use of some of the parameters
or manufacturing trajectories, in addition to the limitations of the technology. The
control and minimization of these defects means that despite the great expansion
of this technique, it still has a significant margin of improvement. Some of the most
characteristic defects are:

- *Bubbles or porosities.* The process takes intrinsic the appearance of bubbles. Working with inadequate parameters causes the appearance of more bubbles, causing an increase in the weakness of the structure. There are numerous studies to reach a density close to 100% or not, but always controlled for many applications [74, 75]. The porosity can be used as an advantage, for example, in medical implantology.
- *Air gap.* It is a defect similar to the previous one, with the only difference being that it is produced by an insufficient overlap between threads, due to an inadequate thickness or flow.
- *Seam/sewing.* It occurs at the beginning and end of each layer due to the change in trajectory that the head must make. It should be taken to an area where it does not affect or can be disguised, for example, in a corner. For cylindrical pieces it can be avoided by making a spiral path, so that the height increases progressively, without a layer change properly.
- *Cracks and crevices.* There is a bad adhesion between layers that may be due to internal tensions that may appear in the piece. If this defect appears only in the first layer it is called "warping" [74] and it causes a curving of the piece due to the internal tensions. It can also be associated with a high coefficient of thermal expansion of materials and is solved by reducing the temperature difference between the environment and manufacturing. If only occurs in the first layer, it can be solved with the use of some type of adhesive or using higher platform temperatures.
- *Cantilever detachment.* In this technology it is essential to create supporting structures when the piece consists of a cantilevered part. Otherwise, the action of gravity causes large deviations to appear in these areas. If adequate use is made of the supports, however, defects may arise due to the extraction of the same.
- *Deformations.* The major deformations appear in small section parts. The material is extruded and, without having time for the layer to cool, the nozzle passes over again placing another layer. This causes the overflow of the layer appearing high dimensional and form deviations. There are also excess of materials deposited in specific areas. Normally this defect is associated with an incorrect extrusion.
- *Residual filament.* Waste of material left lying in empty displacement areas. These rows appear by an insufficient retraction of the thread, and by the temperature to which the head is and the action of gravity.

As in any AM technology, the process of converting the model into STL and subsequent slicing simplifies the geometry losing resolution in most cases, especially when processing circular or small parts [75]. Also the restriction of reaching small details due to the own technology, and to the characteristic defects of the same one. These defects can be reduced with the use of some manufacturing parameters, as well as other system variables. However, the surface finish of the pieces will depend to a greater extent on the extrusion head and the layer height used. Therefore, the poor surface finish and the low dimensional accuracy seem to be a major obstacle against the commercial production of parts, customized or not, by FDM (Figure 2.20) [68].

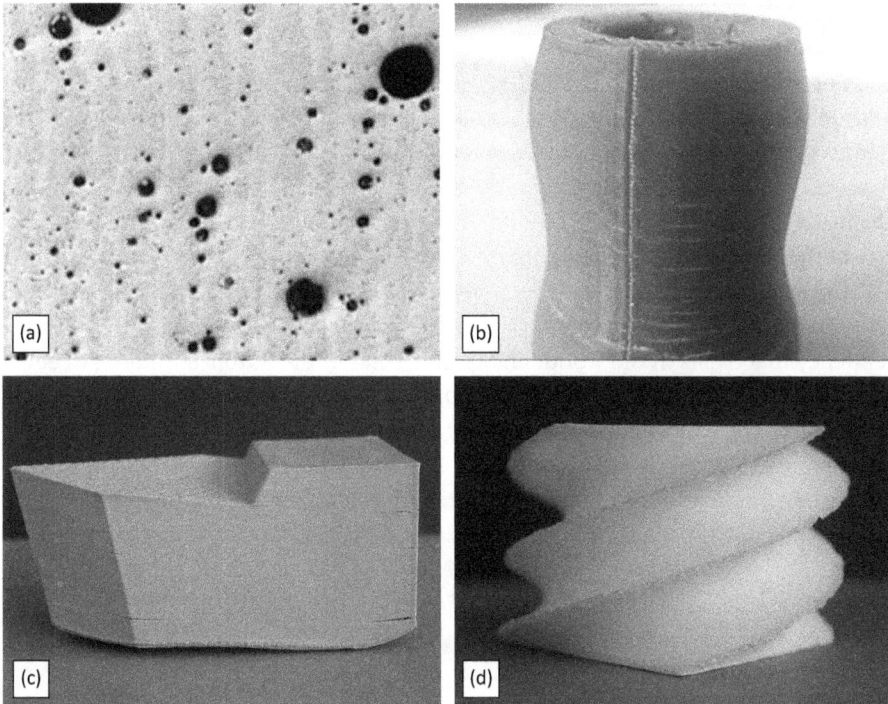

Figure 2.20: Characteristic defects of the FDM technology: (a) porosities; (b) seam; (c) warping; and (d) cantilever.

The techniques used to improve the surface finish are classified into two categories: preprocessing and postprocessing. All surface refinement methods adopted before the manufacture of FDM parts are classified as preprocessing and have to do with the modification in the design and manufacturing parameters. On the other hand in the postprocessing, the pieces are treated after the extrusion is completed under the nozzle [76].

For all this, although there are still some doubts about its applicability in mass production, the use of FDM in the industry is increasing due to the new technological advances that allow to control these defects. Being a technology in development to create objects of precision and great possibility of materials, FDM can offer a way to replace in some cases the conventional manufacturing techniques in the near future [8, 11].

2.2.4 Environmental and ergonomic considerations in FDM

There are some environmental aspects to consider for this technology. It is possible to reduce the consumption of materials, but that does not mean that it is a zero waste technology.

For example, there is material waste, such as support structures. In addition, with the polymeric materials there is a need to renew the material, and there are limited recycling rates. Therefore, the idea that the technology is totally green and clean, especially in the manufacturing phase, is not correct in all cases [77–80]. Likewise, the material can release in its conformation gases harmful to health, as is the case of one of the most used materials in this technology, ABS.

However, every time there is more recycling in addition to the use of biodegradable materials. The use of this type of material should be encouraged due to the tendency that exists in the world, and in particular in the European Union, to eliminate the use of plastics by 2030. This is one of the reasons why the polylactic acid (PLA), a biodegradable material, is taking more and more relevance.

Another aspect to take into account is the ergonomics, since there is a noise emitted by this type of machines that, although it does not exceed the limit by regulations, are usually annoying, like the continuous noise of a robot, especially if there are many machines on a small space.

In short and without a doubt, FDM offers the possibility of changing the very essence of design and manufacturing, from the assembly and transformation processes to the synthesis of advanced materials in the final product. This direct synthesis of materials reduces waste and has the potential to improve the performance and reliability of the product. However, an effort must be made between process control engineers, designers, and environmental specialists to arrive at a fundamental understanding of the possible environmental impacts of AM, change the design if necessary, assess the degree to which they could occur, define their distribution, and establish the regulations that will allow the control and prevention of damage, as well as estimate the costs [80, 81].

2.3 Materials used in FDM

To date, the dominant part of the FDM industry has been based on the printing of a single material [3, 11]. This problem, along with the limited options of materials available and compatible with commercial equipment, offers very limited variations in the physical–chemical properties of the objects obtained with AM. These limitations have led to the development of new materials as well as multipurpose/multiproduct equipment with partial control over the composition and properties of the material, offering composite materials in layers [82, 83].

Some of the most used materials in FDM are listed in Table 2.2, although there are more and more materials and derivatives, commercial or not, with a great variety of specific characteristics for specific applications (conductivity, biodegradability, etc.). On the one hand, not all of these materials have been published in studies related to FDM, although they are mentioned as alternatives and with characteristics that would bring certain advantages. In addition, many of them are common consumables of FDM.

Table 2.2: Main materials used in the FDM process.

Material	Main feature	Acronym	Related studies
Acrylonitrile butadiene styrene	Resistance to impact, abrasion, and chemical agents	ABS	[84, 85]
Polylactic acid	Biodegradable/biocompatible	PLA	[32, 86]
Polycarbonate	Electrical insulation/dimensional stability/impact resistance	PC	[11, 87]
Polycaprolactone	Biodegradable/impact resistance	PCL	[52, 54]
Polyphenylsulfone	Stability against radiation/low ionic impurities/chemical resistance	PPSU/ PPSF	[84, 88]
Polyetherimide	Electrical insulation/mechanical resistance/stability	PEI (Ultem^{TM})	[89]
Polyethylene terephthalate	Biocompatible/chemical resistance	PET	[90]
Polystyrene	Rigidity/fragility/thermal and electrical insulation	PS	[91]
High impact polystyrene	Impact resistance	HIPS	[11]
Polyamide/nylon	Mechanical and wear resistance/ promotes slip/moisture absorption	PA	[72, 92]
Styrene acrylonitrile	Resistance to impact and chemical agents	SAN	[11]
Copolyester	UV and temperature resistance	–	[93]
Acrylonitrile styrene acrylate	High atmospheric and impact resistance	ASA	[94]

Table 2.2 (continued)

Material	Main feature	Acronym	Related studies
Polypropylene	Good adhesion/corrosion resistance	PP	[95]
Polyetherketone	Chemical and hydrolysis resistance	PEEK	[90]
Thermoplastic elastomer	Elasticity/resistance to abrasion and low temperatures	TPE (Filaflex)	[96]
Thermoplastic polyurethane	Resistance to abrasion, O_2, and low temperatures	TPU	[97]
Polyvinyl alcohol	Water soluble/biodegradable	PVA	[98]
Polyacetal/polyoxymethylene	Wear resistance	POM	[99]
Acrylic glass (methacrylate)	Light transmission	PMMA	[100]
Wide variety of reinforced materials (metallic, fibers, etc.)	Different physical–chemical properties	–	[82, 85]
Graphite	Thermal and electrical conductivity	–	[101]
Ceramic (fused deposition ceramics)	Different physical–chemical properties	FDC	[35, 83]
Lead titanate with lead zirconate (piezoelectric ceramics)	Electrical conductivity	PZT	[84]
Wax	Ease of casting (rapid manufacturing of casting models)	–	[83]
Organic/edible materials	Variety of properties/biodegradable	–	[56–59]

On the other hand, the use of fiber-reinforced composites also offers a significant improvement in mechanical properties; however, it requires a complex procedure to be manufactured and is difficult to incorporate into the processing. In addition, advances in the development of AM equipment have allowed the development of premixed materials with fillers such as nanoparticles of multiple materials, carbon nanotubes, fibers, or graphene in order to achieve unique characteristics and capabilities according to the additive used.

However, the addition of fiber to AM materials is not always used to improve the mechanical properties. Fibers, as well as other additives, are also used in the manufacture of intelligent compounds to control the transformation of the structure. This implementation in AM has opened a new field called 4D printing [102, 103]. 4D printing refers to the use of multiple materials with the ability to transform over time or change their shape after printing (Figure 2.21). These structures

Figure 2.21: Series of superimposed photographs showing the automatic folding of a 4D print part [106].

can be programmable and transformed from structures of one or two dimensions to three-dimensional objects [104]. The objective is that a manufactured piece has autonomy to do something by itself: assemble, repair, or respond to some external stimulus among others. That is, according to Skylar Tibbits, director of the 4D printing laboratory at the Massachusetts Institute of Technology [105] 4D printing includes a 3D production programmed so that objects once printed change shape, properties, or even perform computation processes in matters complementary to their environment.

Despite these advances, nonreinforced polymers in particular have been the focus of attention due to the ease of production and availability. The FDM printing industry mainly involves thermoplastic polymers, reinforced or not, with very different characteristics. In this sense, there is a variety of thermoformable plastics available for this technology. However, the tendency that exists to eliminate plastics by the amount of waste they generate will lead to the investigation into the study of biodegradable polymers.

Biodegradable polymers are degraded in physiological environments by splitting the macromolecular chain into smaller fragments and, finally, into simple stable end products [107]. Degradation may be due to aerobic or anaerobic microorganisms, or biologically active processes (e.g., enzymatic reactions) or passive hydrolytic cleavage [108]. All this makes the material as such, and disappear under certain conditions without emitting any type of wastes (harmful gases or pollutants) to the environment.

2.3.1 Polylactic acid

PLA, also called 2-hydroxypropionic acid, is a thermoplastic and biodegradable polymer that is made up of molecules of lactic acid. It is formed by two optically active forms (rotate polarized light in opposite directions) that correspond to two different isomers or enantiomers: L (L-lactide), D (D-lactide), or a mixture of both (meso-lactide) [109]. The commercial PLA for the FDM is usually a mixture, rich in L, of said isomers.

PLA is a natural polymer, derived from renewable sources such as starch, a large carbohydrate that plants synthesize during photosynthesis. Grains such as corn, sugarcane, cassava, potato, or wheat contain a large amount of starch and are the main source for the production of PLA. The bioplastics produced from this polymer have the characteristic of being a resin that can be injected, extruded, and thermoformed [110].

This biopolymer begins with the starch that is extracted, then the microorganisms transform it into a smaller molecule of lactic acid or 2-hydroxy-propionic (monomer), which is the raw material that is polymerized forming chains, with a molecular structure similar to the products of petrochemical origin, which are joined together to form the plastic called PLA (Figure 2.22).

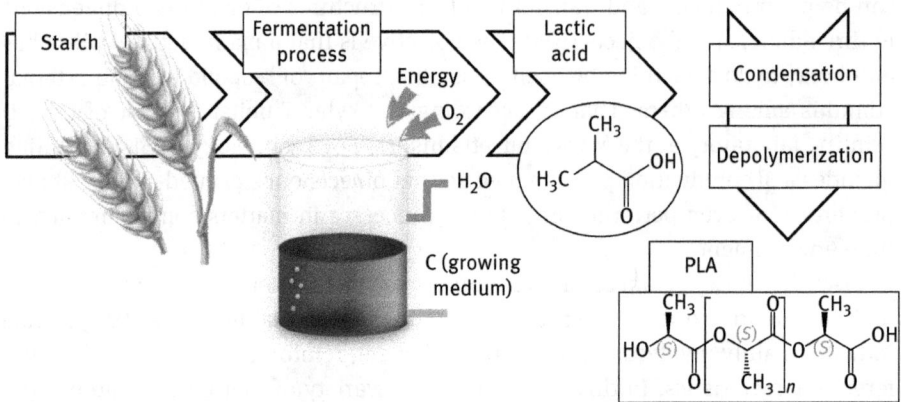

Figure 2.22: Scheme of PLA production.

It has a high average molecular weight (Mw), and is generally produced by polycondensation of lactic acid in the presence of a metallic zinc catalyst (under vacuum and inert atmosphere conditions); and/or by ring-opening polymerization (ROP) [111]. The advantage of ROP is that the reaction can be more easily controlled, thereby varying the characteristics of the resulting polymer in a more controlled manner. NatureWorks LLC is the main producer of PLA, with a capacity of 150,000 tons per year in its facilities. This manufacturer already offers several types of PLA with different qualities, characteristics, and prices [112]. Other companies that produce PLA are Treofan (Germany), Hycail (Holland), and Mitsui Chemicals Inc. (Japan); however, in many cases, the suppliers do not offer data of the producer, so in addition to the great variety of variants of PLA existing without standardizing, there is no traceability.

On the other hand, it is a hygroscopic material and very sensitive to high relative humidity and temperature [113]. This behavior is so important that, during its production, before processing this material, it must be dried to a water content below 100 ppm (0.01% w/w) to avoid hydrolysis (reduction of Mw).

The PLA has played a central role in the replacement of fossil-based polymers for certain applications by being a completely aliphatic polymer. As a polymer, PLA is considered a promising alternative to reduce the problem of municipal solid waste disposal [113]. Due to the great penetration in the market, the worldwide attention, and the increase of PLA production [114], the number of studies and research reports published on the PLA has increased exponentially in the last 25 years, as shown in Figure 2.23. However, currently very few studies have been carried out with FDM technology and this material.

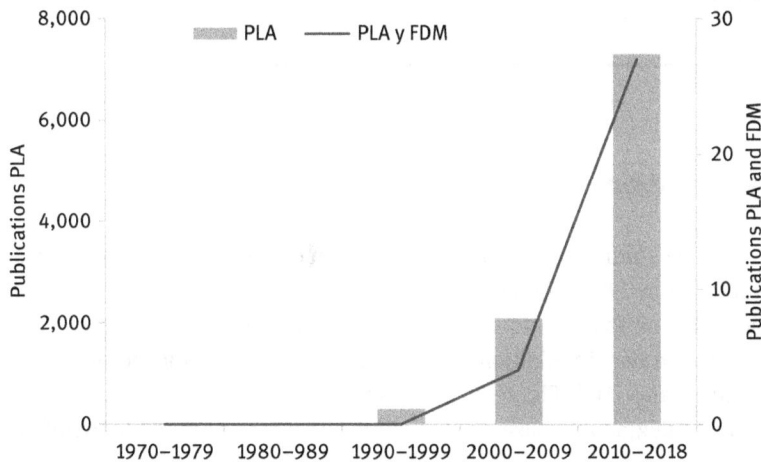

Figure 2.23: Increased interest of the PLA regarding the studies published in *Journal Citation Reports*.

The limiting factors for the processing of PLA are similar to those of fossil-based polymers: degradation in the upper limits of temperature and shear, and poor homogeneity in the lower limits. However, understanding the thermal rheological, crystallinity, and fusion behavior of PLA is very important to optimize its processing and the qualities of the components [115].

Table 2.3 shows a summary of the range of the main parameters recommended and used mostly for PLA. The parameters contemplated are collected by the bibliography and modified in some cases based on experience.

Table 2.3: Main ranges of recommended parameters and used mostly for the PLA.

Parameter	Value	Related studies
Extrusion temperature	190–230 °C	[116]
Layer thickness	0.1–0.5 mm	[63]
Speed	10–40 mm/s	[64]
Flow (ratio)	0.8–1.15	[117]
Retraction	0–3 mm	[118]
Auxiliary structures	Avoid (difficulty in machining)	–
Build platform	No need for heating or adhesive	–
Material diameter	1.75 or 3 mm	–
Nozzle diameter	0.4–0.5 mm	[119]
Fans	0–100%	[37]
Environmental temperature	20–25 °C	[120]
Humidity	20–60%	[45]

2.3.2 PLA applications

PLA is one of the most studied biodegradable plastics in our time and has been available on the market since 1990. The most promising application of the PLA is in containers and packaging for food or other types of substances (Figure 2.24), cutlery, and crockery for disposable use, in addition to the production of films for the protection of crops in primary states [113] (Figure 2.25). However, high fungal growth in materials obtained from biodegradable bases is a negative factor for use in food. Therefore, biopackages are more beneficial for foods that require high respiration and have a short storage life, such as vegetables or bakery products.

Another of its main applications is in living tissues, since PLA is completely depolymerized by chemical hydrolysis. This characteristic makes the PLA widely used in the field of medicine for the production of suture threads, implants or prostheses, capsules for the slow release of drugs, and so on [121, 122]. The biodegradable polymers used in implantology also need meticulous safety evaluations. They need both biodegradability and biocompatibility studies [107, 123, 124]. This means that a compound is compatible with a living tissue or a living system because it is not toxic or harmful and does not cause immune rejection. In addition, the compound is able to support a cell–biomaterial interaction in the tissue in which it is implanted [125, 126]. A polymeric implant is considered biocompatible if it does not

Figure 2.24: Bioplastics for agriculture [128]: (a) mulching agriculture sheets made from PLA and (b) PLA for silage bales.

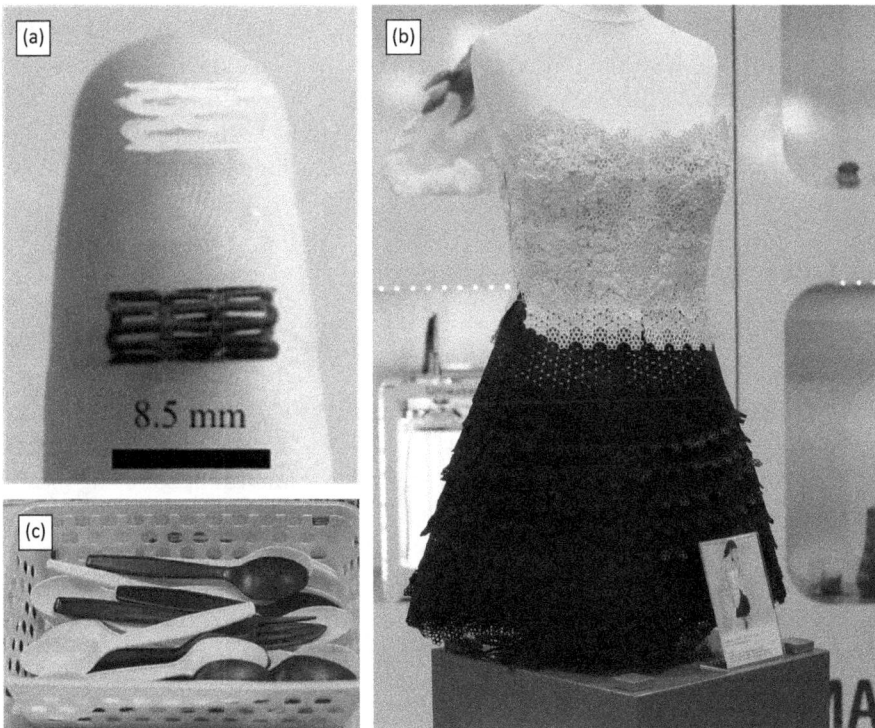

Figure 2.25: Different sectors of PLA application: (a) medicine (stent) [129]; (b) textiles [130]; and (c) polymeric cutlery [131].

lead to the formation of an intolerable inflammatory reaction that is not proportional to its advantageous effects. This means that the implant degrades in the corresponding site until it is completely eliminated, and it is free of microbial and pyrogenic agents or traces of toxic chemicals that are incorporated during the preparation, manufacture, or they are formed during storage or use [127].

On the other hand, the PLA is a very common material in the manufacture of textile fibers due to some of its characteristics. For example, it is a more hydrophilic material than others with similar characteristics such as PET, in addition to having a lower density. In addition, these polymers tend to be stable to ultraviolet light resulting in fabrics with little discoloration [132]. It is also a fireproof and low smoke generation material. All this makes it suitable and very widespread in clothing, upholstery, diapers, and so on.

Sectors such as jewelry, art, and own consumption also have their place for this material. However, it is still a material in FDM of low consumption at industrial level, despite being one of the most used at the domestic level, and hence the little dedication in the current study for this technology, except for certain medical applications. This is mainly due to the fact that it takes one of its main advantages as a drawback, being biodegradable. However, there is a current trend in eliminating or reducing plastic materials in the industry due to the high amount of waste they generate, and this entails some adaptation. In this sense, the PLA is a material increasingly used and whose use should be encouraged. Products such as mobile phone cases, disposable tools, and certain toys can be manufactured with this material.

2.3.3 PLA properties

Due to its biodegradability, barrier properties, and biocompatibility, numerous applications have been found with this biopolymer, since it has a wide range of unusual properties, from the amorphous state to the crystalline state; properties that can be achieved by manipulating the mixtures between isomers, molecular weights, and copolymerization. This, together with the growing interest in not using plastic materials, means that the PLA can be a power in many sectors [133].

The physicochemical, pharmaceutical, and resorption properties depend on the composition of the polymer, its molecular weight, and its crystallinity. The crystallinity can be adjusted from 0% to 40% in the form of linear or branched homopolymers, and as random or block copolymers. This means that their properties vary in very wide ranges.

The processing temperature in FDM is usually close to the glass transition temperature (T_g), and is between 60 and 125 ° C. This depends on the proportion of lactic acid D or L in the polymer. However, the PLA can be plasticized with its monomer or alternatively with oligomeric lactic acid and this allows to decrease the T_g [134, 135].

The PLA also has its mechanical properties in the same range of petrochemical polymers, except for a low elongation [95]. However, this property can be improved during polymerization (copolymerization) or by post-polymerization modifications (plasticizers). Table 2.4 shows the tensile strength of the most common materials in FDM and tested by different researchers. It is observed that the PLA is the most resistant. However, as mentioned earlier, there are different types of PLA, as for the other materials. In this way, increasingly there are more materials of the same and different base of traditional, with improved properties.

Table 2.4: Tensile strength of different thermoplastics manufactured by FDM and tested by different researchers.

Material	Tensile strength (MPa)	Reference
PLA	54	[137]
PEI	40	[89]
PP	32	[95]
ABS	26	[138]
PC	19	[87]

PLA can be as hard as acrylic or as soft as polyethylene; rigid as polystyrene or flexible as an elastomer. It can also be formulated to give a variety of resistances. PLA resins can be subjected to gamma-ray sterilization and is stable when exposed to ultraviolet rays. PLA is also attributed to properties of interest such as softness, resistance to scratching, and wear, so its machinability is usually lower than that of other similar polymers [39, 136].

2.4 Control of surface finish in FDM

FDM process in general lines has a poor surface finish and low dimensional accuracy. This seems to be the main obstacle against the commercial production of parts manufactured by this process [139–141]. As well as the dimensional precision is somewhat remedied with a forecast of possible deviations [64], there are techniques developed to improve the surface finish that are classified into two categories: preprocessing and postprocessing [76]. All the methods of surface improvement adopted in the design and processing phase of the file are classified as preprocessing, while in postprocessing, the parts are subjected to a treatment after the complete extrusion of the piece. Figure 2.26 shows the surface finish improvement techniques for FDM currently studied with only some of its most significant bibliographical references.

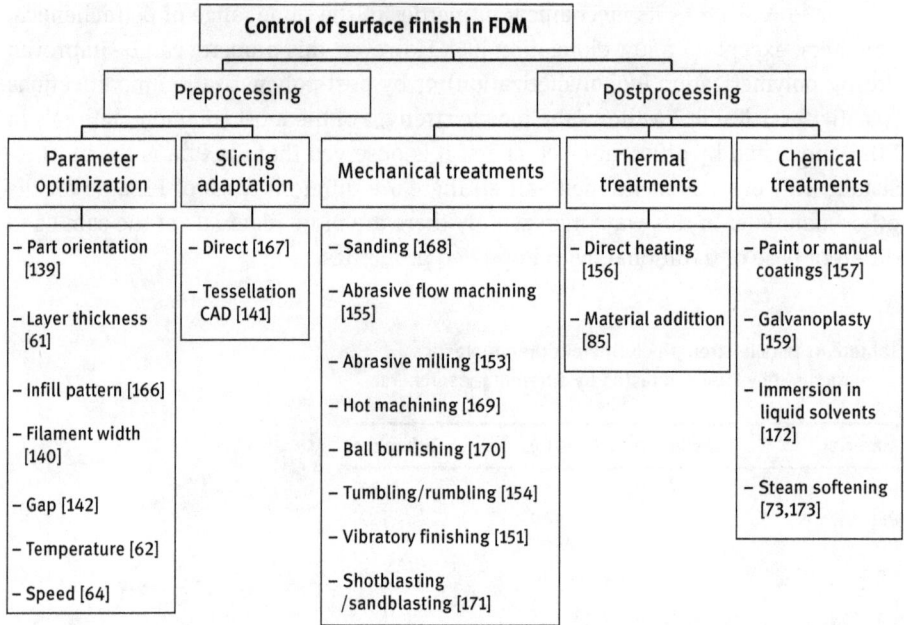

Figure 2.26: Classification of surface finish improvement techniques used in FDM.

Since the inception of the FDM process, extensive research has been conducted to improve the surface finish of FDM parts, but some degree of surface roughness is inevitable. Many researchers have been successful in controlling and reducing surface roughness using different preprocessing and postprocessing techniques. In general, parameters that affect FDM process performance are divided into geometry-specific parameters and process-specific parameters, as machine and material-specific parameters will be restricted [142, 143].

Although surface roughness cannot be eliminated, many researchers have tried to reduce the defect by modifying the process in different parameter groups or by modifying slicing techniques [140, 144–148]. Variation in many of the main process parameters significantly affects surface finish and mechanical properties [62, 149]. Nevertheless, the way to achieve better finishes is through posttreatment application.

2.4.1 Mechanical treatments

These techniques concentrate on the mechanical cutting or under pressure of the peaks of the surface profiles (Figure 2.27). In general, the techniques are reproduced from conventional metal machining or finishing techniques [76]. However, its impact on plastics is drastically different compared to metals, at least for the ABS material, as it is the only one studied.

Figure 2.27: Machining of an FDM part.

On the other hand, some researchers have successfully performed several abrasive finishing techniques for SLA pieces, using shot blasting, tumbling, burnishing, vibratory finishing, and ultrasonic abrasion, achieving an average improvement of 40%, 78%, 81%, 73%, and 60%, respectively, in surface finishing [150, 151]. Submitting parts to an abrasive finish then results in the removal of unwanted material from edges and corners due to the impact of abrasive media, although it is a slow process and in some cases the improvement is insufficient.

Other mechanical treatments studied by other authors to improve the surface finish of parts manufactured by FDM are manual sanding, abrasive fluid machining, or conventional machining even with CNC [152–155]. However, all of this has generally been studied for ABS, and in some cases for other materials, although never for PLA.

In fact, although there is no relevant or high-impact research on the machinability of PLA, from experience and the RepRap community, postprocessing (machining, painting, and above all gluing) for PLA parts is even more complex than for other materials such as ABS, so it will not be considered a viable option in improving the finish. However, perhaps a tensioning, or even a cryogenization of the machined parts can make the use of this type of techniques much easier.

2.4.2 Thermal treatments

Thermal treatments that can be applied on a part manufactured by FDM in general will be dedicated mostly to tasks such as removing excess parts of a part (generally supports) or trying to correct the presence of possible failures in printing, improve the surface finish, or make the union of different parts.

In most cases, these improvement operations will consist of small repairs that give, if possible, the functionality or aesthetics to the piece more than in standardized techniques. Among the typical problems to be solved are the presence of filament burrs, detachments between printing layers, or the breakage of areas of a piece. Given the characteristics of thermoplastics, if the temperature reached in the heating process is not excessive, when the material cools down, it will recover its original properties.

2.4.2.1 Direct heating

By heating the material to the glass transition temperature, it softens and acquires a certain plasticity. If this temperature is exceeded, the material becomes liquid, with all the characteristics that this implies, such as capillarity. This will cause the fluid material to tend to fill the small surface holes in the part.

A variety of tools can be used to heat the material of the part, ranging from professional hot air welding equipment to more affordable options such as hot air paint strippers, hair dryers, tin welders, irons, and use of direct flames (Figure 2.28). However, this type of postprocessing, although evident, has not been studied at a scientific level to demonstrate values with the improvement of qualities and maintenance of the mechanical properties of the parts.

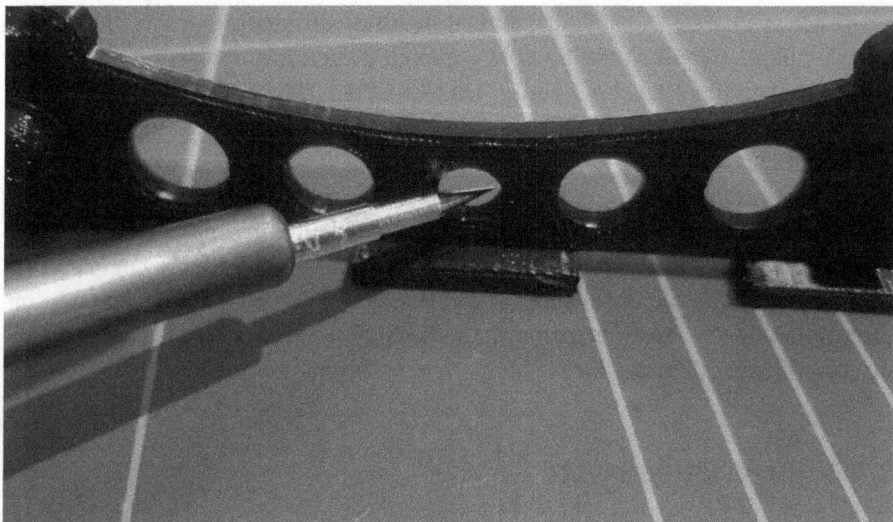

Figure 2.28: Heat treatments on PLA parts manufactured by FDM.

On the other hand, 4D printing, based on intelligent plastics, generally requires heat input to complete the entire manufacturing process, as explained earlier.

2.4.2.2 Fused material addition

These would be the procedures in which there is no direct contribution of thermal energy to the material of the piece, but what is used is a contribution of material, the thermoplastic itself, directly melted. Several techniques derived from the industrial field dedicated to plastic welding can be used. Thus, techniques such as injection welding and extrusion welding can be used.

In the background, this is equipment that applies the same operating principle as the FA equipment itself, in which the same filament with which the piece has been printed can even be used as a filling/gluing material when melting it. These are relatively expensive and specialized equipment, so their use is reserved in practice for advanced users and industrial environments. However, today there are portable FDM equipment with which it is possible to make a small contribution of material in specific areas to cover imperfections or weld.

Friction welding is a technique in which no direct heat is applied and which is another very attractive option for carrying out small repairs and surface welds [85] (Figure 2.29). For this, there are specific tools, spin welders, in which heat is produced by the high-speed rubbing of a piece of thermoplastic against the part, which melts the thermoplastic material of both and gives rise to welding [156], although the great attraction is that generic rotary tools can be used for this purpose.

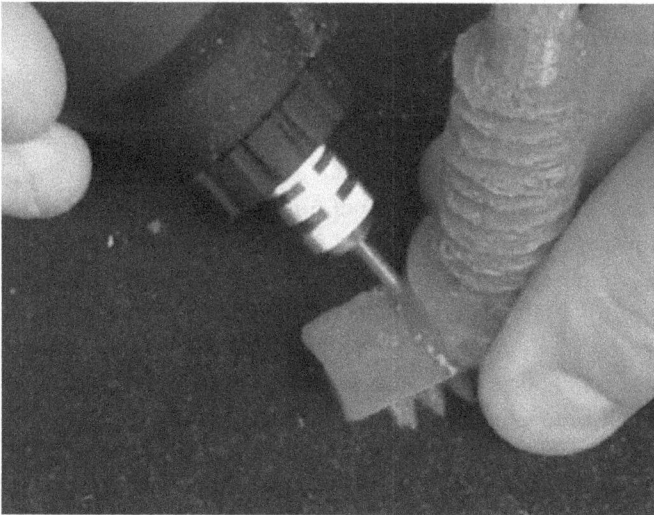

Figure 2.29: Welding of FD fabricated parts.

2.4.3 Chemical treatments

The response of ABS plastic to chemicals, vapors, and coatings has been discussed in different studies; however, PLA is a material whose behavior against chemical agents has not been studied in depth previously. Plastics have a penetrable surface that can be easily modified by chemical action [76]. The main priority of chemical finishing over mechanical finishing techniques is that there is no contact of the tools with the surface, which can guarantee better dimensional and geometric stability.

This section includes different techniques studied to date of chemical treatments applied to 3DP parts, most of them focused on materials such as ABS or ABS matrix.

2.4.3.1 Paint or manual coatings

Few chemical agents such as acetone, heptane, and toluene soften the surface of ABS parts, chemically dissolving surface edges or peaks or bonding layers, and have been used for earlier studies. However, these solvents do not work for PLA, a material that has not been improved or characterized superficially until today by this type of techniques.

Stratasys suggested the use of various primers to clean and polish FDM and SLA manufactured parts. One of the most commonly used finishes today is the manual application of epoxy resin [157]. Although the application of these products saves time and money, manual methods are not uniform and therefore offer an uneven surface finish (Figure 2.30). This can be improved with controlled immersion, pressing, or even projection techniques. There are currently no industrial solutions for the automated application of these products.

2.4.3.2 Galvanoplasty

Strengthening 3D-printed parts through the use of fluids has a significant influence on roughness. However, it is necessary to carefully select the appropriate fluid if roughness is the required dominant property of the model. The best surface roughness for 3D-printed models treated with electroplating is provided by cyanoacrylate, followed by wax and then other fluids [159]. However, there are no fundamental studies on the penetration of the selected fluid, and this directly influences the mechanical properties of the parts. Other researchers study nonmagnetic coatings, for example, based on electrolytic copper. Electroplating copper coatings improve the hardness, corrosion resistance, and tensile strength of plastics [160].

Figure 2.30: Hand-painted 3D-printed tortoise shell [158].

2.4.3.3 Immersion in liquid solvents

Acetone, being a major component of cleaning agents, has good surface polishing characteristics and is widely used for finishing products, usually ABS parts. This is the best-known and most widely used process.

FDM parts are immersed in an acetone solution at different concentrations for a specific time to obtain the desired surface finish [162] (Figure 2.31). Immersion of these parts in acetone solution also improves bending strength, water tightness, moisture, and wear resistance, but slightly reduces tensile strength. On the other hand, they cause an increase in weight, ductility, bending strength, and compressive strength, although a small shrinkage in part dimensions (about 1%) is obtained. Finishing improvements up to the nanometric scale have been achieved using this technique, preserving the geometric characteristics of the part.

FDM parts have also been widely used as sealants involving fluid pressure applications by decreasing the porosity of ABS parts due to this chemical treatment.

Acetone finishing proves to be a highly effective, easy, fast, and economical process. However, there is a risk of eroding and dissolving the small features or details of some areas of the part if exposed to longer periods or using acetone at high

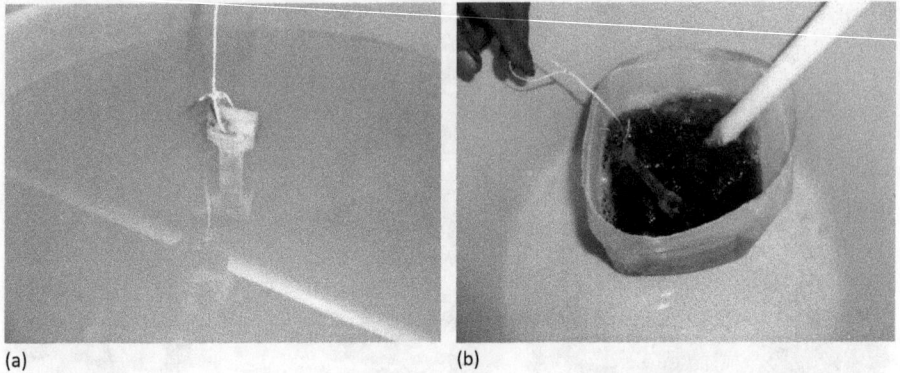

(a)   (b)

Figure 2.31: (a) Silver nitrate-activated sample; (b) nonconductive copper sulfate treatment [161].

concentrations, while the use of a dilute solution can prolong immersion time. However, it is a method that needs to be controlled, automated, and improved in order to achieve a more accurate and systematic finish (Figure 2.32) [76].

(a)   (b)

Figure 2.32: Finishing of part manufactured by FDM: (a) before and (b) after immersion in chemical agents.

2.4.3.4 Steam softening

Finishing techniques have been developed where the surface of the part temporarily refluxes due to exposure to vapors generated by the heating of chemicals in a controlled environment [163, 164]. This has been carried out mainly for the surface finishing of ABS parts. After manufacture, the parts are left to cool for a few minutes and then this steam smoothing is applied to them for different times. A fixed amount of solvent is heated and therefore the (chemical) vapors are elevated to deposit on the ABS part, which is normally hung by a suitable arrangement inside, or dropped onto an easily extractable surface. Vapors condense on the part and penetrate through the surface, flattening the surface [76].

Experimental configurations with forced ventilation have also been developed for the circulation of acetone vapors in a closed chamber, resulting in improvements in the surface finish of FDM parts [165]. Researchers focused on this type of application ensure an improvement in surface finish with an increase in fan rpm, while requiring a longer exposure directly proportional to the size of the part. Tests have also been made using cold vapors (at 20 °C in a closed container) to control the rapid vapor–plastic interaction [166]. Workpiece surface quality is improved with minimal dimensional changes with longer exposures (about 40 min).

Although the use of acetone for PLA is discussed in certain forums, no significant results with the use of this solvent have been found in scientific advances and experience (Figure 2.33).

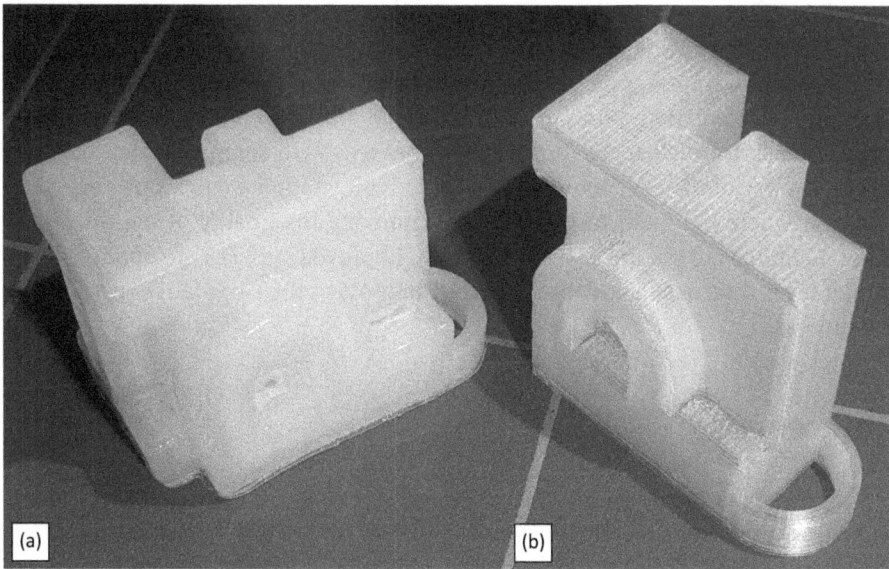

(a) (b)

Figure 2.33: Finishing of part manufactured by FDM: (a) after and (b) before acetone steam treatment.

2.5 Conclusions

Some basic aspects of AM have been reviewed throughout the chapter. To this end, the importance of this group of processes and their evolution has been explained first, in particular the FDM, as it is one of the main processes, the most widespread, and economical, with large margins for improvement.

By this way, the most basic issues associated with this forming process have been shown, which have served as a starting point when it comes to descending toward more specific issues. Among them are the manufacturing phases, the main defects, or the great variety of parameters that govern this process.

One of the most important points is the material to be shaped. Finally, a bibliographic analysis has been carried out of the large number of materials that exist and are continuously suitable for this process. Emphasis is placed on the proposal of biodegradable materials, due to the current trend in the elimination of plastics and the reduction of waste, so that this process can be transferred to a biosustainable industry. For that reason, PLA is proposed as an alternative to petroleum-derived plastics. Despite its great capacity, its implantation in different sectors, and its good characteristics compared to other plastics used in this process, PLA is not one of the most studied at present. Therefore, its processing, its applications, and the great variety of properties it possesses are analyzed.

On the other hand, one of the main restrictions of this process is that it is a slow process compared to others within the AM group. In addition, it has a poor surface finish intrinsic to the process. For that reason, a significant improvement in process performance is required. Hence, a bibliographic analysis has been carried out, where the main techniques for improving the quality of the process in other materials are obtained. Between them, improved surfaces are obtained even if high layer thicknesses are used, thereby improving the two main constraints of the process.

References

[1]   Hashmi, M.S.J. Comprehensive Materials Processing, 13th ed. Elsevier, Netherlands. 2014.
[2]   Gibson, I., Rosen, D. BS. Additive Manufacturing Technologies. 3D Printing, Rapid
        Prototyping, and Direct Digital Manufacturing, 2nd ed. Springer, Germany, 2015.
[3]   Dilberoglu, UM., Gharehpapagh, B., Yaman, U., Dolen, M. The Role of Additive Manufacturing
        in the Era of Industry 4.0. Procedia Manuf [Internet]. The Author(s); 2017;11[June]:545–554.
        Available from: http://dx.doi.org/10.1016/j.promfg.2017.07.148
[4]   ISO 52900:2015. Additive Manufacturing. General principles. Terminology.
[5]   Frazier, WE. Metal additive manufacturing: a review. Journal of Materials Engineering and
        Performance, 2014, 23(6), 1917–1928.
[6]   Vallés, JL. Additive Manufacturing in FP7 and Horizon 2020. Representative from EC Work
        Additive Manufacturing, 2014; Brussels [Belgium].

[7] Mellor, S., Hao, L. DZ. Additive manufacturing: a framework for implementation. International Journal of Production Economics, 2014, 149, 194–201.

[8] Bechthold, L., Fischer, V., Hainzlmaier, A., Hugenroth, D., Ivanova, L., Kroth, K. BR, Sikorska, E. VS. 3D Printing. A qualitative assessment of applications, recent trends and the technology's future potential. Cent Digit Technol Manag. 2015;München.

[9] Lu, Y. Industry 4.0: A survey on technologies, applications and open research issues. Journal of Industrial Information Integration, [Internet]. Elsevier Inc.; 2017, 6, 1–10. Available from: http://dx.doi.org/10.1016/j.jii.2017.04.005

[10] Wohlers, Associates: Fort Collins C. 3D Printing and Additive Manufacturing State of the Industry Annual Worldwide Progress Report. 2017.

[11] Ligon, SC., Liska, R., Stampfl, J., Gurr, M., Mülhaupt, R. Polymers for 3D printing and customized additive manufacturing. Chemical Reviews, 2017, 117(15), 10212–10290.

[12] Shambley, W. Herramental plástico para fundición de impresoras 3D de gran formato. Simple Solutions that work [Internet]. 2017, 62–63. Available from: http://www.palmermfg.com/pdfs/flippingbooks/simple-solutions-2017-04/index.html#1/z

[13] Zelkowitz, M V. A case study in rapid prototyping. Software- Practice and Experience, 1980, 10, 1037–1092.

[14] EFFRA. Factories 4.0 and Beyond. Recommendations for the work programme 18-19-20 of the FoF PPP under Horizon 2020. 2016.

[15] Kruth, J.P. MCL. Progress in Additive Manufacturing and Rapid Prototyping. Annals of ClRP, 1998, 47(2), 525–540.

[16] Levy, G. N., Schindel, R. JPK. Rapid manufacturing and rapid tooling with layer manufacturing (LM) technologies, state of art and future. CIRP Annals – Manufacturing Technology, 2003, 52, 589–609.

[17] Hutmacher, D. W., Tan, K. C., Schantz, J. T., Lim, T. C. CXL. State of the art and future directions of scaffold-based bone engineering from a biomaterials perspective. Journal of Tissue Engineering and Regenerative Medicine, 2007, 1, 245–260.

[18] Dimitrov, D., Schreve, K., de Beer, N. Advances in three dimensional printing – state of the art and future perspectives, Rapid Prototyping Journal, 2006, 12(3), 136–147.

[19] Campbell, I., Diegel, O., Kowen, J. TW. Wohlers Report 2018. Additive Manufacturing and 3D Printing State of the Industry. 2018.

[20] Wohlers, Wong K. 2017 Report on 3D Printing Industry Points to Softened Growth [Internet]. 2017. Available from: http://www.rapidreadytech.com/2017/04/wohlers-2017-report-on-3d-printing-industry-points-to-softened-growth/

[21] Yan, X. PG. A review of rapid prototyping technologies and systems. Computational Design, 1995, 6, 307–318.

[22] ISO 17296:2014. Additive manufacturing – General principles. 2014.

[23] Calleja, A., Urbikain, G., Gonzalez, H., Cerrillo, I., Polvorosa, R., InconelA, Lamikiz A. (R) 718 superalloy machinability evaluation after laser cladding additive manufacturing process. International Journal of Advanced Manufacturing Technology, 2018, 97(5–8), 2873–2885.

[24] Harrysson, O.L.A., Marcelin-Little, D.J. TJH. Applications of metal additive manufacturing in veterinary orthopedic surgery. Journal of the Minerals Metals and Materials Society, 2015, 67, 647–654.

[25] Gurdita, A. The Most Common 3D Printed Prosthetics [Internet]. 2018. Available from: https://all3dp.com/2/the-most-common-3d-printed-prosthetics/

[26] Muthu, S.S., Savalani, MM. Handbook of Sustainability in Additive Manufacturing [Internet]. 1, 2016. Available from: http://link.springer.com/10.1007/978-981-10-0606-7

[27] Touraine, Emmanuel. 3D printed 18K gold limited edition. 2018.

[28]   Touraine, Emmanuel. 3D Printing generatic home collection. 2018.

[29]   Srivastava, VK. A review on advances in rapid prototype 3D printing of multi-functional applications. Science and Technology, 2017, 7(1), 4–24.

[30]   Watkin, H. Wyss Institute Develops Hybrid 3D Printing for Cheap Wearables. 2017.

[31]   Werkheiser, N. NASA Additive Manufacturing Overview [Internet]. 2017 [cited 2018 Jun 21]. Available from: https://ntrs.nasa.gov/archive/nasa/casi.ntrs.nasa.gov/20170001551.pdf

[32]   Valerga Puerta, A.P., Moreno Sanchez, D., Batista M. SJ. Criteria selection for a comparative study of functional performance of fused deposition modelling and vacuum casting processes. Journal of Manufacturing Process [Internet]. Elsevier; 2018, 35, 721–727. Available from: https://doi.org/10.1016/j.jmapro.2018.08.033

[33]   Sireesha, M., Lee, J., Sandeep, A., Kiran, K., Babu, VJ., Kee, BT., et al. A review on additive manufacturing and its way into the oil and gas industry. RSC Advances in Royal Society of Chemistry, 2018, 8, 22460–22468.

[34]   Upcraft, S., Fletcher, R. The rapid prototyping technologies. Assembly Automation [Internet]. 2003, 23(4), 318–330. Available from: http://www.emeraldinsight.com/doi/10.1108/01445150310698634

[35]   Grida, I., Evans, JRG. Extrusion freeforming of ceramics through fine nozzles. Journal of the European Ceramic Society, 2003, 23(5), 629–635.

[36]   Gornet, T. History of Additive Manufacturing [Internet]. 2014. Available from: http://services.igi-global.com/resolvedoi/resolve.aspx?doi=10.4018/978-1-5225-2289-8.ch001

[37]   RepRap Wiki [Internet]. [cited 2018 Aug 2]. Available from: https://www.reprap.org/wiki/RepRap

[38]   Scott, CS., Apparatus and method for creating three-dimensional objects. 1989.

[39]   Farah, S., Anderson, DG., Langer, R. Physical and mechanical properties of PLA, and their functions in widespread applications – A comprehensive review. Advanced Drug Delivery Reviews [Internet], Elsevier B.V.; 2016, 107, 367–392. Available from: http://dx.doi.org/10.1016/j.addr.2016.06.012

[40]   Wong, K. V. AH. A Review of Additive Manufacturing. ISRN Mechanical Engineering, 2012, 38, 208–218.

[41]   Nordin, NAB., Bin, Johar MA., Bin, Ibrahim MHI., Bin, Marwah OMF. Advances in High Temperature Materials for Additive Manufacturing. IOP Conference Series: Materials Science and Engineering, 2017, 226(1), 1–8.

[42]   Fernandez-Vicente, M., Calle, W., Ferrandiz, S., Conejero, A. Effect of Infill Parameters on Tensile Mechanical Behavior in Desktop 3D Printing. 3D Print Additive Manufacturing [Internet], 2016, 3(3), 183–192. Available from: http://online.liebertpub.com/doi/10.1089/3dp.2015.0036

[43]   Rodriguez, J.F., Thomas, J.P. JER. Characterization of the mesostructure of fused-deposition acrylonitrile-butadiene-styrene materials. Rapid Prototyping Journal, 2000, 6(3), 175–186.

[44]   Sood, A.K., Ohdar, R.K. SSM. Parametric appraisal of mechanical property of fused deposition modelling processed parts. Materials and Design, 2010, 31(1), 287–295.

[45]   Valerga, A.P., Batista, M., Salguero, J. FG. Influence of PLA filament conditions on characteristics of FDM parts. Materials (Basel), 2018, 11(1322), 1–13.

[46]   Ceccanti, F., Dini, E., De Kestelier, X., Colla, V. LP. 3D printing technology for a moon outpost exploiting lunar soil. In: Proceedings of the 61th International Astronautical Congress (IAC 2010). Prague (Czech Republic), 2010.

[47]   Khoshnevis, B., Hwang, D. CC. A mega scale fabrication technology. Manufacturing Systems Engineering Service, 2006, 6, 221–251.

[48]   Buswell, R.A., Soar, R.C., Gibbb, A.G.F. AT. Freeform Construction: mega-scale rapid manufacturing for construction. Automation in Construction, 2007, 16, 224–231.

[49] D-Shape. 3D Printed House/Structure Location. New York, US, 2014.

[50] Giannatsis, J. VD. Additive fabrication technologies applied to medicine and health care: a review. International Journal of Advanced Manufacturing Technology, 2009, 40, 116–127.

[51] Billiet, T., Vandenhaute, M., Schelfhout, J., Van Vlierberghe, S. PD. A review of trends and limitations in hydrogel-rapid prototyping for tissue engineering. Jorunal Biomaterials, 2012, 33, 6020–6041.

[52] Ramanath, HS., Chua, CK., Leong, KF., Shah, KD. Melt flow behaviour of poly-ε-caprolactone in fused deposition modelling. Journal of Materials Science: Materials in Medicine, 2008, 19(7), 2541–2550.

[53] Chin Ang, K., Fai Leong, K., Kai, Chua C., Chandrasekaran, M. Investigation of the mechanical properties and porosity relationships in fused deposition modelling-fabricated porous structures. Rapid Prototyping Journal [Internet]. 2006, 12(2), 100–105. Available from: http://www.emeraldinsight.com/doi/10.1108/13552540610652447

[54] Chim, H., Hutmacher, DW., Chou, AM., Oliveira, AL., Reis, RL., Lim, TC., et al. A comparative analysis of scaffold material modifications for load-bearing applications in bone tissue engineering. International Journal of Oral and Maxillofacial Surgery, 2006, 35(10), 928–934.

[55] Wake Forest Institute for Regenerative Medicine print ear, finger bone and kidney structure scaffolds using a 3-D printer [Internet]. 2017. Available from: https://www.jble.af.mil/News/Photos/igphoto/2000833052/

[56] Lipton, J. I., Cutler, M.1, Nigl, F., Cohen, D. HL. Additive manufacturing for the food industry. Trends in Food Science and Technology, 2015, 43, 114–123.

[57] Wegrzyn, T., Golding, M. RA. Food layered manufacture: a new process of constructing solid foods. Trends Food Science Technology, 2012, 1, 66–72.

[58] Severini, C., Derossi, A., Azzollini, D. Variables affecting the printability of foods: Preliminary tests on cereal-based products. Innovative Food Science and Emerging Technologies [Internet], Elsevier Ltd; 2016, 38, 281–291. Available from: http://dx.doi.org/10.1016/j.ifset.2016.10.001

[59] Godoi, FC., Prakash, S., Bhandari, BR. 3d printing technologies applied for food design: Status and prospects. Journal of Food Engineering [Internet]. Elsevier Ltd; 2016, 179, 44–54. Available from: http://dx.doi.org/10.1016/j.jfoodeng.2016.01.025

[60] Dambrāns, K. 3D food printing [Internet]. 2017. Available from: https://www.flickr.com/photos/janitors/21222046631/in/photostream/

[61] Nidagundi, VB., Keshavamurthy, R., Prakash, CPS. Studies on parametric optimization for fused deposition modelling process. Materials Today: Proceedings [Internet]. Elsevier Ltd.; 2015, 2(4–5), 1691–1699. Available from: http://dx.doi.org/10.1016/j.matpr.2015.07.097

[62] Sood, A.K. SSM and RKO. Weighted principal component approach for improving surface finish of abs plastic parts built through fused deposition modelling process. International Journal of Rapid Manufacturing, 2011, 2, (1), 4–27.

[63] Brooks, HL., Rennie, AEW., Abram, TN., McGovern, J., Caron, F. Variable fused deposition modelling – analysis of benefits, concept design and tool path generation. Innovative Developments in Virtual and Physical Prototyping – Proceedings of the 5th International Conference on Advanced Research in Rapid Prototyping [Internet], 2012, [May 2014]:511–517. Available from: http://www.scopus.com/inward/record.url?eid=2-s2.0-84856731282&partnerID=40&md5=9d82ded72946d02d26a0f2d14f7487fd

[64] Valerga, AP., Batista, M., Fernandez, SR., Gomez-Parra, A., Barcena, M. Preliminary study of the influence of manufacturing parameters in fused deposition modeling. Proc 26th DAAAM Int Symp [Internet]. 2016;[2016]:1004–1008. Available from: http://www.daaam.info/Downloads/Pdfs/proceedings/proceedings_2015/141.pdf

[65] Wendt, C., Fernández-Vidal, SR., Gómez-Parra, Á., Batista, M., Marcos, M. Processing and Quality Evaluation of Additive Manufacturing Monolayer Specimens. Advances in Materials Science and Engineering, 2016, 2016, 1–8.

[66] Valerga, AP., Batista, M., Puyana, R., Sambruno, A., Wendt, C., Marcos, M. Preliminary study of PLA wire colour effects on geometric characteristics of parts manufactured by FDM. Procedia Manufacturing, Elsevier B.V.; 2017, 13, 924–931. Available from: https://doi.org/10.1016/j.promfg.2017.09.161

[67] Singh, R., Singh, S., Fraternali, F. Development of in-house composite wire based feed stock filaments of fused deposition modelling for wear-resistant materials and structures. Composites Part B: Engineering, [Internet]. Elsevier Ltd; 2016, 98, 244–249. Available from: http://dx.doi.org/10.1016/j.compositesb.2016.05.038

[68] Nuñez, PJ., Rivas, A., García-Plaza, E., Beamud, E., Sanz-Lobera, A. Dimensional and surface texture characterization in fused deposition modelling (FDM) with ABS plus. Procedia Engineering, [Internet]. Elsevier B.V.; 2015, 132, 856–863. Available from: http://dx.doi.org/10.1016/j.proeng.2015.12.570

[69] Casavola, C., Cazzato, A., Moramarco, V., Pappalettera, G. Residual stress measurement in fused deposition modelling parts. Polymer Test, [Internet]. Elsevier Ltd; 2017, 58, 249–255. Available from: http://dx.doi.org/10.1016/j.polymertesting.2017.01.003

[70] Morgan, HD., Cherry, JA., Jonnalagadda, S., Ewing, D., Sienz, J. Erratum to: Part orientation optimisation for the additive layer manufacture of metal components, 10.1007/s00170-015-8151-6). The International Journal of Advanced Manufacturing Technology [Internet], 2016, 86(5–8), 1689. Available from: http://dx.doi.org/10.1007/s00170-015-8151-6

[71] Ahn, S., Montero, M., Odell, D., Roundy, S., Wright, PK. Anisotropic material properties of fused deposition modeling ABS. Rapid Prototyping Journal [Internet], 2002, 8(4), 248–257. Available from: http://www.emeraldinsight.com/doi/10.1108/13552540210441166

[72] Hashemi Sanatgar, R., Campagne, C., Nierstrasz, V. Investigation of the adhesion properties of direct 3D printing of polymers and nanocomposites on textiles: Effect of FDM printing process parameters. Applied Surface Science [Internet], Elsevier B.V.; 2017, 403, 551–563. Available from: http://dx.doi.org/10.1016/j.apsusc.2017.01.112

[73] Kariz, M., Kuzman, MK., Sernek, M. Adhesive bonding of 3D-printed ABS parts and wood. Journal of Adhesion Science and Technology[Internet], Taylor & Francis, 2017, 31(15), 1683–1690. Available from: http://dx.doi.org/10.1080/01694243.2016.1268414

[74] Too, M.H., Leong, K.F., Chua, C.K., Du, Z.H., Yang, S.F., Cheah, C.M. SLH. Investigation of 3D non-random porous structures by fused deposition modelling. International Journal of Advanced Manufacturing Technology, 2002, 19, 217–223.

[75] Cao, Tong; Kee-Hai, Ho; Scaffold, Swee-Hin T., Design and in vitro study of osteochondral coculture in a three-dimensional porous polycaprolactone scaffold fabricated by fused deposition modeling. Tissue Engineering, 2003, 9, 103–112.

[76] Singh, J., Singh, CR. Pre and post processing techniques to improve surface characteristics of FDM parts: a state of art review and future applications. Rapid Prototyping Journal [Internet], 2017, 23(3). Available from: http://dx.doi.org/10.1108/RPJ-05-2015-0059http://dx.doi.org/10.1108/RPJ-10-2015-0151http://dx.doi.org/10.1108/RPJ-05-2016-0076

[77] Drizo, A. JP. Environmental impacts of rapid prototyping: an overview of research to date. Rapid Prototyping Journal, 2006, 12, 64–71.

[78] Madhavan Nampoothiri, K., Nair, NR., John, RP. An overview of the recent developments in polylactide (PLA) research. Bioresource Technology [Internet]. Elsevier Ltd, 2010, 101(22), 8493–8501. Available from: http://dx.doi.org/10.1016/j.biortech.2010.05.092

[79] Schulze, C., Juraschek, M., Herrmann, C., Thiede, S. Energy analysis of bioplastics processing. Procedia CIRP, [Internet]. 2017, 61, 600–605. Available from: http://dx.doi.org/10.1016/j.procir.2016.11.181

[80] Kellens, K., Mertens, R., Paraskevas, D., Dewulf, W., Duflou, JR. Environmental Impact of Additive Manufacturing Processes: Does AM Contribute to a More Sustainable Way of Part Manufacturing? Procedia CIRP [Internet], 2017, 61[Section 3], 582–587. Available from: http://dx.doi.org/10.1016/j.procir.2016.11.153

[81] Meyer, VB. Prototyping the Environmental Impacts of 3D Printing: Claims and Realities of Additive Manufacturing. Environmental Policy Senior Thesis; 2015.

[82] Parandoush, P., Lin, D. A review on additive manufacturing of polymer-fiber composites. Composite Structures [Internet], Elsevier Ltd; 2017, 182, 36–53. Available from: https://doi.org/10.1016/j.compstruct.2017.08.088

[83] Novakova-Marcincinova, L., Basic, Kuric I. and Advanced materials for fused deposition modeling rapid prototyping technology. Manufacturing Industrial Engineers, 2012, 11(1), 1338–6549.

[84] Novakova-Marcincinova, L., Novak-Marcincin, J., Barna, J., Torok, J. Special materials used in FDM rapid prototyping technology application. In: IEEE 16th International Conference on Intelligent Engineering Systems (INES). 2012.

[85] Kumar, R., Singh, R., Ahuja, IPS. Investigations of mechanical, thermal and morphological properties of FDM fabricated parts for friction welding applications. Measurement: Journal of the International Measurement Confederation [Internet], Elsevier; 2018, 120, [January], 11–20. Available from: https://doi.org/10.1016/j.measurement.2018.02.006

[86] Lanzotti, A., Grasso, M., Staiano, G., and Martorelli, M. The impact of process parameters on mechanical properties of parts fabricated in PLA with an open-source 3-D printer. Rapid Prototyping Journal, 2015, 21(5), 604–617.

[87] Hill, N., Haghi, M. Deposition direction-dependent failure criteria for fused deposition modeling polycarbonate. Rapid Prototyping Journal [Internet], 2014, 20(3), 221–227. Available from: http://www.emeraldinsight.com/doi/10.1108/RPJ-04-2013-0039

[88] Srivastava, M., Maheshwari, S., Kundra, TK. Estimation of the Effect of Process Parameters on Build Time and Model Material Volume for FDM Process Optimization by Response Surface Methodology and Grey Relational Analysis. Advanced 3D Print Additive Manufacturing Technology, 2017, 29–38.

[89] Bagsik, A., Schöoppner, V. Mechanical properties of fused deposition modeling parts manufactured with ULTEM 9085. Proceedings of ANTEC, 2011, 2011, 1294–1298.

[90] Singh, S., Ramakrishna, S., Singh, R. Material issues in additive manufacturing : A review. Journal of Manufacturing Process. The Society of Manufacturing Engineers, 2017, 25, 185–200.

[91] Jia, Y., He, H., Peng, X., Meng, S., Chen, J., Geng, Y. Preparation of a new filament based on polyamide-6 for three-dimensional printing. Polymer Engineering and Science, 2017, 57(12), 1322–1328.

[92] Kamoona, SN., Masood, SH., Mohamed, OA. An investigation on impact resistance of FDM processed Nylon-12 parts using response surface methodology. In: AIP Conference Proceedings 1859 [Internet]. 2017, 1–9. Available from: http://aip.scitation.org/doi/abs/10.1063/1.4990273

[93] For Hla, P. THE, Jackson, A., Hopkins, J. 3D Printing for the HLA laboratory. Human Immunology[Internet]. American Society for Histocompatibility and Immunogenetics, 76, 2015, 66. Available from: http://dx.doi.org/10.1016/j.humimm.2015.07.093

[94] Goyanes, A., Buanz, ABM., Hatton, GB., Gaisford, S., Basit, AW. 3D printing of modified-release aminosalicylate (4-ASA and 5-ASA) tablets. European Journal of Pharmaceutics and Biopharmaceutics, Elsevier B.V.; 2015, 89, 157–162.

[95] Carneiro, OS., Silva, AF., Gomes, R. Fused deposition modeling with polypropylene. Materials Design [Internet], Elsevier Ltd; 2015, 83, 768–776. Available from: http://dx.doi.org/10.1016/j.matdes.2015.06.053

[96] Jang, J., Jung, JW., Pati, F., Huleihil, M. Possible applications of 3D printing technology on textile substrates possible applications of 3D printing technology on textile substrates. IOP Conference Series: Materials Science Engineering, 2016, 141, 012011.

[97] Kim, K., Park, J., Suh, J., Kim, M., Jeong, Y., Park, I. 3D printing of multiaxial force sensors using carbon nanotube (CNT)/ thermoplastic polyurethane (TPU) filaments. Sensors Actuators A Physics [Internet], Elsevier B.V.; 2017, 263, 493–500. Available from: http://dx.doi.org/10.1016/j.sna.2017.07.020

[98] Chen, G., Chen, N., Wang, Q. Preparation of poly(vinyl alcohol)/ionic liquid composites with improved processability and electrical conductivity for fused deposition modeling. Materials Design [Internet], Elsevier Ltd; 2018, 157, 273–283. Available from: https://doi.org/10.1016/j.matdes.2018.07.054

[99] Wang YT., Yeh YT. Effect of Print Angle on Mechanical Properties of FDM 3D Structures Printed with POM Material. In: Bajpai R., Chandrasekhar U. (eds) Innovative Design and Development Practices in Aerospace and Automotive Engineering. Lecture Notes in Mechanical Engineering. Springer, Singapore, 2017.

[100] Street, D., Bergman, J., Messman, J., Kilbey, M. Enhancing thermomechanical properties of PMMA parts fabricated via fused deposition modeling by incorporation of polymer-grafted nanoparticles. In: 255th National Meeting and Exposition of the American-Chemical-Society (ACS) – Nexus of Food, Energy, and Water, New Orleans, LA: 2018, 255.

[101] Dul, S., Fambri, L., Pegoretti, A. Fused deposition modeling with ABS-graphene nanocomposites. Composites PART A, 2016, 85, 1–25.

[102] Sydney Gladman, A., Matsumoto, EA., Nuzzo, RG., Mahadevan, L., Lewis, JA. Biomimetic 4D printing. Nature Materials, 2016, 15(4), 413–418.

[103] Tibbits, S., Cheung, K. Programmable materials for architectural assembly and automation. Assembly Automation [Internet], 2012, 32(3), 216–225. Available from: http://www.emeraldinsight.com/doi/10.1108/01445151211244348

[104] Ge, Q., Dunn, Qi HJ. ML. Active materials by four-dimension printing. Applied Physics Letter, 2013, 103(13), 131901-1,131901-5.

[105] Tibbits, S. 4D printing: Multi-material shape change. Architecture Design, 2014, 84(1), 116–121.

[106] Ge, Qi, Sakhaei, Amir Hosein, Howon Lee, Conner K. NXF& MLD. Multimaterial 4D Printing with Tailorable Shape Memory Polymers. Scientific Reports [Internet], 2016, 6(31110). Available from: https://doi.org/10.1038/srep31110

[107] Mooney, D J., Sano, K., Kaufmann, M P., Majahod, K., Schloo, B., Vacanti, JP. et al. Long term engraftment of hepatocytes transplanted on biodegradable polymer sponges. Journal of Biomedical Materials Research, 1997, 37(4), 13–20.

[108] DW, H. Scaffolds in tissue engineering bone and cartilage. Biomaterials, 2000, 21(25), 29–43.

[109] Saeidlou, S., Huneault, MA., Li, H., Park, CB. Poly(lactic acid) crystallization. Progress in Polymer Science [Internet], Elsevier Ltd; 2012, 37(12), 1657–1677. Available from: http://dx.doi.org/10.1016/j.progpolymsci.2012.07.005

[110] Parada, DC., Laverde, D., Peña, YD., Vazquez, QC. Obtain poly (l-lactic) acid by polycondensation with catalyst metallic zinc. 268 Scientia et Technica XIII, 2007, 36, 267–272

[111] Rudnik, E. Compostable Polymer Materials, Elsevier, Netherlands. 2010.

[112] NatureWorks LLC [Internet]. [cited 2018 Aug 3]. Available from: https://www.natureworksllc.com/

[113] Castro-Aguirre, E., Iñiguez-Franco, F., Samsudin, H., Fang, X., Auras, R. Poly(lactic acid) – Mass production, processing, industrial applications, and end of life. Advanced Drug

Delivery Reviews [Internet], Elsevier B.V.; 2016, 107, 333–366. Available from: http://dx.doi. org/10.1016/j.addr.2016.03.010

[114] Mirabal, A.S., Scholz, L. MC. Market study on bio-based polymers in the word- capacities, production and applications: status quo and trends towards 2020. 2013.

[115] Auras, R., Lim, L.T., Selke, S. HT. Poly(Lactic Acid). Synthesis, Structures, Properties, Processing, and Applications. New Jersey: John Wiley. 2010.

[116] Wendt, C., Valerga, AP., Droste, O., Batista, M., Marcos, M. FEM based evaluation of Fused Layer Modelling monolayers in tensile testing. Procedia Manufacturing [Internet], Elsevier B.V.; 2017, 13[October], 916–923. Available from: https://doi.org/10.1016/j. promfg.2017.09.160

[117] Stewart, Samuel R., Wentz, John E., Allison JT. Experimental and computational fluid dynamic analysis of melt flow behavior in fused deposition modelling of poly(lactic) acid. ASME International Mechanical Engineering Congress and Exposition (IMECE2015). Houston, TX, 2015.

[118] Heshmati, V., Zolali, AM., Favis, BD. Morphology development in poly (lactic acid)/ polyamide11 biobased blends : Chain mobility and interfacial interactions. Polymer (Guildf), Elsevier Ltd; 2017, 120, 197–208.

[119] Gomez-gras, G., Jerez-mesa, R., Travieso-rodriguez, JA., Lluma-fuentes, J. Fatigue performance of fused filament fabrication PLA specimens. Materials Design, Elsevier Ltd; 2018, 140, 278–285.

[120] Tymrak, B.M., Kreiger, M., and Pearce, JM. Mechanical properties of components fabricated with open-source 3-D printers under realistic environmental conditions. Materials Design, 2014, 58, 242–246.

[121] Ramot, Y., Haim-Zada, M., Domb, AJ., Nyska, A. Biocompatibility and safety of PLA and its copolymers. Advanced Drug Delivery Reviews [Internet]. Elsevier B.V.; 2016, 107, 153–162. Available from: http://dx.doi.org/10.1016/j.addr.2016.03.012

[122] Cohn, D., Hotovely Salomon, A. Designing biodegradable multiblock PCL/PLA thermoplastic elastomers. Biomaterials, 2005, 26(15), 2297–2305.

[123] Ramot, Y., Touitou, D., Levin, G., Ickowicz, DE., Zada, MH., Abbas, R., et al. Interspecies differences in reaction to a biodegradable subcutaneous tissue filler: Severe inflammatory granulomatous reaction in the sinclair minipig. Toxicologic Pathology, 2015, 43(2), 267–271.

[124] Satyanarayana, D CP. Biodegradable polymers: challenges and strategies. Journal of Macromolecular Science – Reviews in Macromolecular Chemistry and Physics, 1993, 33(3), 49–68.

[125] Nyska, A., Schiffenbauer, YS., Brami, CT., Maronpot, RR., Ramot, Y. Histopathology of biodegradable polymers: Challenges in interpretation and the use of a novel compact MRI for biocompatibility evaluation. Polymers for Advanced Technologies, 2014, 25(5), 461–467.

[126] Onuki, Y., Bhardwaj, U., Papadimitrakopoulos, F., Burgess, DJ. A Review of the Biocompatibility of Implantable Devices: Current Challenges to Overcome Foreign Body Response. Journal of Diabetes Science and Technology, 2008, 2(6), 1003–1015.

[127] Middleton, JC., Tipton, AJ. Synthetic biodegradable polymers as orthopedic devices. Biomaterials, 2000, 21(23), 2335–2346.

[128] Nicolson, K. Bioplastics for silage bales [Internet]. 2011. Available from: http://www.flickr. com/photos/katynicolson/2697487631/sizes/m/in/photostream

[129] Guerra, AJ., Cano, P., Rabionet, M., Puig, T., Ciurana, J. 3D-Printed PCL / PLA Composite Stents : Towards a New Solution to Cardiovascular Problems. Materials (Basel), 2018, 11, 1679.

[130] Guerra, AJ., Cano, P., Rabionet, M., Puig, T., Ciurana, J. 3D-Printed PCL / PLA Composite Stents: Towards a New Solution to Cardiovascular Problems. Materials, 2018, 11(9), 1679.

[131] Andrea. Example of Cutlery Made from Biodegradable Plastic. Vancouver, Canada, 2009. Avaible from: https://www.thenewslens.com/article/59293

[132] Lim, LT., Auras, R., Rubino, M. Processing technologies for poly(lactic acid). Progress in Polymer Science, 2008, 33(8), 820–852.

[133] Lv, T., Zhou, C., Li, J., Huang, S., Wen, H., Meng, Y., et al. New insight into the mechanism of enhanced crystallization of PLA in PLLA/PDLA mixture. Journal of Applied Polymer Science, 2018, 135(2), 1–7.

[134] Tsuji, H. MS. Poly(L-lactide) 6. Effects of crystallinity on enzymatic hydrolysis of poly(L-lactide) without free amorphous region. Polymer Degradation and Stability, 2001, 71, 415–424.

[135] Cai, H., Dave, V., Gross, RA MS. Effects of physical aging, crystallinity, and orientation on the enzymatic degradation of poly(lactic acid). Journal of Polymer Science Polymer Physics Edition, 1996, 34, 2701–2708.

[136] Mathew, AP., Oksman, K., Sain, M. Mechanical properties of biodegradable composites from poly lactic acid (PLA) and microcrystalline cellulose (MCC). Journal of Applied Polymer Science, 2005, 97(5), :2014–2025.

[137] Song, Y., Li, Y., Song, W., Yee, K., Lee, KY., Tagarielli, VL. Measurements of the mechanical response of unidirectional 3D-printed PLA. Materials Design, 2017, 123, 154–164.

[138] Letcher, T., Waytashek, M., Material Property Testing of 3D-Printed Specimen. Proceedings of the ASME 2014 International Mechanical Engineering Congress and Exposition IMECE2014. November 14–20, 2014, Canada.

[139] Vijay, P., Danaiah, P., Rajesh, D. K V. Critical Parameters Effecting the Rapid Prototyping Surface Finish. Journal of Mechanical Engineering and Automation [Internet]. 2012, 1(1), 17–20. Available from: http://article.sapub.org/10.5923.j.jmea.20110101.03.html

[140] Kumar, GSB. and P. Methods to improve surface finish of parts produced by fused deposition modeling. Manufacturing Science and Technology, 2014, 2(3), 51–55.

[141] Kulkarni, P. and Dutta, D. On the integration of layered manufacturing and material removal process. International Journal of Machine Science Engineering, 2000, 122, 100–108.

[142] Agarwala, M.K., Jamalabad, V.R., Langrana, N.A., Safari, A., Whalen, PJ. and, Danforth, SC. Structural quality of parts processed by fused deposition. Rapid Prototyping Journal, 1996, 2(4), 4–19.

[143] Ahn, D. and Kim, H. and Lee S. Determination of fabrication direction to minimize post-machining in FDM by prediction of non-linear roughness characteristics. Journal of Mechanical Science and Technology, 2005, 19(1), 144–155.

[144] Patel, J.P., Patel, C.P. and Patel UJ. A review on various approach for process parameter optimization of fused deposition modeling (FDM) process and Taguchi approach for optimization. International Journal of Engineering Research and Applications, 2012, 2(2), 361–365.

[145] Patel, R., Patel, S. and Patel J. A review on optimization of process parameter of fused deposition modeling for better dimensional accuracy. International journal of Engineering development and research, 2014, 2(2), 1620–1624.

[146] Ibrahim, D., Ding, S. and Sun, S. Roughness Prediction for FDM Produced Surfaces. In: Proceedings of International Conference on Recent trends in Engineering and Technology. Batam, Indonesia; 2014, 70–74.

[147] Galantucci, GP. and LM. Local-Genetic Slicing of Point Clouds for Rapid Prototyping. Rapid Prototyping Journal, 2008, 14(3), 161–166.

[148] Huang, B., Singamneni, SB. Curved layer adaptive slicing (CLAS) for fused deposition modelling. Rapid Prototyping Journal [Internet], 2015, 21(4), :354–367. Available from: http://www.emeraldinsight.com/doi/10.1108/RPJ-06-2013-0059

[149] Fodron, E. MK and UM. Mechanical and dimensional characteristics of fused deposition modeling build styles. In: Proceedings of 7th Solid Freeform Fabrication Symposium. Austin, USA, 1996, 419–442.

[150] Spencer, J.D., Cobb, R.C., and Dickens PM. Surface finishing techniques for rapid prototyping. In: Technical Paper PE93-169, Rapid Prototyping Conference Dearborn. Michigan, USA; 1993.

[151] Schmid, M., Simon, C. and Levy, GN. Finishing of SLS-parts for rapid manufacturing (RM) – a comprehensive approach. In: Proceedings of 20th Solid Freeform Fabrication Symposium. Austin, USA, 2009, 1–10.

[152] Blair, BM. Post build processing of stereolithography molds. M.S. Thesis, Georgia Institute of Technology, Atlanta, USA, 1998.

[153] Galantucci, L.M., Lavecchia, F. and Percoco, G. Experimental study aiming to enhance the surface finish of fused deposition modeled parts. CIRP Annals – Manufacturing Technology, 2009, 58, 189–192.

[154] Boschetto, A., and Bottini, L. Surface improvement of fused deposition modeling parts by barrel finishing. Rapid Prototyping Journal, 2015,21(6), 686–696.

[155] Williams, R.E., Walczyk, D.F. and Dang HT. Using abrasive flow machining to seal and finish conformal channels in laminated tooling. Rapid Prototyping Journal 2007, 13(2), 64–75.

[156] Kumar, R., Singh, R., Ahuja, IPS., Amendola, A., Penna, R. Friction welding for the manufacturing of PA6 and ABS structures reinforced with Fe particles. Composites Part B: Engineering [Internet], Elsevier Ltd; 2018,132, 244–257. Available from: https://doi.org/10.1016/j.compositesb.2017.08.018

[157] Stratasys. Application Manual 3.0 [Internet]. Document No: FAM3.0. 1997. Available from: http://3d4u.org/MyFDM/wp-content/uploads/2010/12/2000APPL.pdf

[158] Augur, H. "Freddy" Receives Hand-Painted 3D Printed Tortoise Shell. 2016.

[159] Galeta, T., G. Šimunović MM. Impact of strengthening fluids on roughness of 3D printed models. Metallurgy, 2015,1(54), 231–234.

[160] Raja, K., Naiju, CD., Narayanan, S., Jeeva, PA., Karthikeyan, S. Metallization of ABS plastics prepared by FDM-RP process and evaluation of corrosion and hardness characteristics. International Journal of ChemTech Research, 2014, 6(14), 5490–5493.

[161] Kannan, S., Senthilkumaran, D. Investigating the influence of electroplating layer thickness on the tensile strength for fused deposition processed abs thermoplastics. International Journal of Engineering and Technology, 2014, 6(2), 1047–1052.

[162] Rao, A. S., Dharap, M.A., Venkatesh, J. V. L. and Ojha, D. Investigation of post processing techniques to reduce the surface roughness of fused deposition modeled parts. International Journal of Mechanical Engineering Technology, 2012, 3(3):531–544.

[163] Priedeman Jr. W., Thomas Smith, D. US 8123999 B2, 2012. 1–10.

[164] Zinniel, RL. Surface treatment method for rapid manufactured three-dimensional objects. US 2008/0169585 A1, 2008, 1–8.

[165] Kuo, CC., Mao, RC. Development of a precision surface polishing system for parts fabricated by fused deposition modeling. Material Manufacturing Processes, 2016, 31(8), 1113–1138.

[166] Garg, A., Bhattacharya, A., Batish, A. On surface finish and dimensional accuracy of fdm parts after cold vapor treatment. Materials Manufacturing Processes. 2016, 31(4), 522–529.

[167] Yang, S., Zhao, YF. Additive manufacturing-enabled design theory and methodology: a critical review. International Journal of Advanced Manufacturing Technology, 2015, 80(1–4), 327–342.

[168] Kuczko, W., Górski, F., Wichniarek, R., Buń, P. Influence of post-processing on accuracy of FDM products. Advances in Science and Technology Research Journal [Internet], 2017, 11(2),

172–179. Available from: http://www.journalssystem.com/astrj/Influence-of-post-processing-on-accuracy-of-FDM-products,70996,0,2.html

[169] Pandey, P.M., Reddy, N.V. SGD. Improvement of surface finish by stair case machining in fused deposition modelling. Journal of Materials Processing Technology, 2003, 132, 323–331.

[170] Vinitha, M., Rao, A.N. MKM. Optimization of speed parameters in burnishing of samples fabricated by fused deposition modeling. International Journal of Mechanical Industrial Engineering, 2012, 2(2), 10–12.

[171] Hanus, A., Špirutová, N., Beňo, J. Surface quality of foundry pattern manufactured by FDM method – rapid prototyping. Archives of foundry engineering, 2011, 11(1), 15–20

[172] Hambali, RH., Cheong, KM., Azizan, N. Analysis of the influence of chemical treatment to the strength and surface roughness of FDM. IOP Conference Series: Materials Science and Engineering, 2017, 210(1), 1–9.

[173] Lalehpour, A., Barari, A. Post processing for fused deposition modeling parts with acetone vapour bath. IFAC-Papers Online [Internet]. Elsevier B.V.; 2016, 49(31), 42–48. Available from: http://dx.doi.org/10.1016/j.ifacol.2016.12.159

Viktor P. Astakhov, Swapnil Patel

3 Development of the basic drill design for cored holes in additive and subtractive manufacturing

Abstract: One of the practical ways to increase productivity of additive manufacturing (AM) and subtractive manufacturing (SM) operations is a reduction in the volume of the work material in part blanks used than for finishing machining operations. One of the most used ways is to make blanks with cored holes. In AM, it significantly reduces the time needed to make a blank and cost of raw material(s) spent per blank; in both AM and SM, it increases productivity and efficiency of hole-making operations.

In practice, drilling/reaming of cored holes involves a major difficulty; the axis of the cored hole is always shifted (with rather generous tolerance) from that of the finished hole. This chapter presents a step-by-step development of the basic drill design for cored holes. A new design concept was developed; practical drills according to this concept were made, tested, and implemented for drilling cored holes. The test and implementation result showed a two- to threefold increase in tool life with no drill premature failures.

Keywords: cored holes, drill designs, tool life, additive and subtractive manufacturing, force balance, hole accuracy, automotive industry

3.1 Introduction

One of the practical ways to increase productivity of a machining operation is reduction in the volume of the work material removed by machining; as this result in multiple advantages:
1. Reduction of the amount of the work material needed to make a part.
2. As a smaller chip is formed, near-dry, cryogenic, CO_2, and other economical methods of coolant supply can be used.
3. Increased tool life; as tool wear is proportional to the volume of the work material being removed.

Reduction of the volume of the work material removed by machining can be achieved if the near-net-shape blanks are used. It may require redesigning blanks, utilizing different processes to produce these blanks [additive manufacturing (AM), forging, extrusion, cold-forming, die casting, etc.].

https://doi.org/10.1515/9783110549775-003

3.1.1 Cored hole in subtractive manufacturing

In the automotive industry, when die castings are used as blanks, cored holes, that is, holes made in castings with some stock to be removed by semifinishing drilling and then reaming, are commonplace today. Figure 3.1 shows a typical example of a die casting of the valve body of a six-speed automatic transmission. As can be seen, a great number of cored holes are used. The use of such hole often eliminates the need in solid drilling so that the two- or three-pass hole-making operations (two or three cutting tools) can be substituted by single-pass operations with a finishing reamer having a drill point to break a small flash at the bottom of the hole being drilled. However, a number of problems have to be solved to make cored hole machining practical and reliable [1, 2].

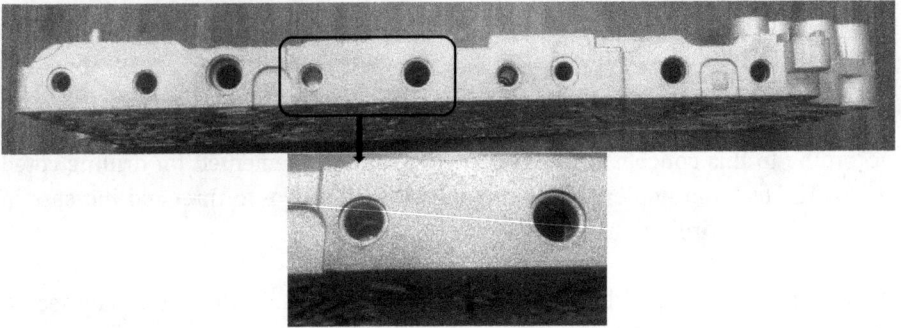

Figure 3.1: (a) and (b) Example of cored holes made in an aluminum die casting.

In practice, drilling/reaming of cored holes involves a major problem. The essence of the problem is that the axis of the cored hole is always shifted from that is required for the finish hole. The older the casting die (known as the cavity), the greater the shift due to wear of core pins. Obviously, casting suppliers are trying to reduce their costs by using casting dies as long as possible due to the significant cost of new dies. The tolerances on the location of the axis of cored holes are rather generous (up to 1.8 mm) so that this axis can be significantly shifted from the actual intended axis on the hole in the manufactured part.

3.1.2 Cored hole in additive manufacturing

The use of cored holes in AM is even more beneficial compared to SM. Significant reduction of process time and considerable savings on the expensive powder used are of prime importance. Figure 3.2 shows an example of an AM part made on a Renishaw RenAM 500M metal AM machine. As can be seen, a number of cored holes are the case.

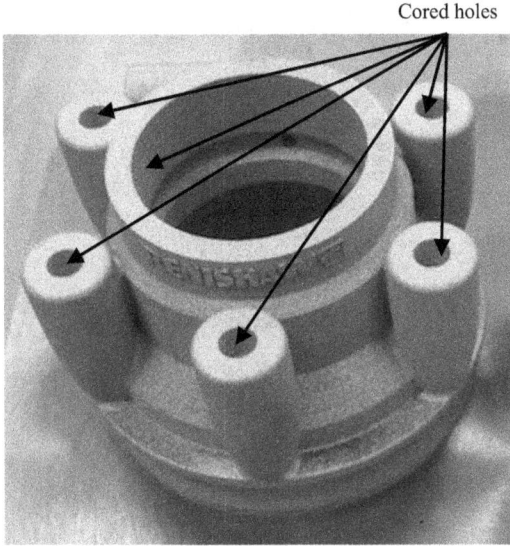

Figure 3.2: Example of cored holes made in an AM blank.

Dimensional inaccuracy of parts produced using AM can be even more problematic [3], particularly when considering a prototype or high value part where the end use is for a component requiring tight dimensional control [4, 5]. The layering process used in AM methods can result in rough surfaces, and possible deviations from the dimensions and tolerances specified the CAD model and/or other geometrical anomalies [6]. Typically, the CAD model is converted to a stereolithography (*.stl) file format, where the designed geometries and surfaces are discretized into geometric meshes. A macrolevel "staircase" effect can occur on part surfaces due to this discretization [7]. In addition, it has been reported that melt pool dynamics have a large influence on sidewall dimensions for AM made parts [8]. The risk of occurrence for curling, waviness, and surface roughness is also all influenced by the previously discussed process and material parameters. To minimize geometrical anomalies, a stable melt pool size/shape is required [8]. This is not easy to achieve in practice so that the so-called Marangoni effect has a strong influence on melt pool size and shape and can introduce anomalies in deposited layers due to its dependence on composition and the local thermal gradients [4].

3.2 Problem with machining cored holes

The problem starts when one attempts to machine these holes using regular-design tools with the results shown in Figure 3.3 – its consequences range from the out-of-size (shape) holes to broken tools. It is already reported by Astakhov

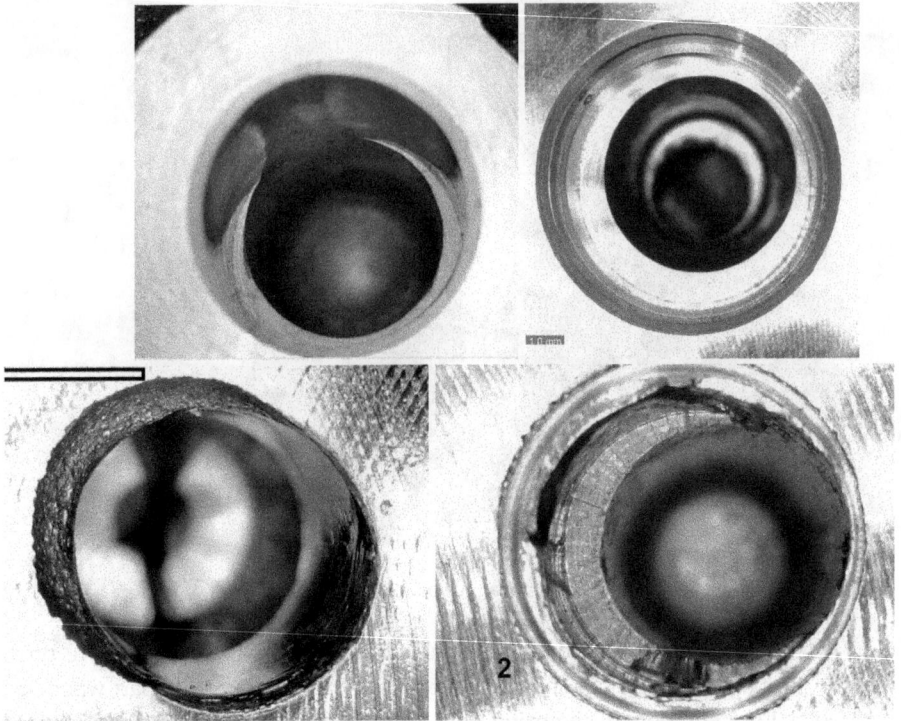

Figure 3.3: Problem with machining of cored holes.

[9], and many problems in machining of cored holes in casting are caused by the use of improper cutting tools. To realize the extent of a problem with the existing drill design in terms of handling cored holes, one should first understand the concept of the cutting edge, number of cutting edges in a common drill shown in Figure 3.4(a), components of the cutting force acting on each cutting edge, and force balance in drilling.

3.2.1 Force balance

Three cutting edges are attributed to each flute of the drill: 1–2 – the major cutting edge; 2–3 – the chisel edge; 1–1′ – the minor cutting edge as shown in Figure 3.4(b). Their orthogonal component of the resultant cutting force should be assigned to each cutting edge. Figure 3.4(b) shows such component applied to the major cutting edge 1–2. The main or power component of the resultant force, F_P, is normally the greatest component. The force in the feed direction is known as the feed or axial force, F_A. The component in the radial direction, F_R, is known as the radial component.

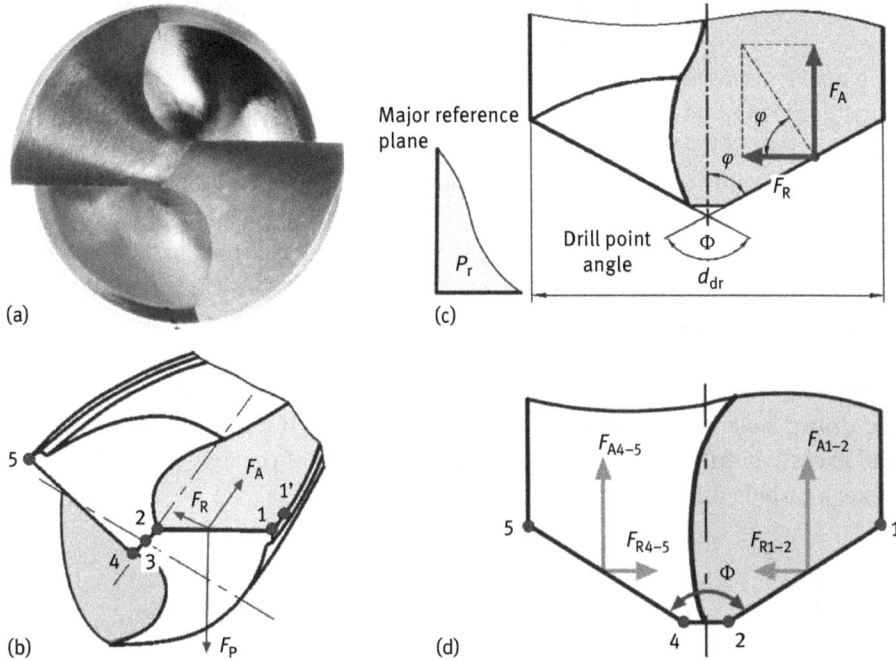

Figure 3.4: Drill and its force balance: (a) common two-flute drill, (b) force components on the cutting edge, (c) axial and radial components as related by the half-point angle, and (c) force balance commonly presented in literature.

One of the major drill geometry parameters is the point angle, Φ, considered in the major reference plane P_r [10] as shown in Figure 3.4(c). In the tool design, however, the half-point angle $\varphi = \Phi/2$ is considered. This is because this angle defines the ratio of the axial and radial force components as

$$\frac{F_A}{F_R} = \tan \varphi \tag{3.1}$$

In drilling, proper performance of a drill of the common design completely relies on the force balance [10]. A force balance usually considered in the literature on drilling [10] is shown in Figure 3.4(d). Engagement of each of the major cutting edges, 1–2 and 4–5, with the workpiece results in the axial forces, F_{A1-2} and F_{A4-5}, and radial forces, F_{R1-2} and F_{R4-5}, respectively. The axial forces are balanced by the feed force applied by the machine/spindle, whereas the radial forces, being theoretically equal and of opposite direction, should counterbalance each other when the drill is designed and manufactured properly, and applied as intended. In other words, the discussed force balance is maintained if both major cutting edges are entering the workpiece simultaneously.

3.2.2 Violation of the force balance in drilling cored holes

The above-discussed proper force balance, which assures intended drill working conditions, may not be the case in drilling of cored holes. As schematically depicted in Figure 3.5(a), when the axis of the cored hole is shifted from the drill rotation axis, the only one major cutting edge (edge 1–2 as shown) is actually engaged in cutting, particularly at the entrance of the hole being machined. As such, the radial force on cutting edge 1–2, $F_{R1–2}$, is not balanced so it tries to bend the drill. Figure 3.5(b) shows that, trying to enter the cored hole, the drill is bent by radial force $F_{R1–2}$. Then the drill rotates (the most common case in modern machining centers), and is subjected to "whirl" motion till point b touches the edge of the cored hole (Figure 3.5(b)). As a result, an excessive bell mouth (the tapered past of the drilled hole at the entrance) [11], oversize hole and out-of-shape conditions, and excessive drill wear are reported. When the shift of the axis (Figure 3.5(a)) is more, a carbide drill simply brakes.

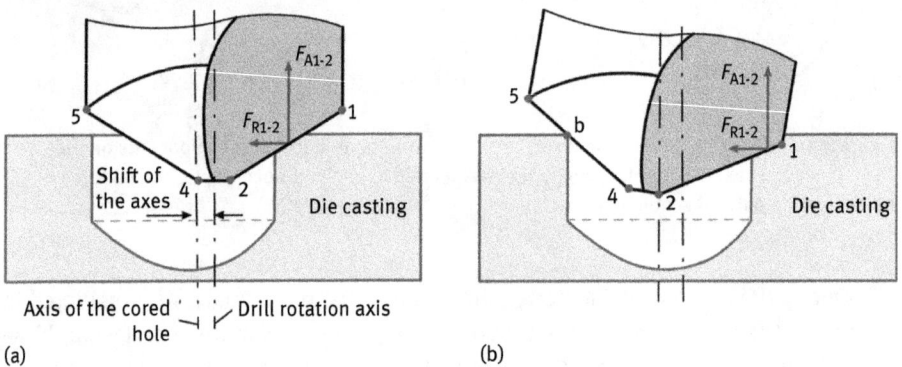

Figure 3.5: A major problem with drilling of a cored hole: (a) the radial force on the cutting edge 1–2 is not balanced and (b) drill banding by the unbalanced radial force.

For years, the discussed problem was simply ignored because the diameter of cored holes was rather small compared to that of the drilled holes, tolerance on the drilled and then reamed holes were not tight, drilling speed and the feed rate were not aggressive as restricted by machine capabilities, and primarily relatively low available spindle rpm's and power. In the modern automotive industry, the diameter of the cored holes has increased in an attempt to cut less work material, tolerances on the hole are tightened at least four times in the last 10 years, and rotational speeds and feed rate are quadrupled over the same time period. All these improvements including significantly more rigid machine spindles and part fixtures brought the discussed problem to the forefront.

3.3 Developing of the problem solution

Developing of the complete problem solution is presented using a special example of cored holes where three worst-case scenario takes place, that is, where the following is the case: (1) the face of the cored hole is not perpendicular to the axis of the drill, (2) cored hole is shifted from this axis, and (3) quality (diametric accuracy and cylindricity) of the drilled hole is of high importance to part performance. The rational is as follows: if a drill to handle these requirements can be developed, then the developed tool can drill efficiently virtually any other cored hole in die castings or in parts made by AM. A drilling operation including all the listed conditions is shown in Figure 3.6, which shows the tool layout, and Figure 3.7 shows the process drawing. As shown in these figures, a rather tight tolerance (for a cored hole) is assigned on the drilled hole and the position of the hole axis is tightly controlled by two position tolerances. Further analysis of this hole reveals that it is a hole for tapping an NPSF 1/8–27 pipe threat subjected to a high-pressure leakage test. That is why the tolerance on the drill diameter is tight.

3.3.1 Initial drill design, its common improvement, and result

The initial drill design assigned at the beginning of the program is shown in Figure 3.8. As shown, its critical first stage is a 120° point angle carbide drill with internal coolant supply, which, as discussed above, is not suitable for the application. As it was found soon enough, it was replaced by a common drill design recommended by leading drill manufacturers to handle cored holes and inclined hole entrances (shown in Figure 3.9). As shown, this drill has a 180° point angle, sufficient clearance angle of the primary and secondary flank faces (SECTION B-B in Figure 3.9). According to the model shown in Figure 3.4(d) and eq. (3.1), the radial forces on the major cutting edges are zero so that there is no drill bending/distortion on the entrance of the cored hole being drilled even if the axis of this hole is shifted with respect to the drill rotational axis. Therefore, the expectation was that the problem should be solved with the implementation of this drill design once and forever.

Unfortunately, the problem was not solved although it was not obvious as related to the drill. Instead, frequently occurring chipping/fracturing of the tap that follows the drill was found to be the case. Two common modes of tap chipping/fracture were found: (a) fracture of the face and (b) fracture of the most loaded cutting tooth are shown in Figure 3.10, which was a surprise for those recommended this drill design for cored holes. Instead of taking a close look at drill design, all efforts were directed toward trying taps made by various tap manufacturers and various tap materials with no or very limited success.

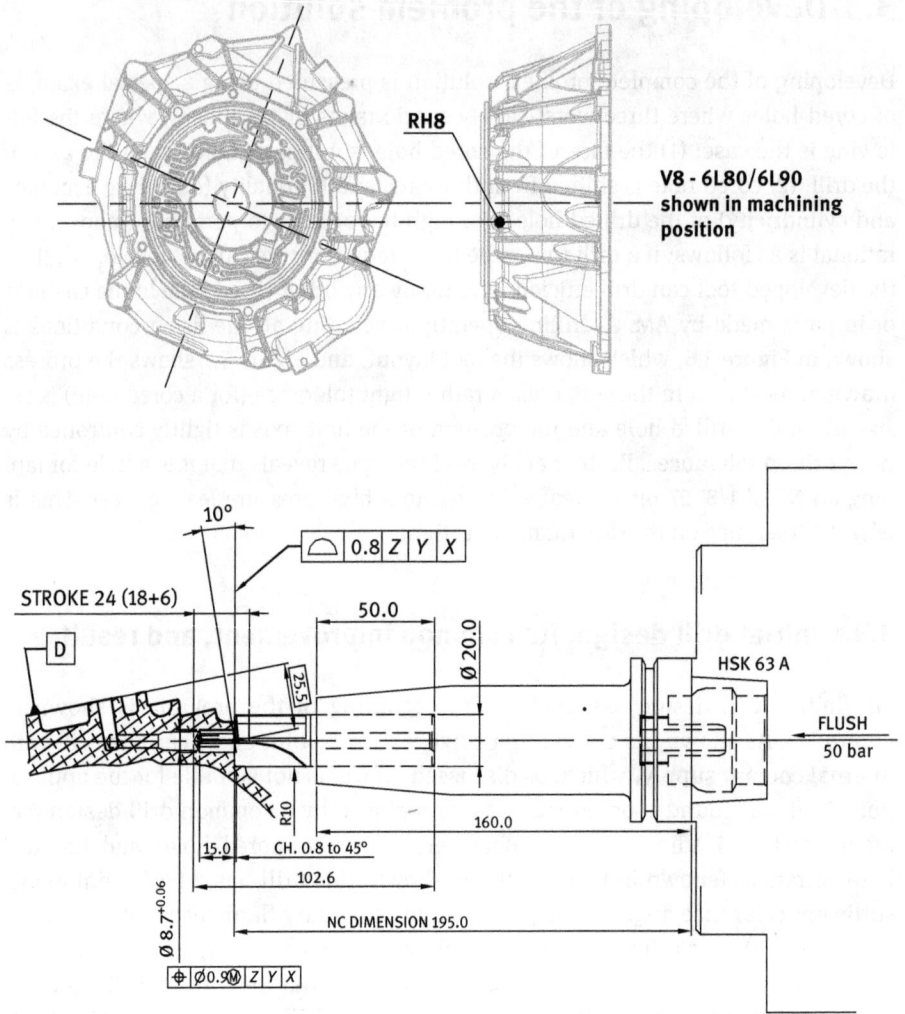

Figure 3.6: Tool layout.

3.3.2 Analysis of the problem and its apparent solution

3.3.2.1 Taping problem

The first step in our analysis of the problem was a graphical analysis of tap chipping/fracture as presented in Figure 3.11. Figure 3.11(a) shows the normal/intended tap entrance when the axis of the drilled hole and tap is coincident (within a tolerance). When this is the case, the face of the tap first enters into the hole, and the first contact between the tap and the hole occurs on the chamfer by the first cutting

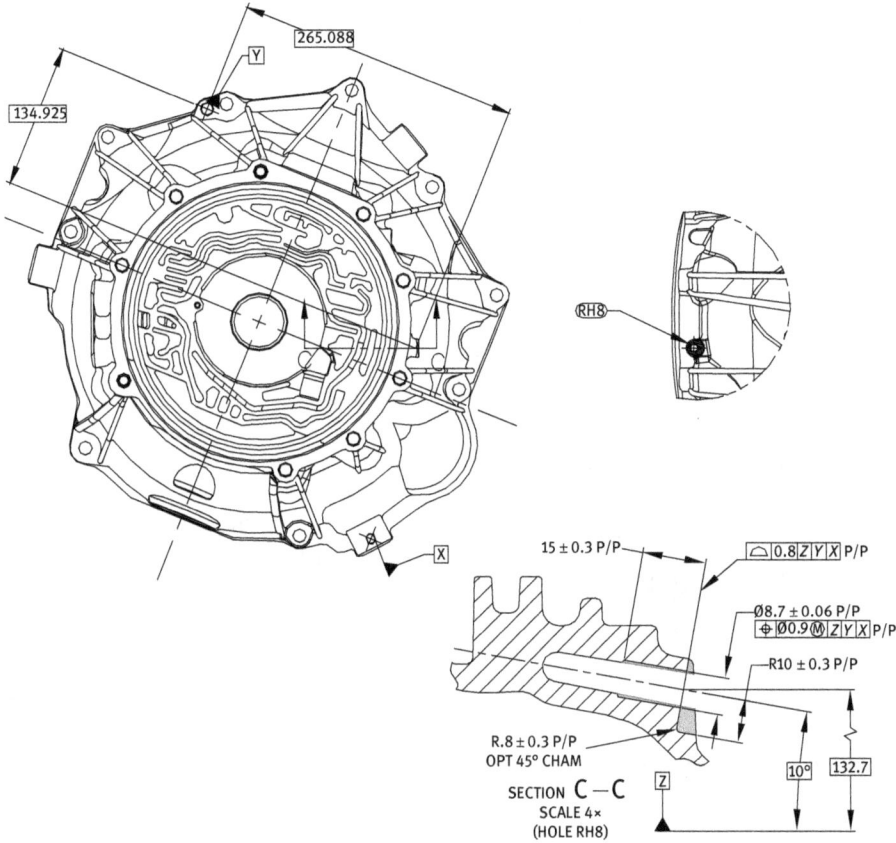

Figure 3.7: Process drawing.

tooth of the tap. As such, a small mismatch (within a tight tolerance) of the axis of the tap and the hole is intended to be compensated by small tap bending so that the load on each cutting tooth of the tap is almost the same. Under these conditions, the tap cuts a normal thread. Figure 3.11(b) shows the abnormal tap entrance when the axes of the tap and the drilled hole are not coincident. As shown, unintended place of contact between the tap and the hole takes place. When the mismatch of the axis is too great, the face of the tap contacts the edge of the drilled hole. As this face has no means to cut, it simply fractured as shown in Figure 3.10(a). When the mismatch is smaller, that is, when the face can enter the drilled hole, but is still greater than the maximum allowable (by tolerance) mismatch, the bending of the tap causes significant force on its cutting teeth. As such, the tooth with the greatest (combined cutting and bending) load fractures in the manner is shown in Figure 3.10(b).

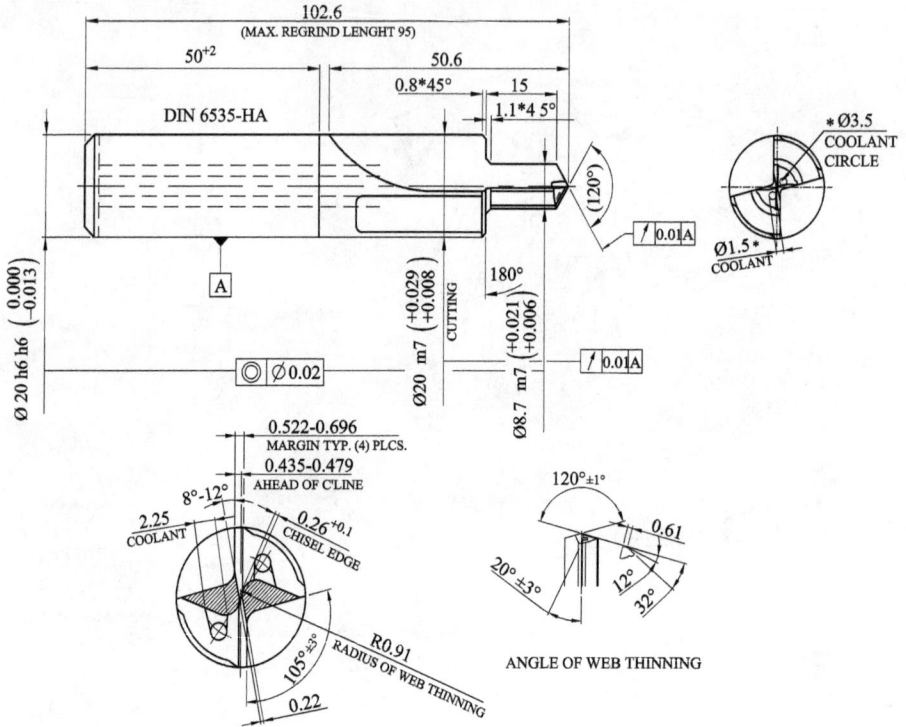

Figure 3.8: Initial drill design.

3.3.2.2 A 180° point angle drill problem

3.3.2.2.1 Entrance problem

The second step of our analysis was to understand why a drill of 180° point angle, commonly recommended by leading drill manufacturers for solving problems with cored holes, drilling, and with nonflat face drill entrances does not actually solve the problem. This problem is explained with the help of Figure. 3.12, which shows a drill of 180° point angle entering a cored hole axis of which is shifted with respect to the drill rotating axis. As can be seen that under such condition, the uncut chip cross-sectional area of the chip cut by the right major cutting edge (shown as 1-2-2'-1') is greater than that cut by the left major cutting edge (shown as 4-5-5'-1'). It is to say that when the shift of the cored hole is relatively small, this drill can make some number of holes. When the shift is greater (still well within the tolerance assigned to die casting blanks), the drill makes holes that cause tap chipping/fracture in the manner shown in Figure 3.10. The drill can make the hole with the intended axis location (coincident with the drill rotating axis), and thus can compensate for the shift of the axis of cored hole only when the side cutting edges, namely 1–1' and 5–5', are

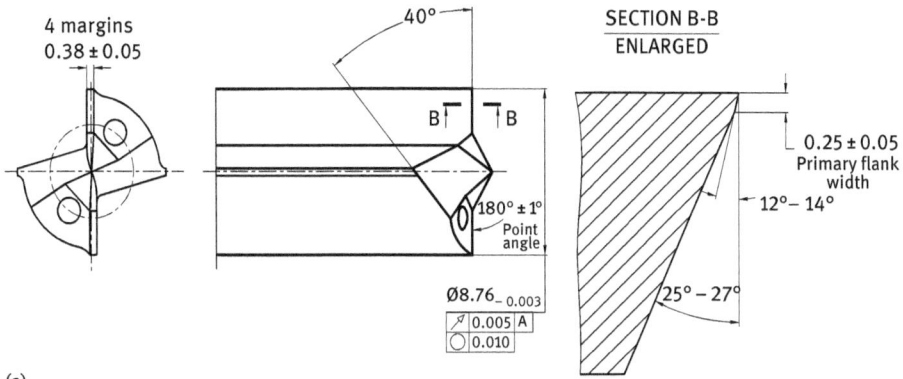

4 margins
0.38 ± 0.05

40°

SECTION B-B
ENLARGED

0.25 ± 0.05
Primary flank
width

12° – 14°

180° ± 1°
Point
angle

Ø8.76₋₀.₀₀₃

25° – 27°

(a)

(b)

Figure 3.9: (a) and (b) Drill design used in the first attempt to solve the problem.

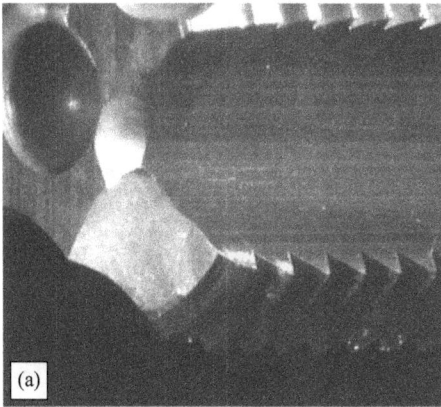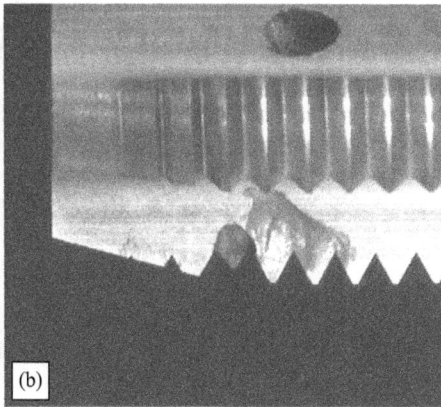

(a)

(b)

Figure 3.10: Two common modes of tap chipping/fracture: (a) fracture of the face and (b) fracture of the most loaded cutting tooth.

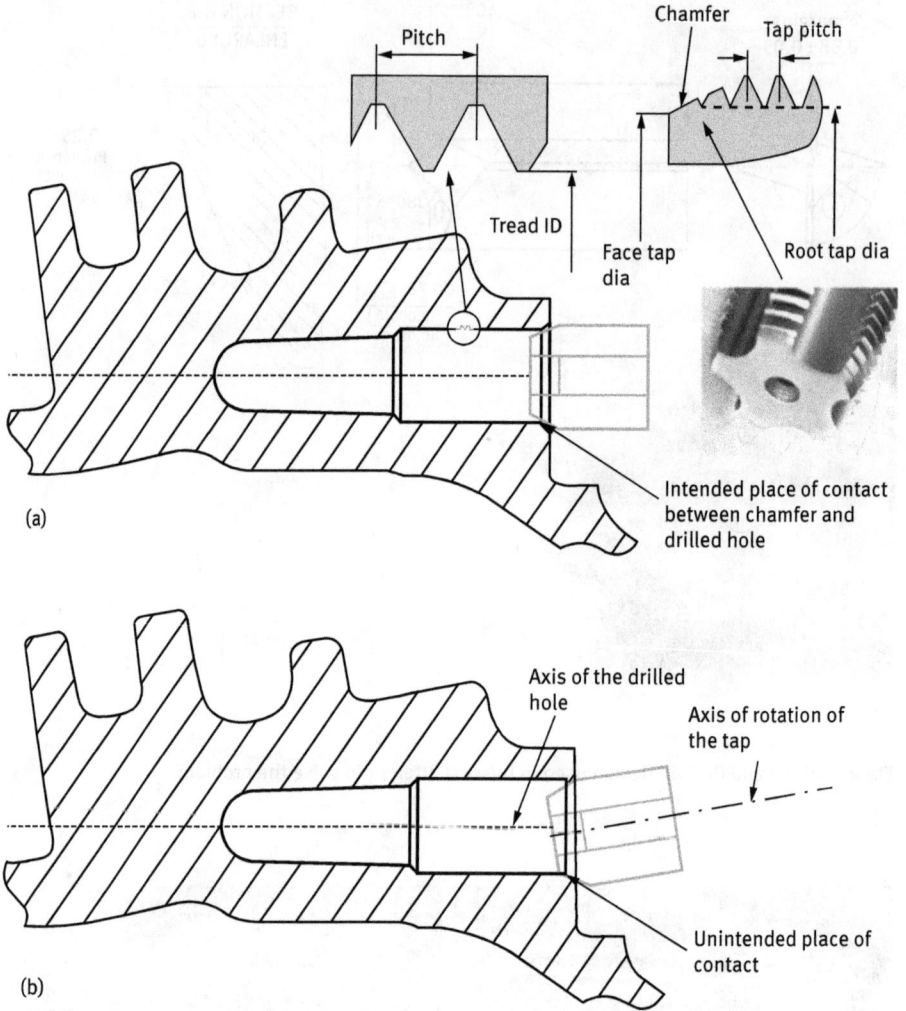

Figure 3.11: Graphical analysis of the tab entrance into the drilled hole: (a) normal/intended tap entrance, and (b) abnormal tap entrance when the axes of the tap and the drilled hole are not coincident.

actually cut similar to the side edges on a plunging end mill tool. The problem is that these edges are not cutting edges as they do not have a positive clearance angle so they can't cut [10]. This is because the margin serving as the flank face of the side cutting edges is cylindrical, that is, has a zero clearance angle. As such, when drill tries to enter the hole in the manner shown in Figure 3.12, a significant axial force F_{cll} arises that bends the drill causing the problem shown in Figure 3.5(b).

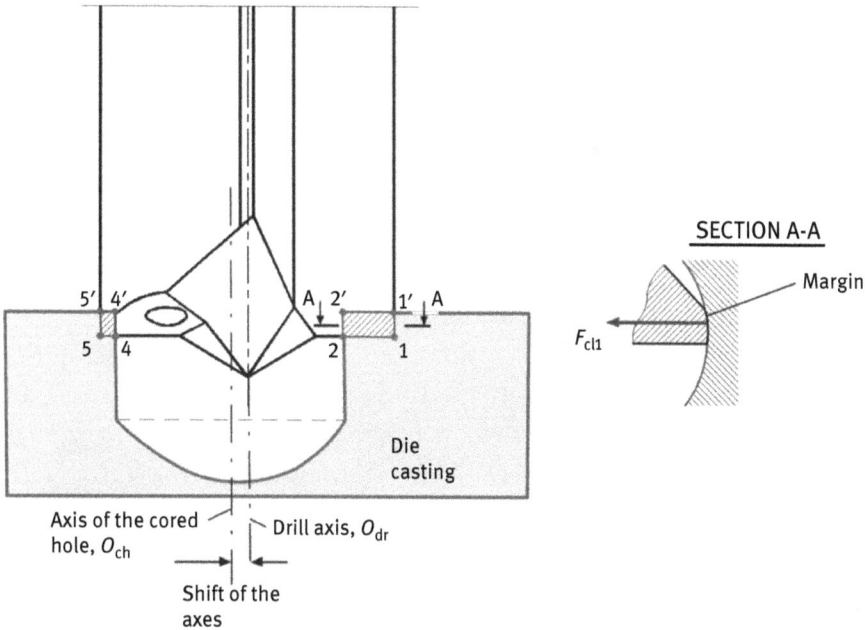

Figure 3.12: The problem with a common drill of 180° point angle in terms of handling of cored holes is shown.

3.3.2.2.2 Drilling problem caused by the feed per tooth

The drilling problem with a 180° point angle drill working with high feed per tooth is in the geometry of the side cutting edge of standard drills. To understand this problem, let's consider a hypothetical drill shown in Figure 3.13. As shown, this drill has a single (major) cutting edge and no minor (side) cutting edge. Figure 3.14 shows the axial cross section of the hole being drilled by this tool. For clarity, the feed per revolution is significantly exaggerated. Although this profile looks odd, Figure 3.15 explains this result. Figure 3.15(a) shows two successive positions of the drill with the standard (for carbide drills) 140° (the half point angle is 70°) point angle. As can be seen, because there is no side cutting edge provided, a part of the work material represented by triangle ABC forms at each drill revolution that yields in the hole profile is shown in Figure 3.14. Figure 3.15(b) shows what happens when the point is increased to 170° (the half-point angle is 85°). As can be seen, much larger area $ABB'C$ is to be cut by the side cutting edge. Obviously, the maximum area to be cut by the side cutting edge is when the port angle is 180° (the half-point angle is 90°), that is, it is the same as in the drill design shown in Figure 3.9. The problem with cutting of this maximum area is the same as described in Section 3.3.2.2.1 – the side cutting edge can't cut as it is not provided with a positive clearance angle. When a point angle is 160° or greater and when the feed per revolution is great as in drilling of

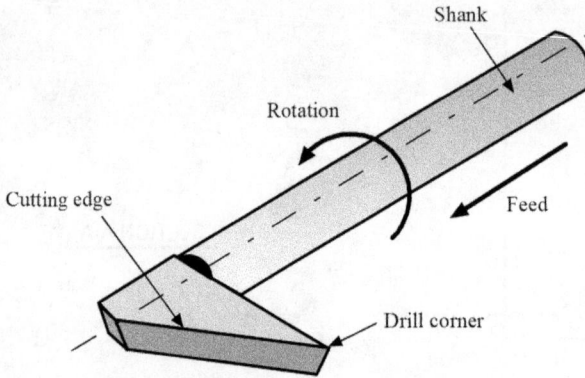

Figure 3.13: A hypothetical cutting tool.

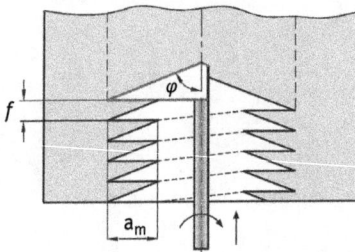

Figure 3.14: Profile of the hole drilled by the hypothetical drill shown in Figure 3.13.

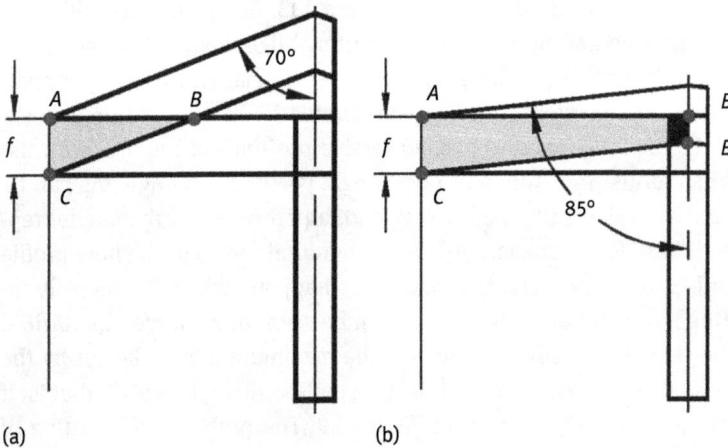

(a)   (b)

Figure 3.15: Explanation of the profile shown in Figure 3.14 and influence of the half-point angle on the uncut chip thickness cut by the side cutting edge.

aluminum alloys in the automotive industry, the absence of the positive clearance angle on the side cutting edge makes drilling unstable and causes premature wear of the drill periphery corners. Figure 3.16 shows a margin of a straight-flute drill for high-speed machining of high-silicon aluminum alloy with a high feed. As can be seen, the aluminum deposit forms on this margin as the side cutting edge adjacent to the drill corner cannot cut.

Figure 3.16: Aluminum deposit of the flank face of the side cutting edge which is the drill margin.

3.3.2.2.3 Drilling problem caused by elastic recovery (springback)

There is another not well-revealed problem particular to all drills but it is of great concern in machining of work materials of low modulus of elasticity, for example, aluminum and titanium alloys. This problem is material elastic recovery (a.k.a. stringback) of the work material. Unfortunately, the nature of this phenomenon and its inherent link to the modulus of elasticity of the work material are not clearly understood by cutting tool and machining process designers/specialists. A need is felt to clarify these items to help specialists to design better drilling tools.

As defined by Astakhov [12] and then supported by many others, for example, by Atkins [13, 14], the process of metal cutting is the purposeful fracture of the work material as physical separation of the workpiece on two portions, namely the machined part and the chip, takes place. To understand the concept of stringback, one should consider the stress–strain diagram of the work material up to fracture. A generalized and thus simplified stress–strain diagram is shown in Figure 3.17. The most important aspects relevant to the foregoing analysis are segments between the

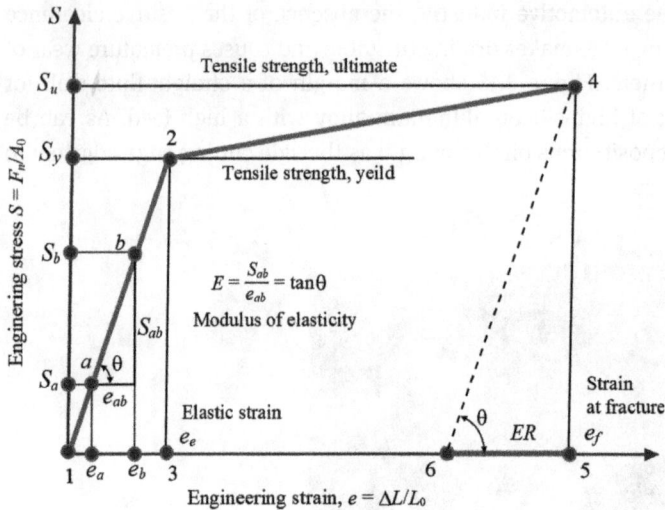

Figure 3.17: A simplified stress–strain diagram.

numbered points. When a solid material is subjected to small stresses, the bonds between the atoms are stretched. When the stress is removed, the bonds relax and the material returns to its original shape. This reversible deformation is called elastic deformation represented by segment 1–2 in the diagram. For most materials, segment 1–2 is linear. Segment 1–3 is known as elastic strain, that is, the maximum elastic deformation the material can be subjected to. The stress at point 2 is known as the yield strength of the material.

The slope of this linear segment (tan θ) is called the modulus of elasticity or Young's modulus, E determined by selecting two points on segment 1–2 (a and b) in the manner shown in Figure 3.17 so that

$$E = \tan \theta = \frac{S_b - S_a}{e_b - e_a} = \frac{S_{ab}}{e_{ab}} \tag{3.2}$$

where S_a, S_b, e_a, and e_b are stresses and strains at points a and b, respectively.

Segment 3–4 in the diagram (Figure 3.17) is called the strain hardening region. If the applied stress exceeds the yield strength, the material exhibits a combination of the elastic (reversible) and plastic (irreversible) deformations. If applied stress can grow further up to point 4 on the diagram then fracture occurs. The strain corresponding to point 4 is known as the strain at fracture. In Figure 3.17, it is represented by segment 1–5 on the strain axis. After fracture, however, the applied stress is released and the permanent strain found in the work material (represented by segment 1–6 in Figure 3.17) is less than that at fracture by the elastic strain represented by distance 6–5 in the stress–strain diagram. As such, the location of point 6 is readily found by drawing a line from point 4 parallel to line 1–2, that is, at angle

θ. Segment 5–6 on the strain axis is known as elastic recovery, ER, which is also known as springback in materials processing.

The next step is to apply the concept of elastic recovery to drilling. Figure 3.18 shows a simple model where no drill runout is considered for the sake of simplifying further considerations. In Figure 3.18(a), the drill corners 1 and 5 separate the chip from the rest of the workpiece forming the wall of the hole being drilled. The diameter of this wall is always equal to that of the drill, that is, d_{dr}. Once the corners 1 and 5 of the drill pass a certain part of the machined hole, the load due to cutting is removed so that the applied stress returns to zero. As a result, the machined hole shrinks due to springback (elastic recovery) because elastic deformation is recoverable deformation. Note that this recovery is not only due to mechanical action of corners 1 and 5 but also due to heating and subsequent cooling of the work material adjacent to the hole being drilled. In other words, the thermal energy due to plastic deformation of the work material and friction between the tool and the workpiece in their relative motion causes thermal expansion of the work material around the drill terminal end. When the corners 1 and 5 advance further, the work material contracts due to cooling by the coolant. As a result of the mentioned mechanical and thermal factors, the diameter of the hole being machined d_{dh} becomes smaller than that of the diameter of the drill d_{dr} by Δ_{er} as shown in Figure 3.18(b).

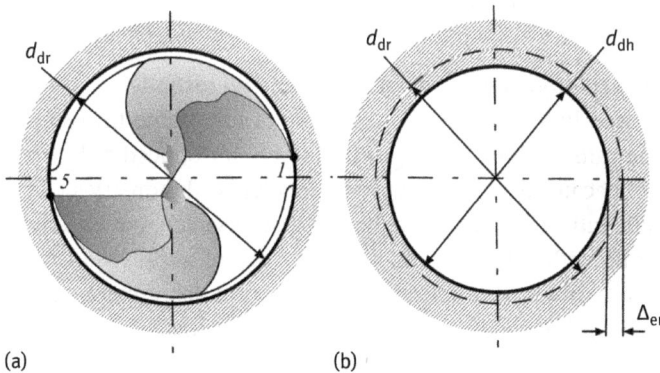

Figure 3.18: Model of elastic recovery in drilling: (a) the diameter of the hole being drilled at the point of cut and (b) elastic recovery behind this point.

Several variables influence the amount of springback. Among others, the stress at fracture (defines the height of the starting point of the unloading line represented by point 4 in Figure 3.17), the modulus of elasticity (defines the slope of the unloading line represented by line 4–6 in Figure 3.17), and thermal expansion are of importance. It is obvious from the diagram shown in Figure 3.17 that the higher the strength of the work material, the greater the springback; the lower elasticity modulus, the higher springback. Consider a few practical examples. As follows from the diagram shown in Figure 3.17, the said springback of a unit volume of the work material can be determined using triangle *456* as the ratio of the ultimate strength of the work material, S_u, and its modulus of elasticity, E, that is, springback = S_u/E.

Consider practical examples:

1. *Elastic recovery*. As the modulus of elasticity is almost the same for wide group of steels (E = 200 GPa), the springback is determined by the strength of the steel. For cold-drawn steel AISI 1012, having S_u = 270 MPa, springback = 0.00135, while for annealed steel AISI 1095, having S_u = 650 MPa, springback = 0.00325. The matter gets worse when titanium or aluminum alloys are drilled. For commonly used annealed titanium alloy Ti–Al6–4 V (Grade 5), S_u = 880 MPa, E = 113.8 GPa, and springback = 0.00772, that is, four times greater than that of medium-carbon steels. This explains known difficulties with drill binding in machining of titanium alloys. For common in the automotive industry, aluminum alloy A380 is used at TTO S_u = 320, MPa E = 71 MPa, and springback = 0.00450, which is greater than that of steels..

2. *Thermal contraction*. The coefficient of thermal expansion for steel AISI 1012 is 12.0 µm/m°K whereas for A380 is 21.8 µm/m°K, that is, it is almost twice greater for steel AISI 1012.

As a result, the combined (springback and thermal) hole shrinkage is much (more than threefold) greater for A380 than for steel AISI 1012.

3.3.2.2.4 Why the discussed problems were not noticeable till recently

The drilling problem caused by the feed per tooth was not noticeable till recently because high penetration rate (high cutting feed) drills were introduced only when the quest for productivity became a mainstream trend in advanced industries where the cycle time was significantly affected by machining time. As such, advanced design of solid carbide drills with internal high-pressure coolant supply and high-speed machines with linear drives were introduced to assure much greater cutting feed in drilling. Increased cutting speed also added to the above-discussed thermo-shrinkage problem.

Although the problem with springback was around since the first drill used, its severity was not of central concern because the system (total) runout of drill caused by runout of the drill itself, drill holder, and machine spindle was significantly greater than springback. At best, the so-called backtaper (a slight decrease in diameter from front to back in the body of the drill), which prevents the drill from jamming in the hole being drilled, was used. As such, backtaper of 0.08 mm/100 mm was used for years.

With the introduction and wide use of high-precision drills, tool holders, and machine spindles, the problem with springback came to the forefront. The authors introduced drills with the backtaper of 3 µm/mm (0.3 mm/100 mm) and reamers with the backtaper of up to 2 μm/mm (0.2 mm/100 mm), that is, more than threefold greater than commonly used in industry. Our implementation experience shown that, as suggested by the above analysis, backtaper should be customized based on particular drilling conditions. However, even with increased backtaper, the margin behind a small side edge adjacent to the drill corner (see edge 1–1′ in Figure 3.4(b)) is still suffering from excessive rubbing due to springback.

The foregoing analysis explains why the drill of the design shown in Figure 3.9 is not suitable for drilling a cored hole, particularly under strict quality requirements. Figure 3.19 shows its wear pattern for the above-described applica-tion. As can be seen, the areas adjacent to the drill corners are severely worn after drilling less than 1,000 holes. Obviously, a new design is needed to deal with the problem.

Figure 3.19: Wear pattern in a common drill of 180° point angle.

3.3.2.3 Solution of the problem

To solve the problem, the role and thus the design of drill margins should be re-thought for high-speed high-penetration rate drilling using modern drilling systems of high accuracy, particularly in drilling cored holes. The solution of the problem was suggested by Astakhov (figure 163 in [10]) and is referred to as Design Rule #1. The major line of thought in the formulation of this rule is to convert a small portion of the margin adjacent into the drill corner (e.g., 1–1′ in Figure 3.4(b)) in a legitimate side cutting edge. It is described as follows with the help of Figure 3.20. The tradi-tional drill/reamer margin is cylindrical as shown in Figure 3.20(a). Its width a_m is selected in the range of 0.3–0.6 mm depending on the drill diameter. This range was

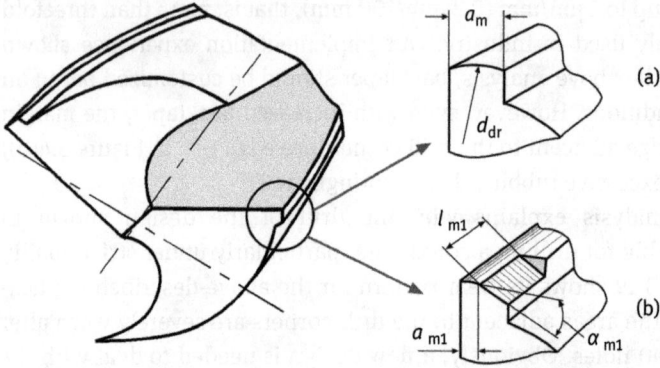

Figure 3.20: Traditional and advanced design/geometry of drill margins: (a) standard margin, and (b) narrowed margin with the clearance angle applied over the short axial distance.

established experimentally in the nineteenth century and hardly changed since then. As discussed earlier, an aggressive backtaper is one of the solutions for high-speed drilling but it only reduces severity and not solves the problem. Much better results, however, can be achieved if this margin is modified in the manner shown in Figure 3.20(b). The narrow margin of width a_{m1} = 0.08 ... 0.10 mm is made over the axial distance l_{m1} equal to approximately three times greater than the feed per revolution of the drill, and the clearance angle α_{m1} = 5° ... 7° is applied as shown in Figure 3.20(b). Such a design significantly improves the cutting conditions of the minor cutting edge, while the narrow margin still maintains drill stability. The side cutting edge of this design can handle: (a) cutting of the portions of the hole shown in Figures 3.14 and 3.15, (b) uneven uncut chip thickness at the entrance of cored holes (Figure 3.12), and (c) machined hole springback (Figure 3.18).

Using the above-described idea, a new drill design was developed for the drill of the considered case (see Figure 3.21). VIEW A shows the extent of the flank face of the margin from the dill corner (0.5 mm). SECTION C-C shows the width of the stabilizing chamfer (0.08$_{+0.02}$) and the clearance angle of the side cutting edge (7°). A drill made according to this drawing is shown in Figure 3.22 with the legitimate side cutting edge made on the drill margin.

The results of testing and implementation of the drills shown in Figure 3.22 showed the following:

1. The above-described tap breakage problem was completely solved.
2. Tool life of 5,000 holes was achieved although some drills do not made this tool life. It was found that this happens when the shift of the axis of cored holes is excessive but still within the tolerance assigned by the die casting drawing. This suggests that the problem was lessened but not completely solved because (a) the developed drill design (Figure 3.21) can't handle cored hole of excessive axis shift, and (b) tool life of 5,000 holes is smaller than

SECTION B-B
ENLARGED

0.25 ± 0.05
PRIMARY
FLANK
WIDTH

12° – 14°

25° – 27°

A

B

B

B

C

180° ± 1°

Point angle

Ø8.76 −0.003

⌀ 0.005	A
◯ 0.010	

VIEW A
ENLARGED

0.5

4 × 0.38 ± 0.05

SECTION B-B
ENLARGED

0.08

7°

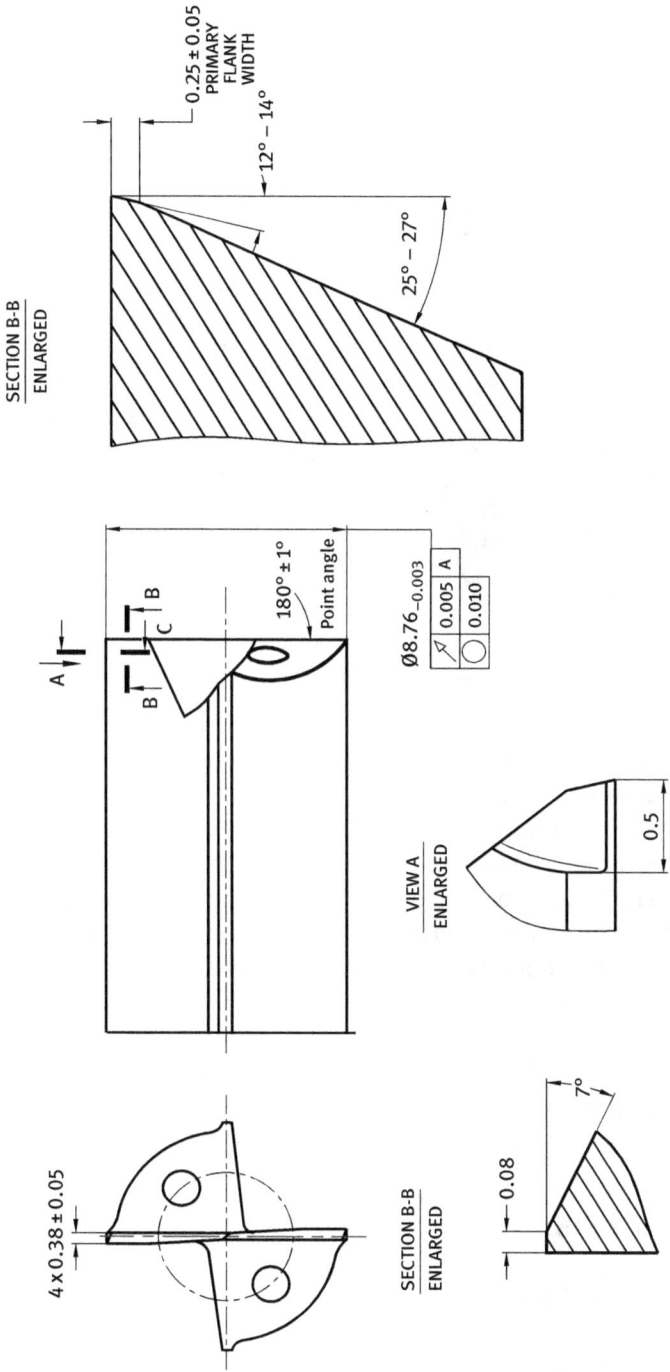

Figure 3.21: Drawing of a drill with advanced margin design.

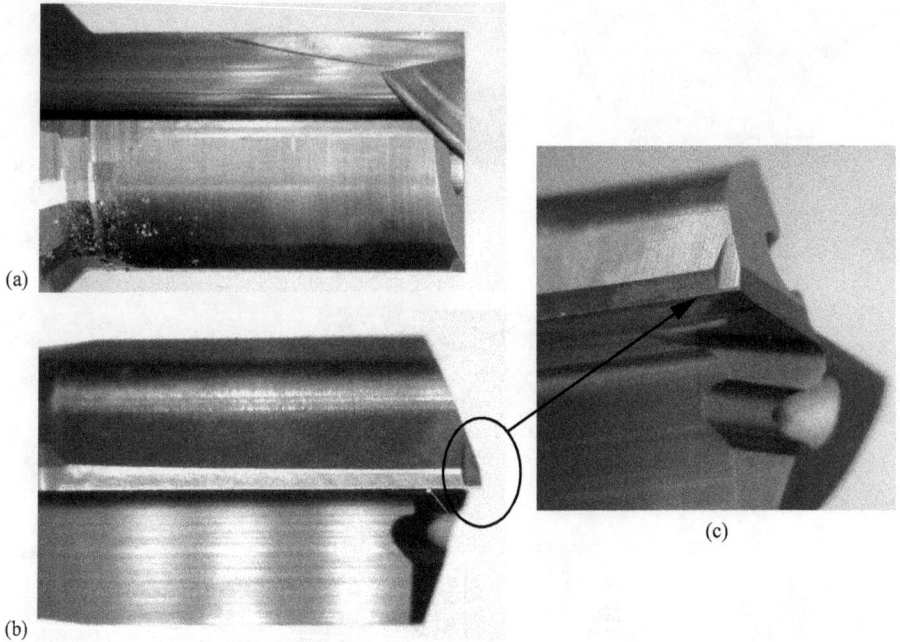

Figure 3.22: Drill made according to the drawing shown in Figure 3.21.

    that of other carbide drills used in drilling of noncored holes for the above-described part (Figure 3.7).

3.   Design Rule #1, that is, converting a small portion of the margin adjacent to the drill corner in a legitimate side cutting edge is to be applied not only to cored hole drills but also for all drills having great ($\geq 140°$) point angles, working with high cutting feeds, or drilling work materials of low modulus of elasticity, or drilling of working material of great coefficient of thermal expansion, that is, for more than a half carbide drill used in industry.

Therefore, further analysis is needed to understand the essence of the problem with tool life, and thus to develop a better drill design.

3.4 Deeper analysis

3.4.1 Complete force balance

To understand the problem with machining cored holes, and thus to provide its complete solution, to a deeper look (compare with that common ones presented in literature [15]) at the drill, force balance was taken. Although in this section the concept of the force balance is discussed for a twist drill as the most common drill type used in industry, its essence and the way of assessment are fully applicable for *any type of drilling tools*.

For convenience of further considerations, the right-hand $x_0y_0z_0$ coordinate system, illustrated in Figure 3.23, is set as follows:

1. The z_0-axis is along the longitudinal axis of the drill, with sense as shown in Figure 3.23, toward the drill holder.
2. The y_0-axis is perpendicular to the z_0-axis in the sense shown in Figure 3.23. The intersection of these axes constitutes the coordinate origin O as shown in Figure 3.23.
3. The x_0-axis is perpendicular to the y_0 and z_0 axes as shown in Figure 3.23.

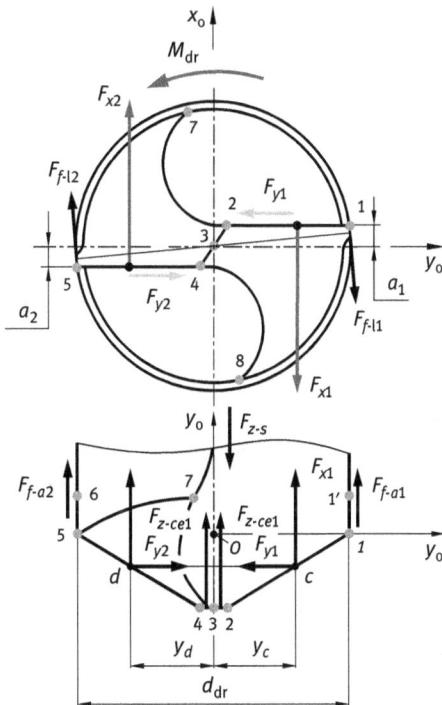

Figure 3.23: Complete force balance for a two-flute drill.

This system should also be regarded as the datum system in drill and drilling machine accessories (drill holder, starting bushing, etc.) design/setting/adjustment.

Complete force balance for a two-flute drill is shown in Figure 3.23. In this diagram, F_{x1} and F_{x2} are resultant power components, F_{y1} and F_{y2} are the radial components, and F_{z1} and F_{z2} are the axial components of the cutting forces acting on the first (1–2) and the second (4–5) major cutting edges (lips), respectively. The power and radial components of the cutting forces that act on the two parts (2–3 and 3–4) of the chisel edge are not shown in Figure 3.23 as these are small, while the axial components (F_{z-cz1} and F_{z-cz2}) shown in Figure 3.23 are significant. Forces F_{t-m1} and F_{t-m2} are tangential and F_{f-m1} and F_{f-m2} are friction forces on cylindrical lands (margins) 1–6 and 5–7, respectively.

The drilling torque applied through the spindle of the machine is calculated as

$$M_{dr} = F_{x1}r_{x1} + F_{x2}r_{x2} + F_{y1}a_o + F_{y2}a_o + F_{t-m1}(d_{dr}/2) + F_{t-m2}(d_{dr}/2) \qquad (3.3)$$

and the axial force applied by the spindle is

$$F_{z-s} = F_{z1} + F_{z2} + F_{z-cz1} + F_{z-cz2} + F_{f-m1} + F_{f-m2} \qquad (3.4)$$

The shown drill is in the static equilibrium in the x_0y_0 and z_0y_0 planes if and only if the following two equilibrium conditions are justified:

In the x_0y_0 plane

$$F_{x1}r_{x1} + F_{y1}a_o + F_{t-m1} = F_{x2}r_{x2} + F_{y2}a_o + F_{t-m2} \qquad (3.5)$$

In the z_0y_0 plane provided that arms of forces F_{z-cz1} and F_{z-cz2} are negligibly small

$$F_{z1}r_{z1} + F_{z-cz1} + F_{f-m1}(d_{dr}/2) = F_{z2}r_{z2} + F_{f-a2}(d_{dr}/2) \qquad (3.6)$$

As such

$$F_{y1} = F_{y2} \text{ and } F_{m1} = F_{m2} \qquad (3.7)$$

Equations (3.3)–(3.7) establish the full force balance, known as the theoretical or intended force balance. Unfortunately, this balance is not the case in practical drilling operations as many addition factors tend to disturb this balance [10]. Assigning tight tolerances on the drill geometry parameters (e.g., lip-height variation, flute spacing, chisel edge centrality, and runout of the lands (margins)), one can minimize the deviation from the intended force balance in "normal" drilling. Moreover, standard drills are provided with lands (margins) to prevent the side cutting of the drill due to possible force unbalance.

Unfortunately, this is not the case in cored drilling. It is discussed above that an unbalanced radial force in cored hole drilling can be avoided by using a drill with a 180° point angle and, moreover, drill bending can be minimized if the side cutting edge is provided with a clearance angle. However, as pointed out above, these measures can only lessen but not solve the problem. This is because the complete force balance shown in Figure 3.23 should be conserved instead of the simplified force balance shown in Figure 3.4(d).

When one considered the complete force balance (Figure 3.23), he or she should notice that when a drill enters a cored hole in the manner shown in Figure 3.5(a), not only the radial force (F_{R1-2} in Figure 3.5a and F_{y1} in Figure 3.23) cause drill bending but also the unbalanced power component, F_{y1} causes drill bending in the $x_0 y_0$ plane because force F_{y2}, which should be generated in cutting edge 4–5 does not exist as this edge is not engaged in cutting. When one uses a drill with a 180° point angle with the side cutting edge, the radial unbalanced force does not exist, at least theoretically. As a result, there is no bending of the drill at the hole entrance in the $y_0 z_0$ plane in the manner shown in Figure 3.5. However, the unbalanced resultant power component F_{y1} does exist, which is the cause of drill bending in $x_0 y_0$ plane. Moreover, the resultant power component F_{y1} is much greater than the radial force so that its effect on drill bending is much greater than any unbalanced radial force that presents a major problem. Unfortunately, this issue has never been discussed in literature on drilling so that it was never addressed in any known drill design.

Because the bending of a drill even with an 180° point angle in the $x_0 y_0$ plane by force F_{y1} is revealed as the root cause of the problem, a new drill design should be developed to minimize this bending. As clearly shown in Figure 3.23, there is no one available drill design components, which can restrict the discussed bending. Two closest are the tops of the heels shown as points 7 and 8 in Figure 3.23. However, in the common drill design these points are located on the diameter which is smaller than the diameter of periphery points 1 and 5. Such a drill is shown in Figure 3.24(a).

Auxiliary circular land (margin)

(a)   (b)

Figure 3.24: Drills: (a) common design of the top of the heel, and (b) auxiliary circular land (margin) is made on the top of the heel.

The problem with drill bending in the $x_0 y_0$ plane at the entrance of the hole being drilled is known for last 100 years even in drilling of noncored holes because the technology of drill manufacturing did not allow to grind the drill point perfectly (technically) symmetrical. Particularly, the major cutting edges suffered from a significant so-called lip-height variation (see Ref. [10] for a detailed explanation), which almost always caused an unbalanced force in the x_0y_0 plane. An enlarged and tapered entrance of the hole being drilled often occurred as a result. This is known as the bellmouth in industry. The introduction of CNC drill-point grinding machines first allowed improving drill point symmetry and then to grind drills of more complex design to reduce drill bending at the entrance.

Figure 3.24(b) shows a drill design in which the auxiliary circular lands (margins) are made on the top of the hills with the intent to prevent drill wandering at the entrance of the hole being drilled. This can significantly reduce the extent of the bellmouth, and improve roughness and staginess of drilled holes. Our implementation experience of drill with the auxiliary margins shows that these advantages can be gained if and only if the drill is designed properly. The latter is explained as follows.

Although the problem with bellmouth formation was reduced, it has been never solved as no adequate explanation to the bellmouth formation was provided, and thus the role of the auxiliary margins was explained. With the introduction of new carbide submicron grades having much greater flexural strength, also known as modulus of rupture, or bend strength, or transverse rupture strength, drills with the advanced planar flank face became commonplace. Moreover, the primary clearance angle applied to these drills is increased to its optimal value ($10°$–$14°$ compared to $6°$–$7°$ used in the past). As a result, the secondary clearance angle was increased up to $35°$, particularly for drills with internal coolant supply to provide better cooling and lubrication conditions at the flank face, that is, where these two actions are mostly needed. The problem with great clearance angle is that the locations of the top of the heels where the auxiliary circular lands (margins) start (points 7 and 8 in Figure 3.23) were thrown far back in the axial direction (along the z_0 axis) from the cutting edges (point 8 in Figure 3.25). This resulted in three issues: (1) When the power force component, F_{y1}, is unbalanced, a great arm of this force l_{Fy} (Figure 3.25) causes significant bending moment = $F_{y1} \cdot l_{Fy}$. (2) As points 7 and 8 are too far in the

Figure 3.25: Unbalanced force F_{y1} and it arm l_{Fy}.

axial direction from the cutting edge, the auxiliary circular lands (margins) do not provide any help over the drill entrance where their actions are mostly needed. (3) A drill is provided with backtaper, that is, its diameter decreases along z_0 axis. As a result, the diameter at point 8 is smaller than the drill diameter so that even when point 8 enters the hole being drilled, it can't provide adequate support to the drill.

The problem with drill entrance stability has become even greater when various so-called split-point designs (an example is shown in Figure 3.26a) were developed and made as a standard (gallery) feature of modern CNC drill point grinding machine. This advanced design aims at solving the long-standing problems: (1) reduction of a high axial force due to the chisel edge, and (2) improvement of drill self-centering. According to this concept, the chisel edge is ground with a neutral rake face (a zero rake angle) which makes it able to cut compared to that in common drills where the rake angle of the chisel edge reaches $-70°$ so that such a chisel edge extrudes the work material. In the drill shown in Figure 3.26(a) and in the model in Figure 3.26(b), the split of the chisel edge is achieved by grinding an additional secondary flank plane(s). In the planar design of the drill flank faces, which is common today, the chisel edge is split by a tertiary flank plane as shown in Figure 3.27 which makes the axial location of points 7 and 8 even further from the major cutting edges. It is to say that having solved the problems with the chisel edge via a wide introduction of the split-point design, specialists/tool designers did not pay attention to the location of points 7 and 8. Figures 3.26(b) and 3.27 clearly show that the drill heel extends too far back with respect to the major cutting edges so that even when a drill is provided with auxiliary circular lands (margins), they provide no help neither on the drill entrance nor while drilling.

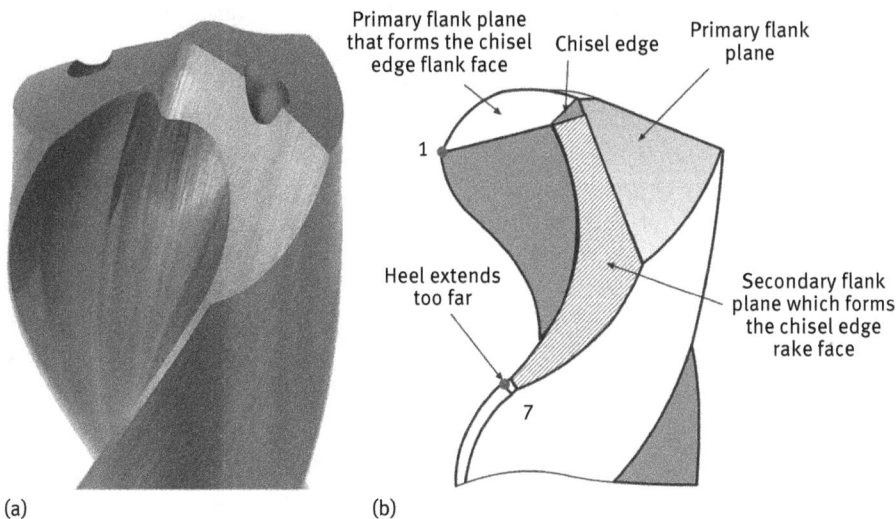

(a)   (b)

Figure 3.26: Modern design of a split-point drill: (a) drill and (b) model.

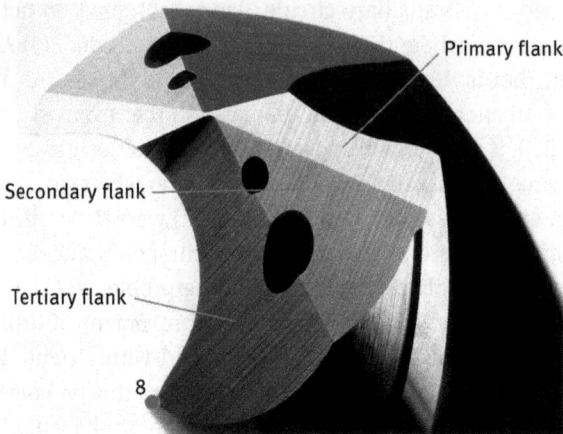

Figure 3.27: An example of split-point design drills with a tertiary flank.

The foregoing analysis reveals that most modern drill designs can't provide drill stability at the entrance of the hole being drilled due to incomplete understanding of drill entrance condition. Therefore, a need is felt to develop a new design concept based on the principle of drill entrance stability (Design Rule #2) developed by Astakhov [16]. Applying this concept to a current drill design, one can formulate a modified design rule (Design Rule #2.1): Drill-free penetration in the feed direction while assuring the maximum entrance stability is achieved by the location of points 7 and 8 (Figure 3.23) at a distance of $l_{Fy} = f/4$ from the drill corners, where f is the cutting feed, mm/rev. Obviously, due to the drill manufacturing tolerances and setting (in the holder, machine, etc.) inaccuracies, this theoretical distance should be increased. A modeling test result showed that $l_{Fy} \approx 1.5\,f$ can be used.

The next step was to develop a practical drill design to satisfy Design Rule #2.1. As per Figure 3.7, if a drill is provided with an optimal 10° primary angle and 25° clearance angle, then the location of point 8 seems to be geometrically fixed. The experience with the development of the flank faces [17] of drill shows that it is not so. The development of a new design concept is shown in Figure 3.28. As shown, the flank face of a drill is designed as consisting of three facets. The first one is the primary flank plane and the second one is the secondary flank plane. Both planes are provided with the optimal flank angles as discussed above so the optimal work performance of the drill is not compromised. The third facet is the tertiary flank plane. It starts at certain distance l_s from the rake face (cutting edge) as shown in Figure 3.28 and is positioned so that the distance between the drill corner and point 8 $l_s \approx 1.5 \cdot f$, that is, as shown by the result of the modeling test.

The next step was to assure that the proposed concept can be realized practically on modern multiaxis CNC drill point grinding machines, that is, to develop a practical procedure of its manufacturing. To do that, modeling of possibility of

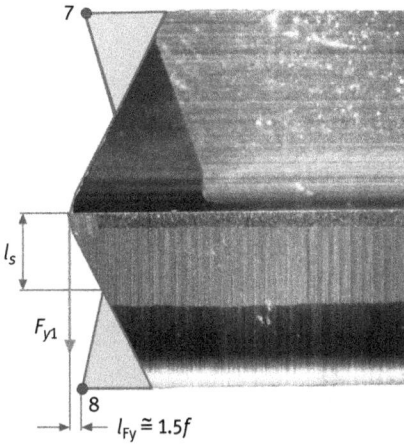

Figure 3.28: Development of a new design concept.

grinding a special secondary flank plane face using the Helitronic Tool Studio software on a modern Walters Helitronic Vision CNC drill point grinding machine was carried out as shown in Figure 3.29. Figure 3.29(a) shows a grinding wheel positioning to assure that the second flank lane can be ground independently of the geometry/position of the first and tertiary flank planes. As such the proper grinding wheel shape and its diameter were also selected to achieve the intended result. Figure 3.29(b) shows the result of grinding. As shown, the proposed design concept can be fully realized using modern CNC drill point grinding machines.

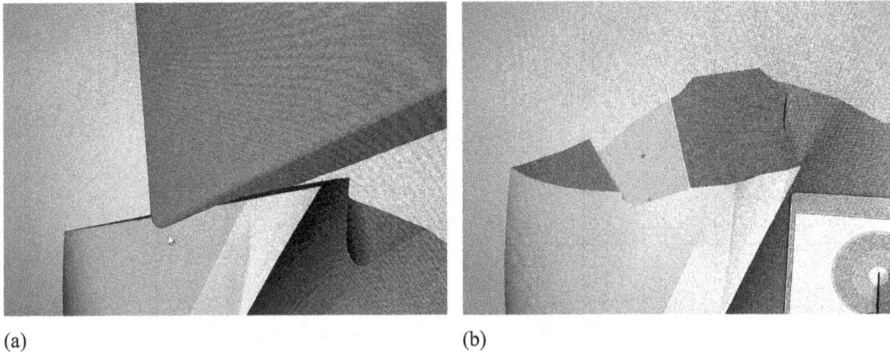

(a) (b)

Figure 3.29: Modeling of possibility of grinding a special secondary flank plane face using the Helitronic Tool Studio software on a modern Walters Helitronic Vision CNC drill point grinding machine: (a) grading wheel positioning and (b) result of grinding.

Figure 3.30 shows the most relevant extract from the developed tool drawing where the proposed concept is realized. SECTION B-B shows the location of the front ends of the auxiliary circular lands (margins) with respect to the drill corners/major

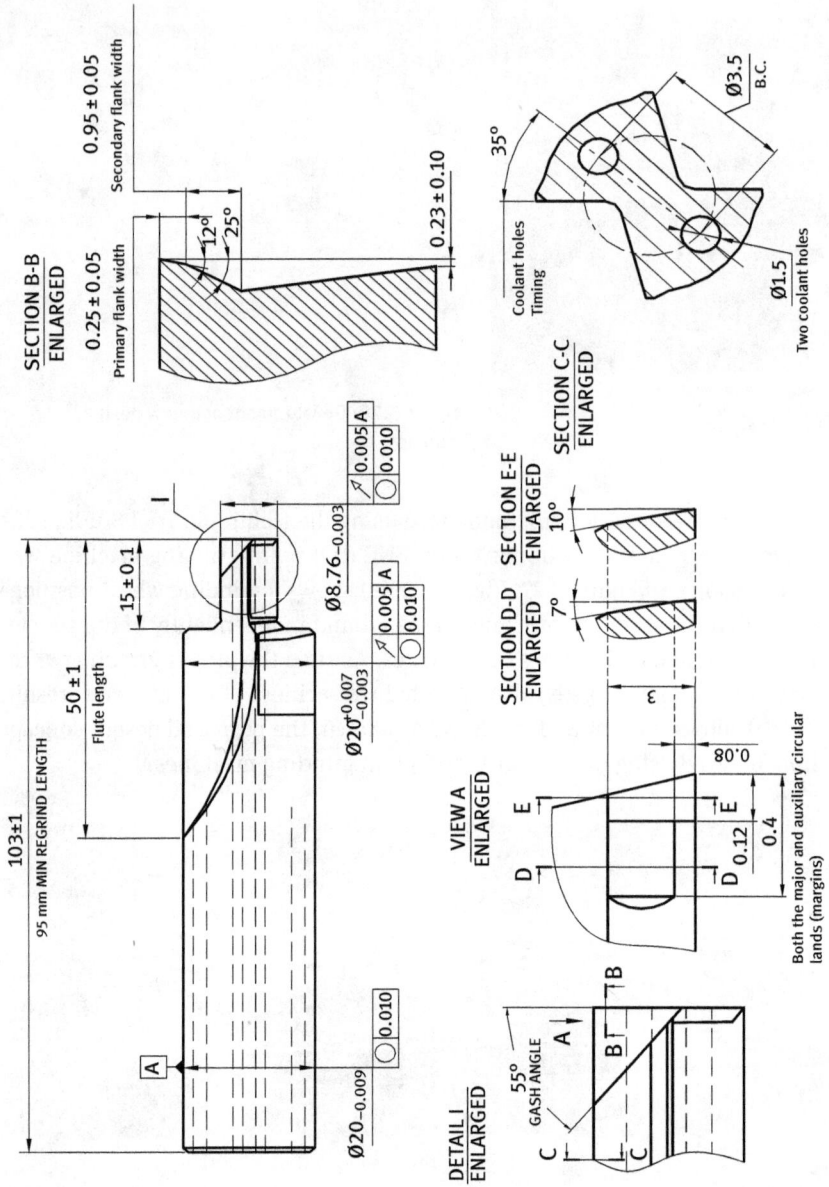

Figure 3.30: The most relevant extracts from the developed tool drawing.

cutting edges (as the drill has an 180° point angle) – the location of point 8 as per Figure 3.23 is actually shown. It is clear that the axial location of point 7 is the same as the drill is symmetrical. VIEW A shows the geometry of the side cutting edge (edges 5–6 and 1–1′ are shown in Figure 3.23). Compared to the previous design (see Figures 3.21 and 3.22), the flank face of this edge include two parts: the first part (SECTION E-E in Figure 3.30) is ground as it is usually made on end mills, that is, ground up to a sharp edge; the second part (SECTION D-D in Figure 3.30) is provided with a thin (0.08 mm) circular land (similar to that in previous shown in Figure 3.21) for drill stability. Note that the front portions of the auxiliary margins are provided with the same geometry.

The drill was made to the drawing, inspected, and set into a manufacturing cell. Its tool life was tripled compared to the previously used drill (Figures 3.21 and 3.22). Figures 3.31 and 3.32 show particularities of the new design and wear patterns after 10,000 holes. Figure 3.31(a) shows the face view of the drill, and Figure 3.31(b) and (c) shows two "3D" views where particularities of the rake faces, auxiliary circular lands (margins), and location of the gash forming the rake faces can be seen. Figure 3.31(d)

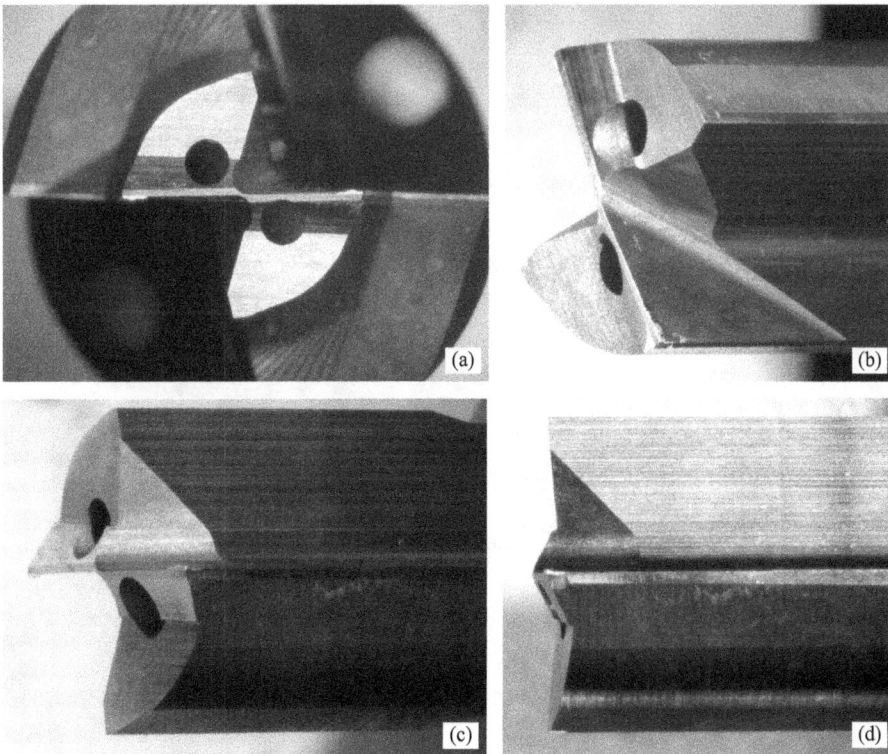

Figure 3.31: Particularities of the new drill design.

shows the side view of the drill where the result of the proposed concept (see Figure 3.28) can be clearly seen due to "uncommon" positions of the tertiary flank plains. Figure 3.32 depicts the advantages of the developed design concept. The first advantage is obvious – the drill doubled tool life of the current drills with no single problem reported. Figure 3.32(a) shows the rake face having a small built-up edge in the vicinity of drill corner. Figure 3.32(b) shows this built-up edge from the face. Figure 3.32(c) shows the wear of the drill corner and side cutting edge on the circular land (margin). As shown, both the corner and cutting edge on the circular land are not worn so that tool life was further extended to 15,000 holes, that is, the drill easily tripled the maximum tool life of the previously used drill. Figure 3.32(d) shows the wear pattern on the face of the auxiliary circular land (margin). As shown, this wear is insignificant as only small trace of the aluminum built-up edge is observed in the very corner. Such a small wear on the drill corners and the front ends of the major and auxiliary circular lands (margins) suggests that the number of drill regrinds can be substantially unstressed till min. regrind length (95 mm per Figure 3.30) is reached.

Figure 3.32: Wear pattern of the developed drill after a 10,000 cycles (holes) run.

Figure 3.33 shows a section of the machined hole with the tapped thread. As shown, the proper thread was machined even when the axis of the cored hole is significantly shifted from that of the drill/thread.

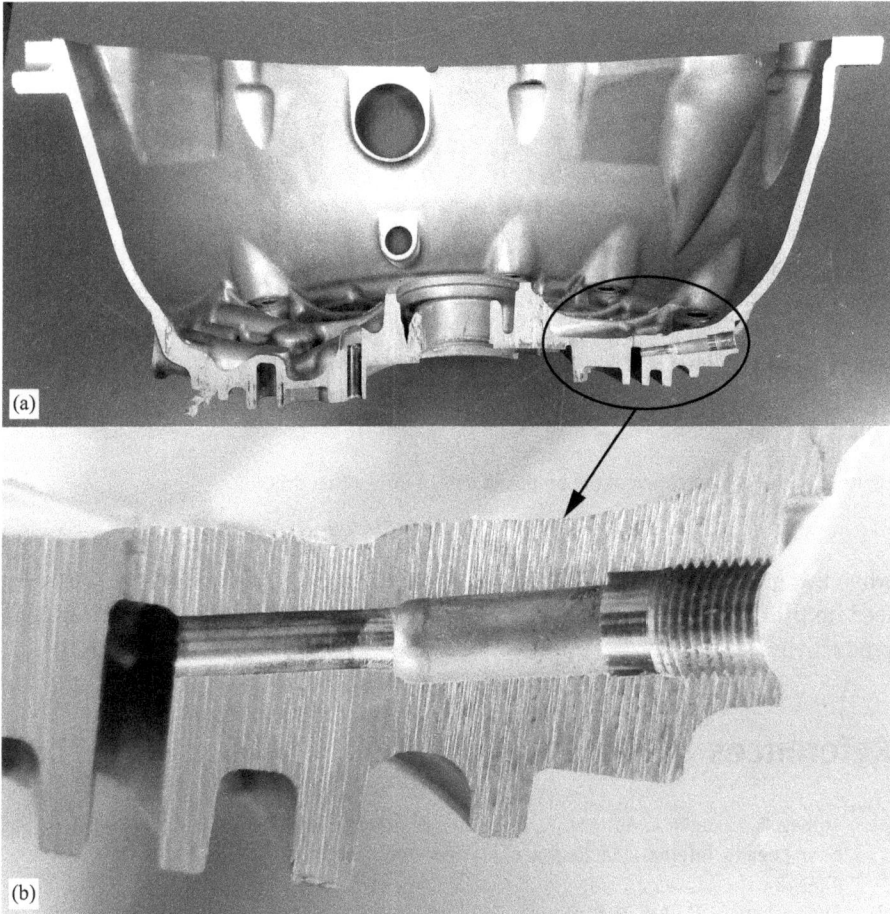

Figure 3.33: (a) and (b) The proper thread was machined even when the axis of the cored hole is significantly shifted from that of the drill/thread.

To prove that the developed basic design of the drill for cored holes is not particular for the specific drilling conditions, it was implemented in other drills. One of these is shown in Figure 3.34. Its implementation was resulted in doubling the tool life.

As a conclusion, it can be stated that the basic design of drill for cored hole was developed, tested, and implemented. Moreover, as with the new design, drill entrance stability increases so that the quality of the entrance of the hole being drilled is improved, and the proposed design can be used for drills (both straight and twist)

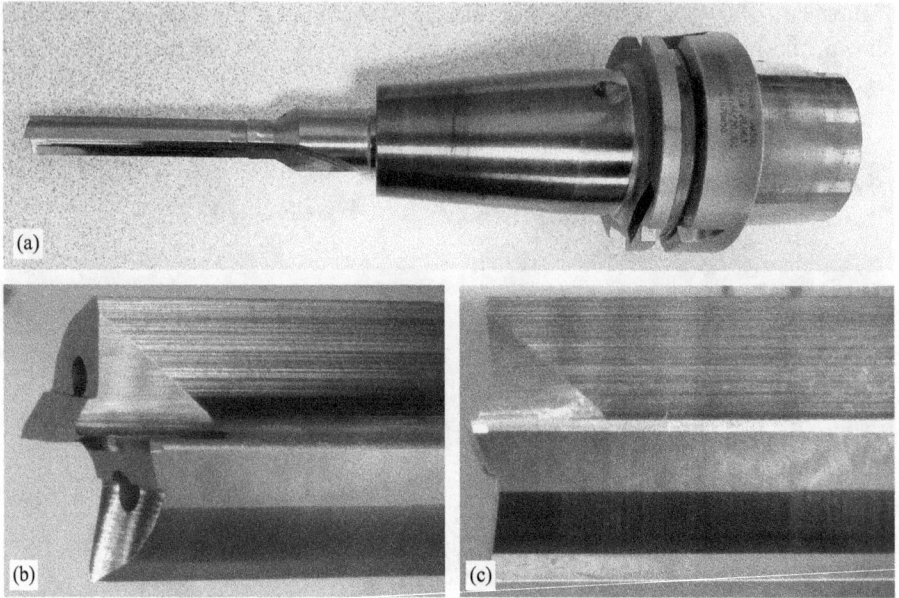

Figure 3.34: (a–c) Realization of the proposed design for another drill.

when this quality is of some concern. As such, there is no need to reduce the cutting feed on drill entrance as it is common in drilling, which can result in high drilling productivity.

References

[1]   Atabey, F., Lazoglu, I., Altintas, Y., Mechanics of boring processes – Part II – multi-insert boring heads. International Journal of Machine Tools and Manufacture, 2003, 43(5), 477–484.

[2]   Bhattacharyya, O., Jun, M.B., Kapoor, S.G., DeVor, R.E., The effects of process faults and misalignments on the cutting force system and hole quality in reaming. International Journal of Machine Tools and Manufacture, 2006, 46, 1281–1290.

[3]   Taheri, H., Shoaib, M.R.B.M., Koester, L.W., Bigelow, T.A., Powder-based additive manufacturing – a review of types of defects, generation mechanisms, detection, property evaluation and metrology. International Journal of Additive and Subtractive Materials Manufacturing, 2017, 1(2), 172–209.

[4]   Smith, C.J., Derguti, F., Hernandez Nava, E., Thomas, M., Tammas-Williams, S., Gulizia, S., Fraser, D., Todd, I., Dimensional accuracy of electron beam melting (EBM) additive manufacture with regard to weight optimized truss structures. Journal of Materials Processing Technology, 2016, 216(229), 128–138.

[5]   Farrell, S.P., Deering, J., Analysis of seeded defects in laser additive manufactured 300 M steel. Materials Performance and Characterization, 2018, 7(1), 300–315.

[6]   Zhang, B., Yongtao, L., Bai, Q., Defect formation mechanisms in selective laser melting: a review. Chinese Journal of Mechanical Engineering, 2017, 30, 515–527.

[7]   Moroni, G., Syam, W.P., Petro, S., Towards early estimation of part accuracy in additive manufacturing. Procedia CIRP, 2014, 21, 300–305.

[8]   Lee, Y.S., Farson, D.F., Surface tension-powered build dimension control in laser additive manufacturing process. International Journal of Advanced Manufacturing Technology, 2015, 85(5), 1035–1044.

[9]   Astakhov, V.P., Improving sustainability of machining operation as a system endeavor, in Sustainable Machining, J.P. Davim, Editor. 2017, pp. 1–29, London: Springer.

[10] Astakhov, V.P., Drills: Science and Technology of Advanced Operations. 2014, Boca Raton, FL: CRC Press.

[11] Astakhov, V.P., The mechanisms of bell mouth formation in gundrilling when the drill rotates and the workpiece is stationary. Part 1: The first stage of drill entrance. International Journal of Machine Tools and Manufacture, 2002, 42, 1135–1144.

[12] Astakhov, V.P., Metal Cutting Mechanics. 1998/1999, Boca Raton, USA: CRC Press.

[13] Atkins, A.G., Modelling metal cutting using modern ductile fracture mechanics: quantitative explanations for some longstanding problems. International Journal of Mechanical Science, 2003, 43, 373–396.

[14] Atkins, T., The Science and Engineering of Cutting. 2009, Amsterdam: Butterworth-Heinemann.

[15] Klocke, F., Manufacturing Processes 1: Cutting. 2011, London: Springer-Verlag Berlin Heidelberg.

[16] Astakhov, V.P., Geometry of Single-Point Turning Tools and Drills. Fundamentals and Practical Applications. 2010, London: Springer.

[17] Astakhov V.P., G., V.V., Osman, M.O.M., A novel approach to the design of self piloting drills. Part 1. Geometry of the cutting tip and grinding process. ASME Journal of Eng. for Industry, 1995, 117, 453–463.

Ke Huang, Tianxing Chang, Yandong Jing,
Xuewei Fang, Bingheng Lu

4 Additive manufacturing of magnesium alloys

Abstract: Significant progress has been made in understanding the mechanisms involved during additive manufacturing (AM) processes of high-performance metallic materials, which leads to their wide applications in aerospace, medical, and transportation industries. However, little investigation has been conducted on AM of lightweight magnesium alloys due to the difficulty in processing. In this chapter, the characteristics of magnesium alloys and their application areas are first introduced. Two typical AM methods to process magnesium alloys, namely, wire arc AM and selective laser melting, are discussed in terms of their principles, processing parameters, and their relation to the microstructures and properties, as well as the encountered defects and their controlling. Because the field is rapidly evolving, emerging or less commonly used AM methods such as ultrasonic AM, friction stirring AM, and cold spray AM are also briefly presented. Focus is placed on the process–microstructure–property relationship of the deposited magnesium alloy parts. Various approaches to eliminate defects inside the fabricated parts are also considered to improve their properties. The current status of AM of magnesium alloy, existing difficulties, and gaps in the scientific understanding and the research trends for the further expansion of AM of magnesium alloy are all illustrated.

Keywords: Additive Manufacturing, Selective Laser Melting, Wire arc additive manufacturing, Microstructure, Defects, Magnesium Alloys

4.1 Introduction

The investigation of additive manufacturing (AM) of metallic materials has been mostly focused on titanium alloys, steels, aluminum alloys, Ni-based superalloys, and so on, driven by the increasing demands in aerospace and medical industries. Further investigation related to AM is now being actively carried out for other emerging metallic materials, such as high entropy alloys and shape memory alloys. However, little investigation has been conducted on AM of lightweight magnesium alloys due to the difficulty in processing [1], even though it is expected to play a big role in manufacturing components with complex geometry that are difficult to be realized by traditional processing methods due to their intrinsic characteristic such as low plasticity. A chapter focusing on the latest development of AM of Mg alloys, as well as the related difficulties and corresponding solutions, is therefore of high interest.

https://doi.org/10.1515/9783110549775-004

The characteristics and application areas of Mg alloys, as well as a brief review of AM of metallic materials are first presented in Section 4.1. According to the available literature on AM of Mg alloys, two AM methods are considered in detail. Wire arc additive manufacturing (WAAM) of Mg alloys is first discussed in Section 4.2, while Section 4.3 is devoted to selective laser melting (SLM). Finally, other less common AM methods that can be used to process Mg alloys is given in Section 4.4. Each section is provided with a brief introduction, as well as a list of latest references, which will enable interested readers to skip certain unrelated topics, as well as delve further into a particular subject.

4.1.1 The characteristic of Mg alloys

Magnesium and magnesium alloys exhibit a wealth of valuable properties, making them of great interest for use across a wide range of fields [2]. Magnesium is a silvery light alkali earth metal; its key properties are listed in Table 4.1. Its weight, as the lightest practical metal, is almost two-third of aluminum, and one quarter of iron. Magnesium is widespread in the nature and it is one of the essential elements toward human body. It was not discovered, however, until in the nineteenth century. Thus, compared to other alloys, magnesium alloy is an emerging material in the modern industry; however, the consumption of magnesium alloy is skyrocketing at 20% a year.

Table 4.1: Basic properties of magnesium.

Character	Unit	Value	Character	Unit	Value
Atomic coefficient		12	Specific resistance	nΩ·m	47
Valence		2	Specific heat	J/(mol·K)	1.03
Relative atomic mass		24.3	Recrystallization temperature	°C	150
Density(room temperature)	g/cm^3	1.74	Standard electrode potential	V	−2.37
Melting point	°C	649	Heat conductivity	W/(mol·K)	156
Boiling point	°C	1,107	Coefficient of thermal expansion	10^{-6}/K	26

4.1.1.1 Advantages of magnesium alloys

1) Low density:
   Mg has the lowest density in the present engineering metal. Its density is one-fifth of steel, a quarter of Zn, and two-third of Al. While casting magnesium alloy has similar stiffness as that of aluminum alloy, its density is much lighter than Al alloy.
2) High toughness and good damping properties:
   Mg alloy deforms when subjected to external forces. However, the energy it absorbs is 1.5 times of aluminum under the same impact load. Mg alloy is an ideal

material in high-frequency and multifrequency vibration environment on account of its high damping capacity.

3) Low heat capacity, fast solidification rate, and good die casting performance:
Mg alloy is a good die-casting material with good flowability and rapid solidification rate. It can fabricate parts with both flat surface and clear edge. In addition, its production can avoid the dimension error caused by excessive contraction. Due to its low heat capacity, the production efficiency of Mg alloy is 40% higher than Al alloy castings with identical geometry.

4) Excellent machinability:
Among all commonly used metals, Mg alloy is considered to be relatively easy to process. In addition, high surface finish can be achieved without further grinding or polishing.

4.1.1.2 Disadvantages of magnesium alloys

1) Inflammability
Mg has an affinity for oxygen. When it is in solid state under high-temperature condition, it will react easily with oxygen and give off a lot of heat. Its production, magnesium oxide, has poor thermal conductivity, which hinders the heat emission and further promotes the oxidation reaction.

2) Low plasticity
Mg has a hexagonal close-packed crystal structure. It only has one principle slip plane and three slip systems at room temperature. Its plastic deformation, as a result, mainly depends on the coordination of slip and twinning. But when the temperature is over 250 °C, the additional slip systems will be activated and the plastic deformation capacity of Mg alloy will improve.

3) Poor corrosion resistance
Factors influencing corrosion resistance of magnesium alloys include the following two aspects. First, because the oxide film on the surface of magnesium alloy is usually thin and porous, the corrosion resistance of magnesium alloy is not significantly improved. Second, magnesium is easy to be corroded in the presence of other metal elements (especially Cu, Fe, and Ni) due to its low corrosion potential.

4.1.2 The typical application area of magnesium alloys

4.1.2.1 Transportation

There is no need to reiterate here the importance of reducing fuel consumption in the transportation industry. The observed increase in the use of aluminum and its alloys in automobiles is a direct consequence of the pursuit for weight reduction in the past decades. As good candidate materials to replace Al alloys, Mg alloys are getting more and more attention. Since Mg alloy is ~35% lighter than Al alloy, it can realize extra weight reduction in automotive structures, especially when it is used in thicker sections. Mg alloy has been used extensively, nowadays, in automobile industry, such as wheel, gearbox body, instrument panel, car door, and engine block [3].

Magnesium alloys have also been used on many occasions in high-speed trains all around the world. The ICE high-speed rail train made by Siemens and TGV Duplex double deck high-speed train in France are both equipped with Mg alloy seats. The seats, stands, armrests, and ground mats in the Japanese Shinkansen N700 series are all made of Mg alloys. The KTX express train in Korea also changed its Al Alloy seats to the Mg alloy seats, in order to reduce its weight and cost. Mg alloy models commonly used in rail transit equipment includes AZ31B, AZ61A, AZ91D, ZK60, and AM60B.

4.1.2.2 Medical industry

The conventionally used metallic materials in medical industry today are stainless steels, Co-based alloys, titanium alloys, and so on. However, all of them often exhibit defects such as stress shielding and metal ion releases. Secondary surgical operation is usually inevitable to prevent long-term exposure of body with the toxic implant contents. Mg alloy is revealing its advantages as the development of biodegradable metals progresses [4, 5]. Magnesium can dissolve in the body fluid; implanted magnesium, as a result, can degrade gradually without any debris if it is well controlled. Mg alloy is also biocompatible; excess magnesium element will be discharged through urine. In addition, the properties of Mg alloy is also similar as those of human bone.

The main applications of Mg alloys in medical industry include the following three aspects.
1) Bone fixation material: Due to their similar properties toward human bone, excellent biodegradability, and favorable biocompatibility, Mg alloys are considered as one type of promising material in orthopedics.
2) Porous scaffold bone repair material: Porous magnesium, as a biodegradable biomaterial, is far superior to biodegradable polymer materials such as polylactic acid, which have been applied in clinical practice. Moreover, it has biological activity and can induce the growth of bone cells and blood vessels.
3) Cardiovascular stent: Magnesium-based materials can be made into degradable cardiovascular scaffolds due to their degradability and proper mechanical properties.

Biodegradable magnesium and magnesium alloys biodegradable stents are perhaps the largest improvements in the clinical application of biodegradable biomaterials. Biotronik Company in German has fabricated the first WE43 Mg alloys stent using laser-engraving technology, and testified its favorable degradability and properties in the clinical trial.

4.1.2.3 Aeronautics and Astronautics

Lightweight design is one of the directions of the future development for aerospace materials. The application of Mg alloy, practically the lightest structural metallic materials, can bring huge weight reduction and significant increase in aircraft performance. In the aerospace domain, Mg alloys are widely applied in the manufacture of airplanes, missiles, aircrafts, and satellites [6]. Germany has increased investment in the research of Mg alloys. ZIM FLUGSITZ GmbH Company in Germany fabricated Mg alloy aeronautical seats, made of Elektron®43 Mg alloy produced by Magnesium Eletron Company in America, and succeeded in reducing 25% weight. In 2007, Raytheon Company designed the AGM-154C for the USN, and the stern room, wing skeletons, and the equipment box body of this weapon are made of AZ91E Mg alloy and AZ91D Mg alloy. Besides, the EU has also prompted a series of studies of Mg alloys in the FP4, FP5, and FP6 since 1996, and Ostrovsky I predicted, according to the result of the research, that during 2015 and 2020, 10%–15% of civil aircraft parts will be made of Mg alloy.

4.1.3 Overview of metal additive manufacturing

The AM process of metallic materials fall into two categories defined by ASTM standard F2792 [7] as directed energy deposition (DED) and power bed fusion (PBF). The difference between the two methods is that DED utilizes a powder flow/wire feedstock system to direct material onto the build substrate while simultaneously tracing out the build layer. But for the PBF, the powder is already within the build chamber and the energy source need only trace out each layer.

The heat source of DED includes laser, electron beam, plasma arc, and gas metal arc. In all of these DED processes, a three-dimensional (3D) part is fabricated in a layer-by-layer manner following the input of a digitized geometry from a computer-aided design (CAD) file. The distance between the focused beam and the build surface is maintained by a synchronized multiaxis movement of the fixture that holds the substrate and the heat source during layer-by-layer deposition. Supporting structure maybe needed in the parts with overhanging features, to prevent distortion of hot overhangs induced either thermally or under their own weight. Secondary machining is required after the deposition process to achieve the desired surface quality.

The heat source of PBF includes laser and electron beam. It begins with a solid or surface CAD model, orienting it within a build volume to include support structures, slicing into planer layers, defining a scan path, and building file based upon a prespe-cified set of material specific parameters and the specific machine configuration. The part forms by spreading thin layers of powder and fusing pass-by-pass and layers upon layer of this powder, under computer control, within an inert chamber, incre-mentally lowering the Z-axis after each layer. Solid powders are often reused in PBF to avoid wastage, but this can also lead to poor surface finish and mechanical proper-ties of the final part.

Quite a few excellent review papers [1, 8–16] focusing on AM of metallic materi-als are available in the literature; however, a comprehensive review on AM of Mg alloys is still missing.

4.1.3.1 Selective laser melting (SLM)

To avoid the defects in selective laser sintering (SLS) [14], such as low density and poor performance, due to the lack of metallurgical bonding process, Fraunhofer Institute of Laser Technology in German first came up with the SLM, technology in 1995. SLM is a laser-based PBF approach. In contrast to SLS, the part produced by SLM is nearly completely dense. Its mechanical property and internal microstruc-ture, therefore, are improved a lot. However, due to its complex phase transition, from solid to liquid to solid, the volume and temperature of the fabricated parts are both subjected to huge cyclic change, which leads to the increase of internal stress.

4.1.3.2 Electron beam melting (EBM)

EBM was developed by Arcam Company. It is an electron beam-based PBF ap-proach. With the heat source of electron beam, metal powders are melted layer by layer in a vacuum to form completely dense parts. It is suitable for metals that oxi-dize easily and react easily with air, such as Al and Ti. The part produced by EBM has higher density than that by SLS. In addition to this, EBM is a fast metal AM technology, up to 15 kg/h. But there are also some defects in EBM. For example, it has a high cost on account of the vacuum environment and heat source of electron beam. Beyond that, its surface quality, around 0.13–0.20 mm, is slightly inferior to SLM. Some key features of SLM and EBM are listed in Table 4.2.

4.1.3.3 Wire arc additive manufacture (WAAM)

WAAM is a technology that uses arc as heat source, adds wire simultaneously, and prints out metal parts. WAAM can be used to fabricate large-scale parts with rela-tively low cost and it can process Al alloys and Cu alloys, which have high reflectivity

Table 4.2: The comparison of SLM and EBM.

Item	SLM	EBM
Surface quality	Ra: 9–12 μm	Ra: 25–35 μm
Residual stress	High	Low
Heat treatment requirement	Need stress relief annealing or HIP	Stress relief annealing or HIP are not necessary
Minimum wall thickness	0.2–0.5 mm	0.6–1.0 mm
Minimum- value aperture	Φ0.5 mm in vertical direction Φ0.8 mm in horizontal direction	Φ0.2 mm–Φ0.5 mm
Working allowance	0.1–0.5 mm	0.5–2.0 mm

to laser. However, the surface quality of parts produced by WAAM is often very poor and subsequent machining is inevitable.

4.1.3.4 Laser engineering net shaping (LENS)

Optomec combines SLS with laser cladding to propose laser engineering net shaping (LENS) technology in 1988. LENS is a laser-based DED approach. The laser and powder feeding are synchronized in this process. The laser is used to melt the metal, and the powders with different components and properties are sent into the molten pool and then deposited on the substrate. LENS is usually used to produce stainless steel, Ti, and super alloy. Its application includes Aerospace large-component manufacturing and repair of parts.

In contrast to SLM, LENS has higher efficiency. And compared with spraying, electroplating, and surfacing welding, it has the characteristics of high bonding strength between coating and matrix metallurgy. Besides, it can also be used in repairing welding.

4.2 Wire arc additive manufacturing of Mg alloy

4.2.1 Introduction and the equipment of WAAM

WAAM is a wire-based DED approach that uses an electrical arc as a source of fusion to melt the wire feedstock and deposit a part preform, layer by layer, as shown schematically in Figure 4.1.

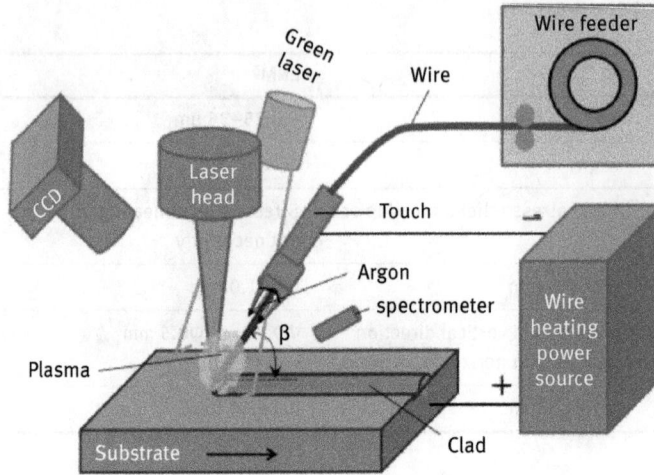

Figure 4.1: Schematic graph showing the principle of WAAM [17].

The origin of the WAAM process can be traced back to the 1920s when Baker proposed to use an electric arc as the heat source with filler wires as feedstock materials to deposit metal ornaments. Since then, consistent progress has been made in the development of this technology, particularly in the last 10 years. A major benefit of the WAAM process relates to the low investment, as the components of a WAAM machine are easily accessible in the welding industry. The processing characteristics may also make the WAAM process preferable compared to the alternative fusion sources. For example, WAAM does not need a vacuum environment to operate as required in electron-beam-based methods. Besides, WAAM has other obvious advantages such as high deposition efficiency and high material utilization rate, which make it possible to fabricate huge Mg alloy components efficiently at low cost. The comparison of various WAAM technique is shown in the Table 4.3.

4.2.1.1 Key components of WAAM equipment

An integrated WAAM hardware system consist of power source, wire feeder, industrial robot, robot control tank, substrate, and welding positioner. The torch was installed on the robot arm moving in the designated path to fabricate metal parts. Its software system is centered on a high-performance computer, equipped with CNC machining software and robot offline simulation software. A typical WAAM equipment is depicted in Figure 4.2.

Table 4.3: Comparison of various WAAM technique [18].

WAAM	Energy source	Features
GTAW based	GTAW	Nonconsumable electrode; separate wire feed process
GMAW based	GMAW	Consumable wire electrode; typical deposition rate 3–4 kg/h; poor arc stability, spatter
	Cold metal transfer (CMT)	Reciprocating consumable wire electrode; typical deposition rate: 2–3 kg/h; low heat input process with zero spatter, high process tolerance
	Tandem GMAW	Two consumable wires electrodes; typical deposition: 6–8 kg/h; easy mixing to control composition for intermetallic materials manufacturing
PAW based	Plasma	Nonconsumable electrode; separate wire feed process; typical deposition rate 2–4 kg/h; wire and torch rotation are needed

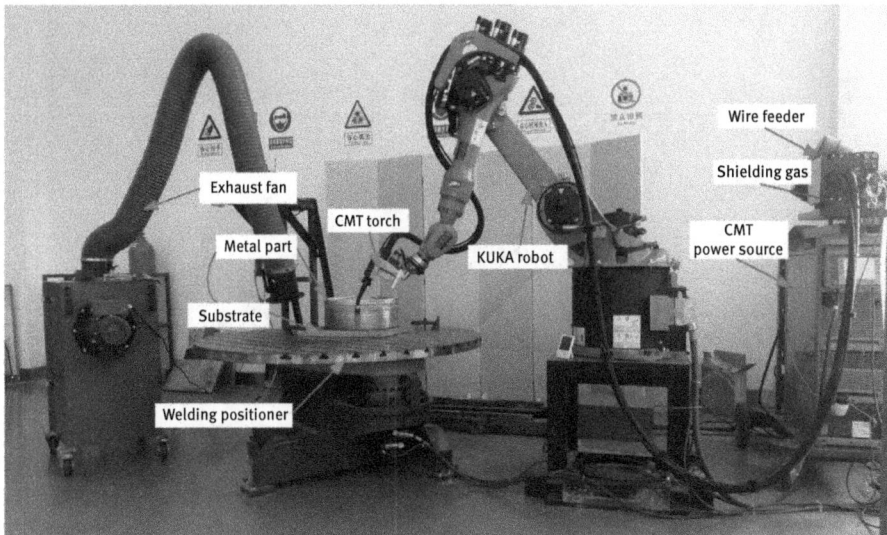

Figure 4.2: The equipment of wire arc additive manufacture (the National Innovation Institute of Additive Manufacturing in China) [19].

4.2.2 The effect of process parameters on microstructures and properties

4.2.2.1 Microstructure

The processing parameters have been confirmed to influence the final microstructure of the fabricated parts in various ways:

Pulse frequency: With the enhancement of pulse frequency in WAAM of AZ31 Mg alloy, grain size first decreases and then increases, as shown in Figure 4.3. This is related to the following reasons. First, increasing pulse current can cause the stirring of the molten pool, leading to refined grains. Second, high pulse current can cause the fast cooling rate of the molten pool, which also results in refining grains. During the base current, the heat input was suddenly reduced and therefore the cooling rate of the molten pool is high causing heterogeneous nucleation to refine grains [20, 21], as shown in Figure 4.3(c). However, with the further increase in pulse frequency, the influence of heat input is greater, leading to higher temperature and lower cooling rate [22], which favors coarser grain size. As a result, the optimum pulse frequency is found to be 5 and 10 Hz for AZ31 [23].

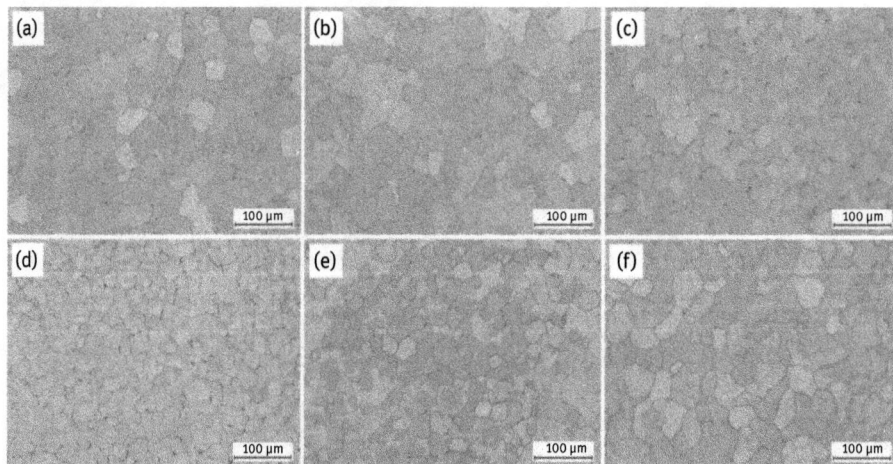

Figure 4.3: Microstructures of the samples deposited by different pulse frequencies [23]: (a) 500 Hz; (b) 100 Hz; (c) 10 Hz; (d) 5 Hz; (e) 2 Hz; and (f) 1 Hz.

Current: As the current increases from 80 to 120 A, the average grain diameter increases from 28 to 32 μm, and the grain aspect ratio also changes in the range of 0. 95–1.19. This is because with the increase of the current, the arc heat input and the temperature of the molten pool also increase, which then lead to slower cooling rate in the molten pool, as well as longer time for the growth of the grains. A coarser grain size is thus observed at higher current, as illustrated in Figure 4.4. Although the heat input increases with increasing current, the cooling rate of molten pool is also high due to the high thermal conductivity. Thus, the balanced effect of current change on the final grain size is actually not significant [24].

(a)80 A   (b)100 A   (c)120 A

Figure 4.4: Microstructure of AZ31 alloy by wire arc additive manufacturing with different current [24].

Torch feed: The study of AZ31B Mg alloy shows that with the increase of the torch feed speed, finer grain structure was observed as shown in Table 4.4. The reason for this is that as the torch feed speed increases, the amount of heat input per unit length decreases, thereby suppressing the temperature rise within the fabricated object. This, in turn, causes an increase in the cooling rate [25].

Table 4.4: Grain size of AZ31 alloy by wire arc additive manufacturing with different pulse frequency and torch speed [23, 25].

Pulse frequency (Hz)	Grain size (µm)	Torch feed speed (mm/min)	Grain size (µm)
1	31	400	50
2	23	600	46
5	21	800	42
10	21		
100	39		
500	37		

4.2.2.2 Macrostructure

In the study of AZ31 Mg alloy by Guo et al. [23], different pulse frequencies were used as variate to investigate the relationship between pulse frequencies and dimensional accuracy of AZ31 Mg alloy. The results show that as the pulse frequency increases, the layer thickness first decreases and then increases until it is stabilized, while the maximum width first increases and then decreases to a constant value. And it was found that the sample fabricated at the frequency of 10 Hz has the smallest layer thickness (1.7 mm) and the largest maximum width (7.3 mm). Meanwhile, the weld ripples, which are obvious at low frequency, become continuous when increasing the pulse frequency. The wall surface is also smoother at higher pulse frequency as shown in Figure 4.5 [23].

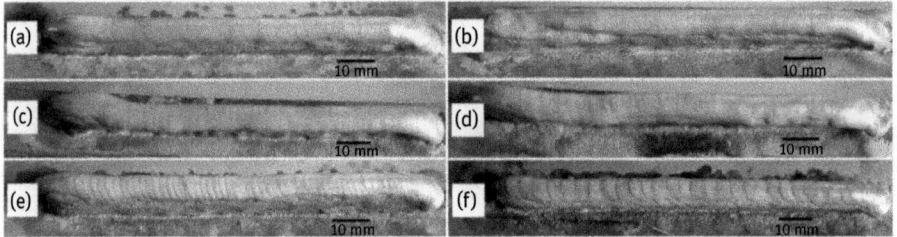

Figure 4.5: Surface morphologies of the samples deposited by different pulse frequency [23]: (a) 500 Hz; (b) 100 Hz; (c) 10 Hz; (d) 5 Hz; (e) 2 Hz; and (f) 1 Hz.

The increase of current seems to deteriorate the surface quality. As can be seen from Figure 4.6, the weld ripples become thicker and the intervals become larger. The reasons are that the molten drop, formed by the melt of welding wire, becomes larger as the current increases. Therefore, when it drops, it forms a larger weld ripple. On the other hand, with the rise of heat input, the fluidity of liquid metal also increases, which results in stronger stirring effect of the arc on molten pool, that is, greater fluctuation [24]. Therefore, higher alternating current contributes to rougher sample surface.

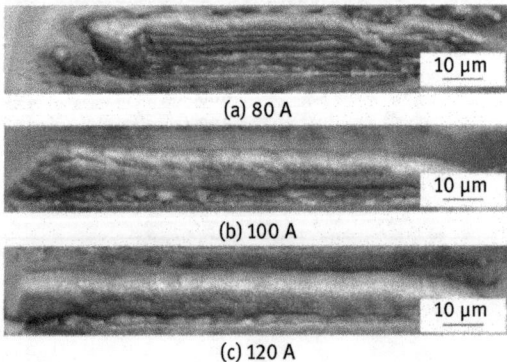

(a) 80 A

(b) 100 A

(c) 120 A

Figure 4.6: The surface morphology of the AZ31 alloy fabricated by wire arc additive manufacturing under different current [24].

Another study by Takagia et al. [25] on AZ31B Mg alloy focused on the combined effect of torch feed speed and welding current on the appearance of the fabricated object. As the torch feed speed decreases, the width and height of the fabricated parts improve. The increase in welding current will also cause an increase in the width and height, as shown in Figures 4.7 and 4.8 [25].

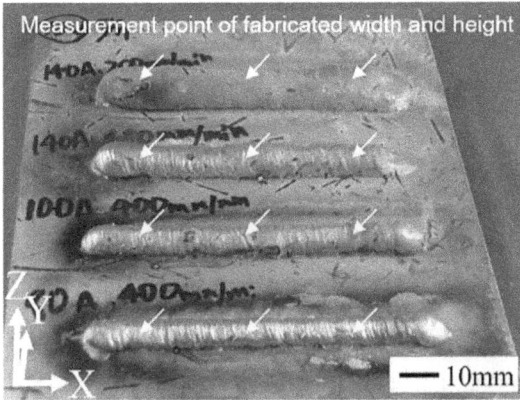

Figure 4.7: Surface morphologies of the AZ31B samples deposited by different current and torch speed [25].

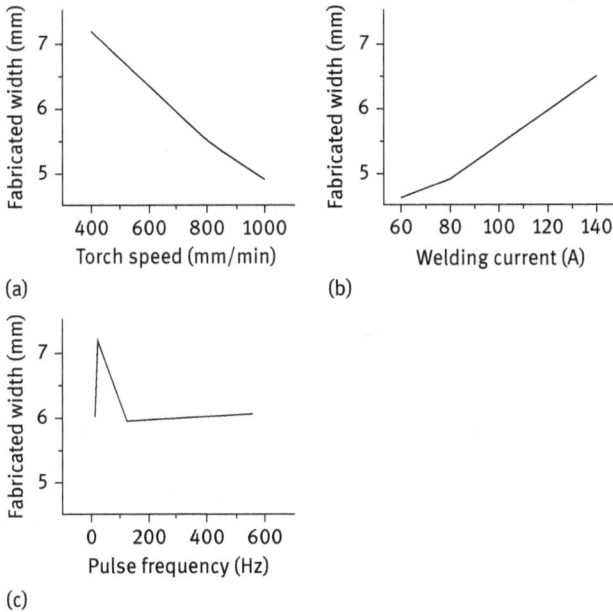

Figure 4.8: The microstructure of AZ31 alloy by wire arc additive manufacturing with different pulse frequency, current, and pulse frequency [23, 25]: (a) torch speed, (b) welding current, and (c) pulse frequency.

It follows that only a few layers were deposited in the above-mentioned experiments; it is unclear why components with larger size were not tested considering that WAAM is an AM technology that is suitable for fabricating large-sized components. Recently, thin walls of AZ with height larger than 60 mm were fabricated at the National Innovation Institute of Additive Manufacturing in China by the current authors, as shown in Figure 4.9, demonstrating that larger Mg components can indeed be fabricated by WAAM.

Figure 4.9: AZ31 walls fabricated at the National Innovation Institute of Additive Manufacturing in China.

4.2.2.3 Mechanical properties

Tensile testing was conducted to investigate the effect of processing conditions on mechanical properties. At present, the influence of pulse frequency and current on the tensile properties of AZ31 Mg alloy have been well documented. The tensile properties include ultimate tensile strength (UTS), yield strength (YS), and elongation (EL). As seen from Table 4.5, with the increase of pulse frequency, the UTS and YS first increase until it reaches the peak near 10 Hz, after which it declines with further increase of pulse frequency. The influence of pulse frequency on El seems to be weaker if the standard errors are considered.

On the contrary, there seems to be no obvious effect of current change on the tensile properties. It is noteworthy that the samples fabricated at 5 and 10 Hz (100 A) exhibit higher UTS (260 MPa) and YS (102 MPa), which are similar to those

Table 4.5: The mechanical properties of AZ31 alloy by wire arc additive manufacturing with different pulse frequency and current [23, 24].

Pulse frequency (Hz)	Current (A)	UTS (MPa)	YS (MPa)	EL (%)
1	100	229±3	90±1	26.3±0.3
2	100	232±1	89±1	26.7±1.2
5	100	258±5	100±3	25.6±3.1
10	100	263±5	104±5	23.0±3.7
100	100	221±11	82±6	23.4±5.8
500	100	231±5	79±2	27.3±0.3
400	80	230±1	78±3	31.0±3.9
400	100	236±1	89±2	27.5±2.1
400	120	234±6	85±7	24.9±2.8

of the forged AZ31 alloy whose UTS and YS are 234 MPa and 131 MPa, respectively, according to ASTM standard B91-12 [26]. Moreover, the elongations of all samples are above 23%, which indicates good ductility [24]. It is clear from all fracture surfaces that dimple rupture is formed, as illustrated in Figure 4.10. Compared with the other samples, the fracture surfaces of the tensile specimens manufactured at 5 and 10 Hz contain a lot of much finer dimples, which indicate relatively higher tensile strength. On the contrary, tensile specimens manufactured at other currents, as shown in Figure 4.11, have similar fracture surfaces. Therefore, it can be concluded that pulse frequency has a remarkable influence on the microstructure and mechanical properties of WAAM AZ31 Mg alloy [23], while the effect of current on microstructure and mechanical properties is not obvious.

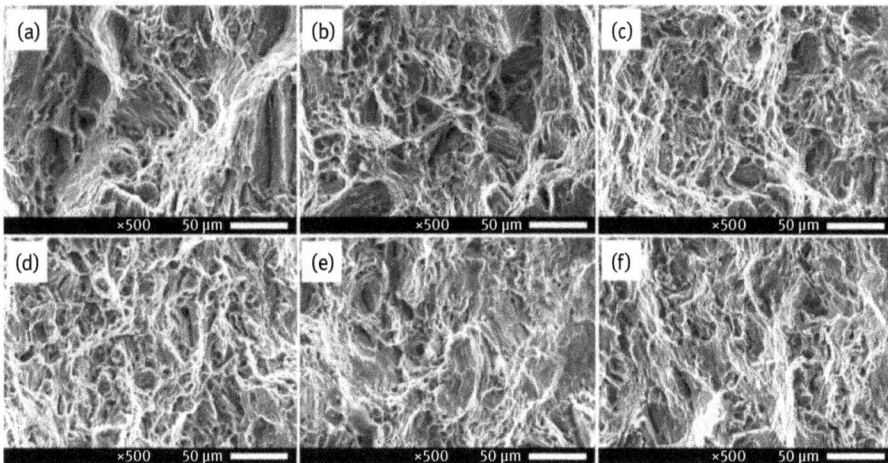

Figure 4.10: SEM fracture graphs of tensile samples deposited by different pulse frequencies [23]: (a) 500 Hz; (b) 100 Hz; (c) 10 Hz; (d) 5 Hz; (e) 2 Hz; and (f) 1 Hz.

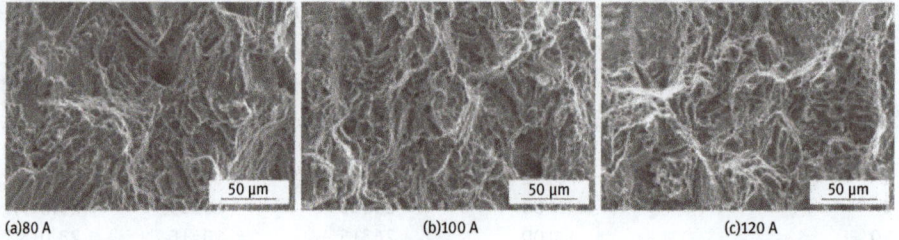

(a)80 A   (b)100 A   (c)120 A

Figure 4.11: SEM fracture graphs of tensile samples deposited by different current [24].

4.2.3 Defects and process control

4.2.3.1 Blowhole and crack

Blowhole defect is one of the major defects in metal AM. It has a significant effect on the quality of the fabricated parts. How to control and even eliminate the blowhole is the key technology in metal AM. Takagia et al. [25] studied the microstructure of AZ31B Mg alloy at different torch speed. They found that at a torch feed speed of roughly 800 mm/min, no blowhole defects measuring 0.156 mm or larger were observed. The defects generated in the samples fabricated at 400 and 600 mm/min are concentrated on the boundary between the first layer and the second layer, as shown in Figure 4.12.

David et al. [27] studied the AZ91D Mg alloy fabricated by WAAM technique. It was found that most of the cracks appeared at the bottom of the specimens, only a few were seen at the top. The defects grew inward, even though not necessary toward the center of specimens in the majority of the cases. The average crack length is about 220 μm, and the deepest crack reaches to 288 μm. But the central portion of the printed parts exhibit no defects. Therefore, all the cracks can be easily removed by a secondary machining process.

4.2.3.2 Distortion and residual stress of WAAM

There is a lot of heat input during the process of WAAM, and the distribution of temperature field is very complex. Research found that the distribution of temperature field is asymmetrical and the difference among the cooling solidification rate in each point is the main reason causing the residual stress. Another reason is the local phase transition [28]. At the time of writing, there is unfortunately no detailed investigation on the residual stress in WAAM of Mg alloys. The residual stress in WAAM of other metallic materials will probably shed light on this relatively new material. Szost et al. [29] compared the microstructure and residual stress of Ti-6Al-4V alloy parts

Defect volume (mm³)

0.50

0.25

0.00

z
x y

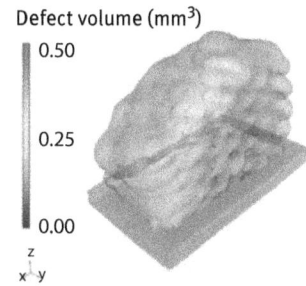

(a) 100 A, 10 V, 400 mm/min

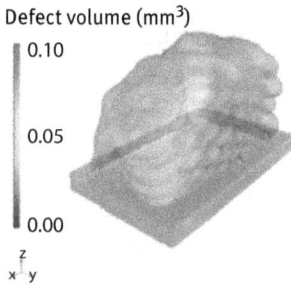

Defect volume (mm³)

0.10

0.05

0.00

z
x y

(b) 100 A, 10 V, 600 mm/min

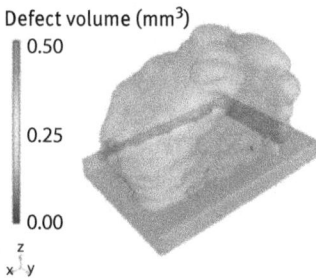

Defect volume (mm³)

0.50

0.25

0.00

z
x y

(c) 100 A, 10 V, 800 mm/min

Figure 4.12: Results obtained via CT scans of fabricated objects [25].

manufactured by WAAM and laser additive. The results show that the residual stress of WAAM forming parts is large in all directions, and the maximum residual stress appears at the bottom of the forming layer.

Various approaches have been developed to improve the residual stress in parts fabricated by WAAM. The controlling of residual stress is particularly important for WAAM since it is usually used to produce large-sized parts. The first category of these approaches refers to the tailoring of the process parameters, which include welding current, welding voltage, feeding speed, ambient temperature, shielding gas flow rate, and so on. There is still lack of systematic investigation on controlling defects through optimized selection or manipulation of these parameters. Path planning also improves the distortion and residual stress evolution in WAAM process [30]. If appropriate deposition path design is used, it will help in the significant improvement in these defects, especially in large-metal fabrication. At present, the residual stress is tailored mainly by some post-treatment methods, which are described in the following sections.

Interlayer rolling is another commonly used method to control residual stress after WAAM. Pual et al. [31] showed that high-pressure rolling has a good stress improvement in wire and arc additively manufactured steel parts. Their results show that both distortion and the peak longitudinal residual stress have been reduced by the rolling process. They concluded that there are two possible causes for the reduction in distortion with the rolled samples: (i) first, an overall reduction in the inherent strain and resulting stresses and modification of the geometry; and (ii) second, rolling reduced the wall height, which will also affect the distortion even where the longitudinal stresses remain similar. Hönnige et al. [32, 33] investigated the method of controlling residual stress of Ti-6Al-4V and aluminum in wire and arc manufacture. In the research of Ti-6Al-4V, they found that interpass rolling can eliminate the strong crystallographic texture of as-deposited material, but has few effect on the elimination of residual stress. On the contrary, thermal stress-relieving significantly reduced the residual stress. For aluminum, postdeposition side-rolling is very effective for controlling residual stresses and distortion in aluminum parts manufactured with WAAM and it increases the hardness by work hardening. Li et al. [34] proved that surface residual stress and microhardness of ultrasonic surface-rolled sample at 160 °C were significantly improved in HIP Ti-6Al-4V alloy.

The laser shock peening (LSP) is also a promising method to tailor the residual stress during WAAM, which then improves the fatigue life of the treated parts. Sun et al. [35] investigated the effect of LSP on WAAM 2319 aluminum alloy. Their results show that LSP can significantly refine the microstructure, with average size decreased from 59.7 to 46.7 μm after LSP; microhardness is also significantly improved due to high density of dislocations generated by LSP. In addition, tensile properties of YS was remarkably enhanced by 72%. Guo et al. [36] studied the influence of LSP on laser additive manufactured Ti6Al4V titanium alloy. It was found that LSP of Ti6Al4V titanium alloy modified the tensile residual stress into a compressive residual stress (CRS) with the maximum value of about 200 MPa and an affected depth of 700 μm. Recently, 3D LSP, combining SLM and LSP processes by applying the LSP treatment on every few SLM layers was developed by Kalentics et al. [37]. It aims at increasing both the CRS magnitude and depth compared to a conventional surface LSP process, with therefore an expected further improvement in fatigue life. Kalentics et al. [37] found that 3D LSP increases both the magnitude and depth of CRS, and with a reduced spot size and pulse energy, 3D LSP can produce deeper CRS than those induced by a conventional LSP treatment with a larger spot size and pulse energy.

4.2.4 Application of WAAM in Mg alloy

Compared with other AM technologies, WAAM has the characteristics of less equipment investment, lower operation cost, high deposition efficiency, and high material utilization rate. It would be a good choice to fabricate some workpieces, when the adoption of traditional processes results in low material utilization, such as Mg alloy wheel hub and seat. It must be noted that there is currently no wide application of WAAM Mg alloy parts in real industry. In fact, the industrial application of Mg alloy is still limited, regardless of its processing method, as discussed earlier in Section 4.1.3. The characteristics of WAAM lend itself to fabricate large-sized component used in transportation, aeronautics, and astronautics industries, whereas it is not likely to be widely used in medical industry where small-sized components with high dimension accuracy are usually required.

WAAM takes continuous "line" as the basic forming unit, which is suitable for rapid forming of large internal frame, reinforcing bar, and wall plate structure. At present, the application of large integral titanium, aluminum, and magnesium alloy structure in aircraft is increasing. Although large integrated structural parts can significantly reduce the weight of the structure, but it also brings great difficulties as compared to the material reduction by conventional methods in terms of material processing and manufacturing. For example, the main bearing member of F35 in the United States still needs to be pressed and formed by a hydraulic press of tens of thousands of tons. In the later stage, a lot of tedious milling, grinding, and other processes are required, and the manufacturing cycle is long. So now, more and more aircraft manufacturer such as Airbus, Bombardier, BAE systems, Lockheed Martin and Astrium, and so on all use WAAM to directly form large structural parts of titanium alloy and high-strength steel material, which shortens the development cycle of large structural parts.

In 2015, the Institute of Metals, Chinese Academy of Sciences, signed a research cooperation agreement with Airbus. Both parties have decided to start two projects of AM powder/silk lean process and magnesium alloy surface protection. Therefore, it can be believed that the applications of WAAM technology in the research and development and manufacturing of magnesium alloy aircraft components will also be a direction of WAAM development in the future.

In terms of maintenance, the RAMLAB set up in Rotterdam, the Netherlands, in 2016. Its purpose is to offer rapid maintenance of large ships in Rotterdam port with WAAM technology. Thus, long-distance transportation is avoided, delivery time is shorted, inventory is reduced, and cost is reduced.

It can be seen that WAAM has huge advantages in parts repair and specific parts replacement. For some parts that cost too much to repair and the cost of processing a single part by traditional machine is also very high, WAAM would be a good alternative.

4.2.5 Future trends in WAAM of Mg alloys

Although it has been a long time for the study of WAAM technique, the development of the WAAM is slow with the restriction of the development of welding technique itself. It was brought into focus again during the last decades. The future development trend may include the following points:

1) Since WAAM is more suitable to produce large parts, it is important to control its residual stress to ensure dimensional accuracy. More research should be conducted on quantitative control of the residual stress during WAAM.

2) The surface quality of parts produced by WAAM is relatively poor, and subsequent machining is inevitable to get desired surface quality. Because of the distortion caused by residual stress, the size of the deposited part will be different from the predefined model, which brings great difficulties to the subsequent processing and demands for 3D measurement of the dimensions of the fabricated parts.

3) At present, the main method of monitoring the process is through visual sensing system. Due to the processing speed constraints, the limitation of spatial layout, and the interference of light path, the technology is not mature enough. So new in situ monitoring system should be developed to control the geometry and quality of the fabricated parts.

4) Current research work on the quality of WAAM components mainly focuses on process optimization and dimension control; the microstructure control of the components has not been equally addressed. Tailoring of internal microstructure to meet the desired properties for different application environment will most likely to receive more attention.

5) With the further requirements for process integration and further reduction in production cost, the emergence of highly capable WAAM machines that combine other manufacturing processes to efficiently transform raw materials into final parts is a key direction for WAAM.

6) Due to the low melting point, high linear expansion coefficient, and thermal conductivity of Mg alloy, as well as strong affinity with oxygen and nitrogen, it is easy to form inclusion and brittle phase after welding and easy to generate welding deformation and thermal crack. It may be necessary to consider using inertial gas environment during WAAM of Mg alloy; it will also be of great help to develop suitable WAAM wires that meet the performance requirements of the final component.

7) WAAM of dissimilar metallic materials, which combines the advantages of each individual material, could overcome the limitation of traditional manufacturing process and extend the wide application of WAAM.

4.3 Selective laser melting of magnesium alloys

4.3.1 Introduction

SLM is a widely used AM technology. Among them, "laser" means that the main energy source of the process is laser, and selective melting means that only the powder of a specific region is melted during the process of melting the powder. A complete SLM system typically includes a laser generator, an automated powder system, a flat subtract, a computer control system, and some auxiliary equipment (e.g., inert gas protection system and auxiliary powder paving equipment) [38].

The main process includes: (i) layer and laser path planning of the model with specific software based on the 3D model of the target workpiece; (ii) fixing and leveling the substrate for processing the powder and the workpiece on the platform and filling the working chamber with protective gas; (iii) spreading a thin metal powder layer on the substrate by a powder laying system, the thickness of which shall be determined according to the thickness of the workpiece slice layer processed by the computer; (iv) the high-energy laser beam scans the powder bed in a predetermined path so that the shape formed by sintering the layer is consistent with the computerized 3D model processed layered pattern; (v) lowering the platform on which the substrate is placed by a predetermined height, and repeating the laser scanning and lowering the platform for the continuous powder layer until the required components are fully built [39]; (vi) after the parts are processed, separate the workpiece from the substrate (e.g., electrical discharge machining) and, if required, move to postprocessing steps. The whole process is shown in Figure 4.13 [40].

Metal powder materials commonly used in laser AM processes include iron alloys, titanium alloys, cobalt chromium alloys, nickel-based alloys, aluminum alloys, copper alloys, zinc alloys, tungsten alloys, gold, and so on. It is noted that there are currently only sparse studies focusing on magnesium alloys due to the technical difficulties related to this type of alloys, which will be detailed in later sections. The properties of the powders, such as the size distribution, shape, surface morphology, composition, and flowability of the powers, greatly affect the performance of the fabricated parts [8]. The characteristic of SLM is shown in Table 4.6.

4.3.2 Application of SLMed magnesium alloys

SLM can be used to manufacture parts with extremely complex shapes. The advantages of magnesium alloys such as good processability, high damping capacity, good casting properties, low density, low melting point, and degradability make them an ideal choice for automotive, aerospace, and medical implants. However, to the best knowledge of the authors, the SLMed Mg alloy components have not been widely used in industry yet, their potential applications are instead introduced here.

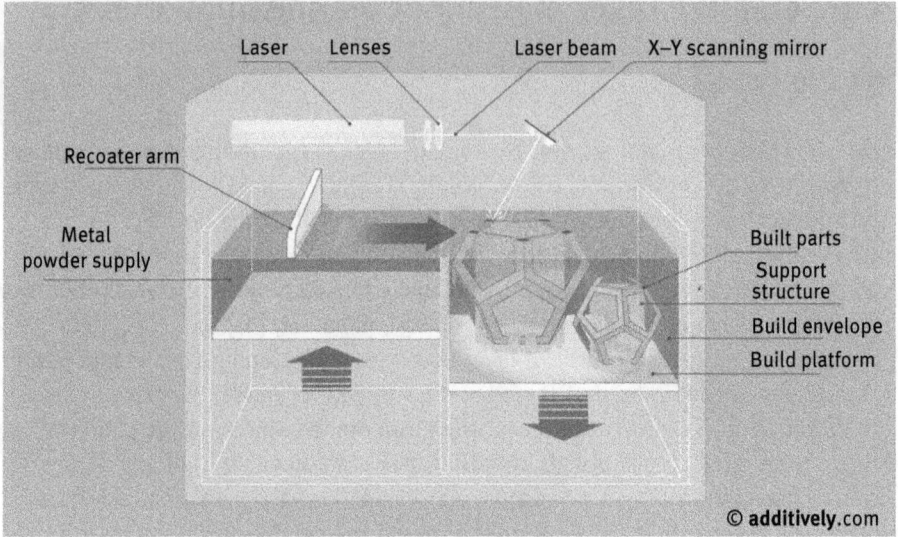

Figure 4.13: The whole SLM process.

Table 4.6: Characteristics of SLM.

Advantages	Disadvantages
– Reduce assembly time and increase material utilization. – Shorter time to market and more flexible production processes. – Ability to manufacture complex structures that are difficult to manufacture with traditional processes. – Better product quality.	– The price is relatively expensive, and the cost of use is high. – Poor surface quality. – Lower material utilization.

Medical implants made of magnesium alloys can degrade themselves in patients' body, thereby avoiding a second procedure to remove the implant after the tissue has healed. At the same time, magnesium alloys have similar elastic modulus to human bones, which can control the stress on the bones in the human body much better [41, 42]. Compared with titanium, cobalt, chromium, and other alloys, magnesium alloys can avoid stress shielding and looseness of implants [43]. At the same time, polymers and bioactive composites are limited by their low load carrying capacity, making magnesium alloys unique in human implants [44]. Recently, the first commercial magnesium alloy named Magnezix screw orthopedic device in Europe has begun production. Magnesium alloys are used to produce bone plates, nails, and animal wound closures [45]. Researchers found that magnesium ions also contribute to the rapid growth of bone tissue [45]. In conclusion, the use of magnesium alloy AM products has superior surface quality, and its biodegradability and low stress shielding make magnesium alloys the perfect choice for medical implant applications. The strength and fatigue life of the magnesium alloys produced by SLM technology are satisfactory. Figure 4.14 shows the complex gyroid cellular structures manufactured by SLM [46].

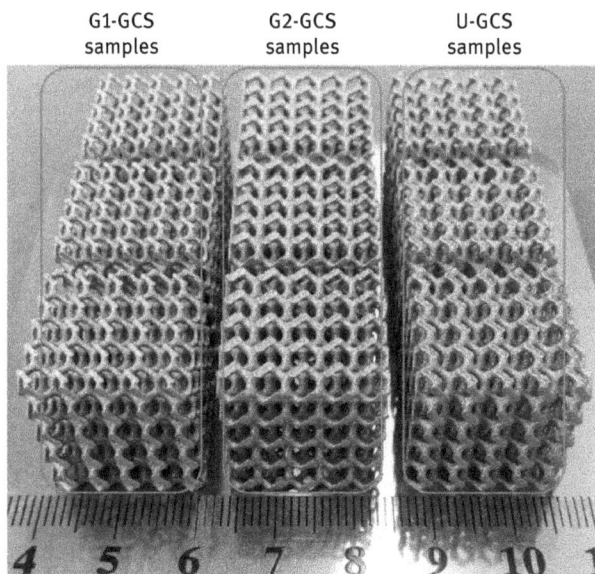

Figure 4.14: The complex gyroid cellular structures manufactured by SLM.

4.3.3 Mechanical properties of magnesium alloy components manufactured by SLM

The performance observed for different SLMed magnesium alloy powders is summarized in Table 4.7. We have got a lot of detailed data from many experiments, for example, Zhang et al. [47] used Mg9 wt%Al in SLM, using a laser power of 15 W and scan speed of 20 mm/s and their production's hardness value reached 75 HV, which is better than casted AZ91 alloy. Figure 4.15 [39] is a comparison of a magnesium alloy test piece manufactured by SLM technology with a cast part and a forged part (Figure 4.13). It can be seen that the hardness of magnesium alloy specimens manufactured by SLM technology has similar values to forged and casted counterparts, but its Young's modulus is clearly lower. However, its lower Young's modulus is closer to Young's modulus of human bones (3–20 Gpa), which makes it more suitable to produce human bone implants. Figure 4.16 [39] is a comparison of tensile

Table 4.7: The performance observed for different SLMed Mg alloy powders [38].

Materials system	General microstructure/ intermetallic phase	Hardness (HV)	Young's modulus (Gpa)	Surface roughness (µm)
Mg [49, 50] *	Equiaxed α-Mg grains, precipitates of MgO along grain boundaries	60–97 (from Gpa)	20.8–38.2	–
Mg [51] *	–	66–74 (from Gpa)	29.9–33.1	19–33
Mg [52]	–	44.7–52.4	–	38.6–51.8
Mg-9Al [53]	Equiaxed α-Mg grains/ $Mg_{17}Al_{12}$, MgO, Al_2O_3	66–85	–	–
AZ91D [54]	Equiaxed α-Mg grains/ $\beta - Mg_{17}Al_{12}$, Al_8Mn_5	85–100	–	–
ZK60 [55]	Oriented dendrites/ MgZn, Mg_7Zn_3	78	–	–
ZK60 [56]	Dendritic/columnar α-Mg; Mg_7Zn_3	70.1–89.2	–	–

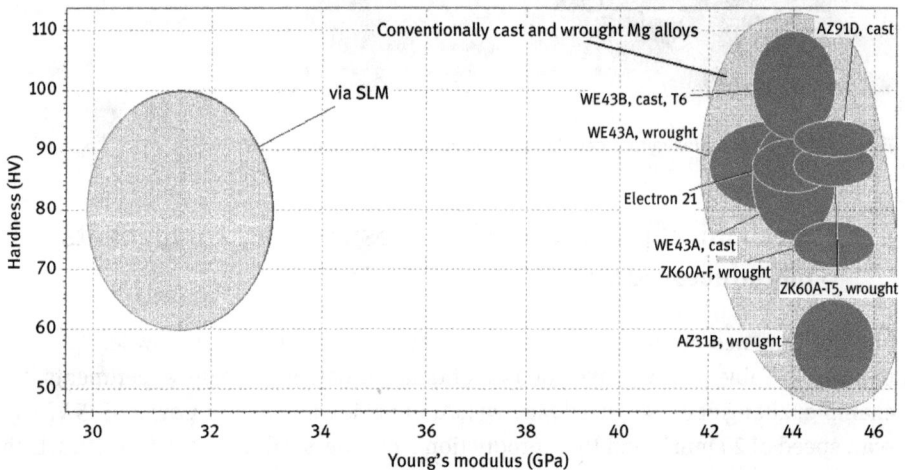

Figure 4.15: The comparison of SLMed test piece with a cast part and a forged part.

properties of SLM processed parts with conventionally cast and wrought magnesium alloys. On the basis of the experimental result from Wei et al. [48], their AZ91D magnesium alloy specimen prepared by SLM technology has a lower elongation than the die cast AZ91D; the details are listed in Table 4.8 [48].

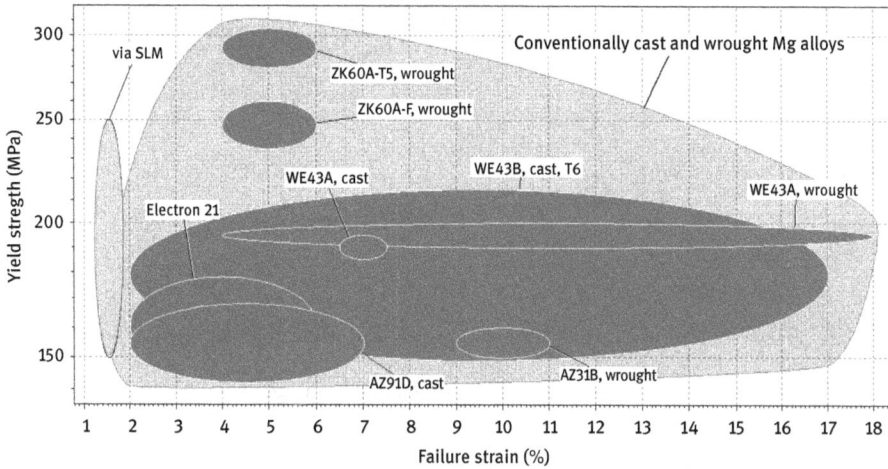

Figure 4.16: The comparison of tensile properties between different materials and methods.

Table 4.8: The comparison of AZ91D's mechanical properties.

Sample	Ultimate strength (MPa)	YS (MPa)	Elongation (%)	Microhardness (HV)
AZ91D (SLMed)	294–298	251–256	1.68–1.99	90–108
Die-cast AZ91D	230	160	3	61.3–63.7

4.3.4 Microstructure

The microstructure of SLMed metallic components is closely related to the fast heating and cooling rates (up to 10^5–10^6 K/s) during the SLM process. Once the investigated material is fixed, it is the process parameters such as laser power and scanning speed that determine the heating and cooling rate, which eventually controls the final microstructure. Since the microstructure of the SLMed components is affected by many parameter variables, its prediction is still a huge challenge. According to the existing literature, three forms of microstructures can be obtained depending on the process parameters: columnar (elongated grain morphology), columnar-plus-equiaxed, and equiaxed (isotropic grain morphology). In Table 4.7 [39], the microstructure characteristics of different magnesium alloys are summarized. Figure 4.17 [50] is the comparison of microstructure manufactured by continuous and pulsed laser. In which the part (a) shows the microstructural of SLMed magnesium under continuous wave irradiation at $1.27*10^9 \mathrm{J/m^2}$ and the part (b) is manufactured under a pulsed wave irradiation at $1.13*10^{12} \mathrm{J/m^2}$. These two pictures are obviously different; the continuous laser leads to a full recrystallization and complete grain growth, when the incomplete growth of the α-Mg phase occurred using the pulsed laser. Figure 4.18

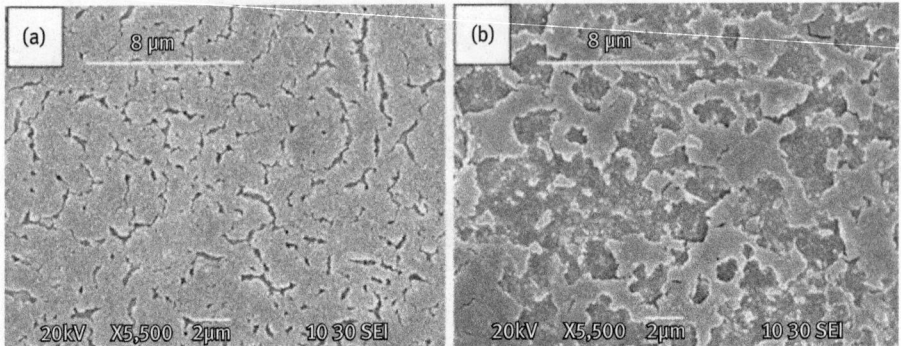

Figure 4.17: The comparison of microstructure manufactured by (a) continuous and (b) pulsed laser.

[56] shows characteristic crystalline structures of ZK60 prepared at different scanning speed and energy density. The highest scanning speed created extreme fine dendrites with articular shape. In part (b), the microstructure changed into disorder column and in part (c) the orderly dispersed equiaxed structure is observed. Finally, at the energy density of 750 J/mm^3, the significantly coarsened equiaxed grains were obtained. Figure 4.19 [50] is the comparison of (a) cast pure Mg ingot and (b) laser-melted Mg at longitudinal cross-section. The SLMed magnesium has a much finer microstructure than conventional casting specimen, which is because of the high solidification rate.

4.3.5 Factors affecting the performance of magnesium alloy specimens manufactured by SLM

The ideal form of SLM manufacturing process is the ability to produce 100% density metallic parts with superior mechanical properties. However, due to the absence of mechanical pressure during processing, and because its fluid dynamics are mainly driven by gravity and capillary forces and thermal effects, it is still practically difficult to reach fully dense (i.e., zero porosity) SLMed components. In addition, discontinuous melting of the melting trajectory and formation of pores could also take place [57, 58]. Figure 4.20(a) [56] shows the porosity and unmelted areas of the ZK60 sample from the SLM production process. These materials undergo different degrees of thermal fluctuations during processing, which may cause residual stresses in the cladding layer during rapid solidification, resulting in the formation of hot cracks and delamination as shown in Figure 4.20(d).

Rapid heating and cooling rates also create a narrow heat affected zone around the melten pool. The presence of such areas can result in changes in the composition and microstructure of the material, affecting the quality and nature of the specimen. Thermal behavior during SLM can be effectively controlled by adjusting processing parameters such as laser power, scan rate, scan spacing, layer thickness, and scan

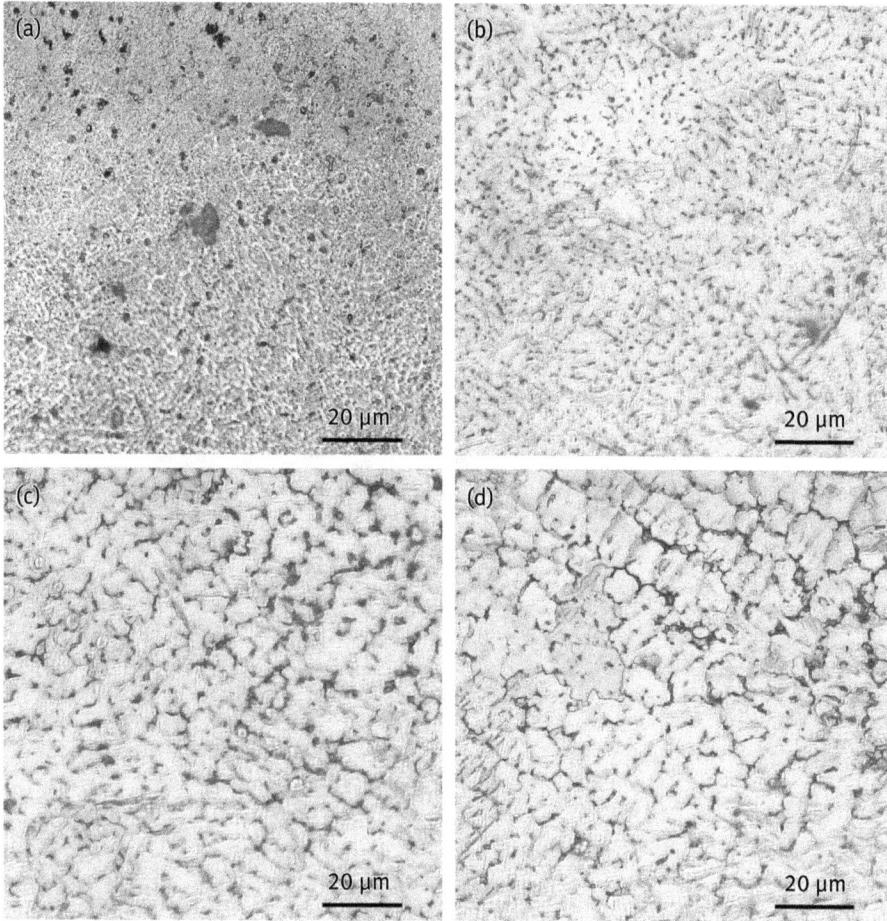

Figure 4.18: Characteristic crystalline structures of ZK60 prepared at different scanning speed and energy Density: (a) 700 mm/min, 420 J/mm³; (b) 600 mm/min, 500 J/mm³; (c) 500 mm/min, 600 J/mm³; and (d) 400 mm/min, 750 J/mm³.

pattern for each layer (which can be seen in Figure 4.21). The main factors affecting the performance of the test piece during the SLM process are listed in Table 4.9.

Throughout the whole process, one needs to carefully adjust these parameters to avoid defects such as oxidation, balling, loss of metallic elements, slag formation in the molten pool, delamination, and other issues [11, 61, 62].

i. Influence of process parameters on properties of the workpiece

Laser is the main source of energy for the entire process. For SLM, the role of the laser is to provide a certain amount of energy, so that the metal powder can melt

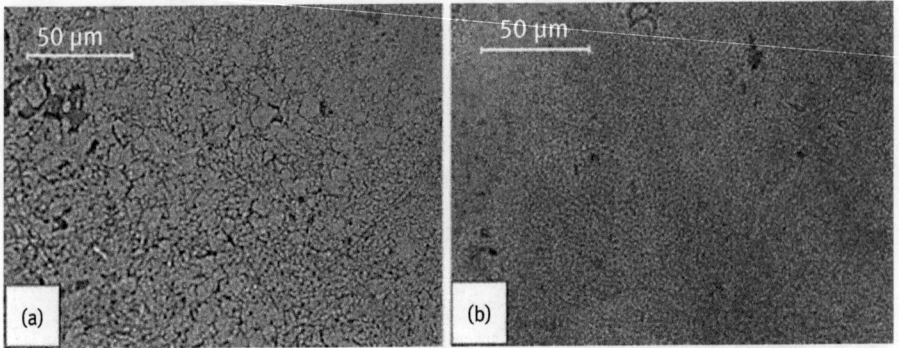

Figure 4.19: The comparison of (a) cast pure Mg ingot and (b) laser-melted Mg.

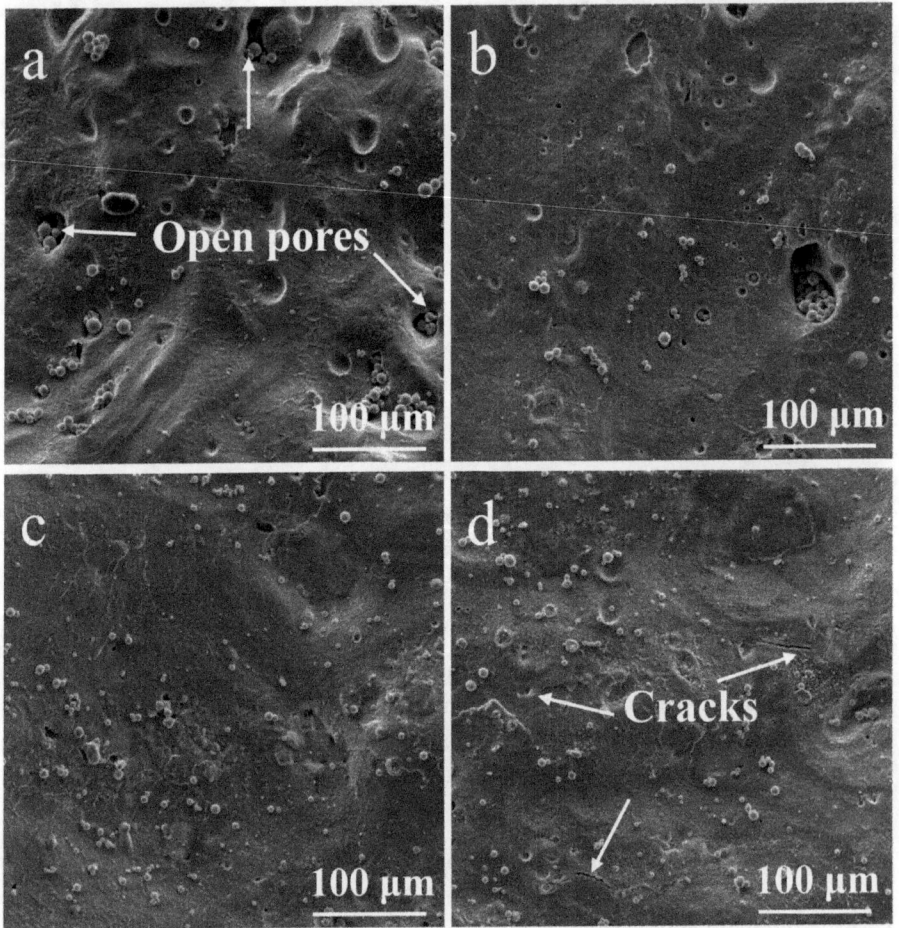

Figure 4.20: The surface of SLMed ZK60 prepared at different energy density (a) 420J/mm^3; (b) 500J/mm^3; (c) 600J/mm^3; (d) 750J/mm^3.

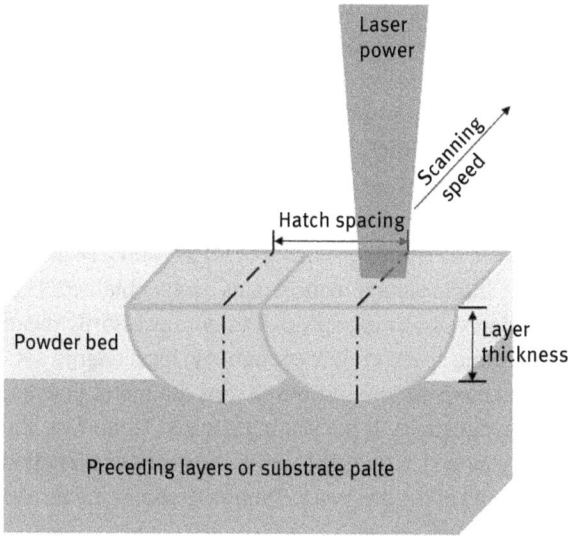

Figure 4.21: The main parameters of SLM process [59].

Table 4.9: Variables influencing the processing and densification mechanism of SLMed parts [60].

SLM processing parameters	Material properties
Laser type	Viscosity
Laser power	Surface tension
Mechanical layering of power	Thermal conductivity
Atmospheric control	Specific heat
Gas flow	Absorptivity/reflectivity
Heaters (bed temperature)	Emissivity
Scan radius	Particle size distribution
Scan vector length	Particle shape
Scan spacing	Melting temperature
Scanning time interval	Chemical composition
Scan rates	Boiling temperature
Thickness of layers	Oxidation tendency

quickly. The parameters related to the laser mainly include laser power, laser energy density, scanning rate, scanning spacing, and so on. Suitable laser parameters, which depend mainly on the chemical composition of the alloy, need to be tailored for each investigated material [53]. In order to obtain a suitable range of parameters, a large number of simulation analyses and experiments are required for each metallic material.

Taking AZ91D magnesium alloy as an example, Wei et al. [54] studied the effect of scanning rate and scan spacing on the density of magnesium alloy workpieces processed by SLM. Under the premise of constant laser power (200 W), the laser energy density can be increased by reducing the scanning rate and reducing the scan spacing, which enables the powder to be better melted. Increasing the scanning spacing results in the reduction in the overlap of the two parallel scan paths, which may cause the adjacent melt to not fully fuse and increase the internal porosity. The experiment achieved a maximum workpiece density of 99.52% at a scanning spacing of 90 μm and a scanning speed of 0.33 m/s. In the study of ZK60, similar conclusions were obtained [55]. However, for the whole process, the slower scanning speed does not necessarily mean the better performance. When the scanning speed is slow enough, the local energy density of the laser on the powder causes evaporation and burning of the metal powder, and leaves burnt on the surface of the substrate. If the scanning speed is too fast, the powder particles will not be completely melted, resulting in voids between the unmelted powders, causing a decrease in the density of the test piece. Table 4.10 shows the results of the experiment for AZ91 [39].

Table 4.10: Relative densities of AZ91 parts formed with varying scanning speed and spacing at laser power = 200 W.

Scanning Speed (mm/s)	Relative density (%) at different scan spacing			
	70 μm	90 μm	110 μm	130 μm
333	99.4	99.5	99.2	98.8
500	99.2	99.3	99	98.4
667	99.1	98.8	93.5	89.1
833	97.4	95.9	84.4	77.2
1,000	91.8	59	76.5	73.4

In addition to the conditions of the laser itself, the layer thickness of the powder also has a significant effect on the quality of the fabricated part. In the case of constant laser parameters, excessive layer thickness will affect the fusion between the powder layers. Compared with the lower layer thickness, the excessive layer thickness requires higher energy density; otherwise, parts with substantial porosity will be formed. In the pursuit of a good balance between fabrication efficiency and part quality, it is necessary to find the most suitable layer thickness.

ii. Influence of powder quality on properties of the workpiece

Powder is the raw material in SLM processing, the qualities of the powder include shape, size, surface morphology, composition, internal porosity (which may result in the gas-induced porosity), fluidity, and apparent density. Among them, the surface morphology of the powder will affect the processing flowability, powder filling characteristics, layer thickness and surface roughness of the workpiece [63], and so on. The use of spherical particles improves the flowability of the powder compared to irregular particles, resulting in higher deposition speed and precision. The metal powders used in the SLM process are typically very fine (10–60 μm), mutual adhesion due to the presence of water should be avoided as much as possible. At the same time, the powder should be as uniform as possible, the size of the powder particles should be similar, and the smaller particles tend to aggregate while the larger particles tend to separate, which leads to instability during the process. It has been found in experiments that finer powder particles lead to greater workpiece density than coarse powder particles [64]. The characteristic material properties of Mg alloys include low absorptivity to the laser beam, low boiling point elements, high thermal conductivity, high coefficient of thermal expansion, tendency to form low melting point eutectic phases, and low viscosity [62].

The laser absorption rate determines the energy efficiency of the laser, which means that at the same laser power, a higher absorption rate can melt the powder better, and a lower absorption rate may cause the powder to be insoluble. Compared to titanium alloys, magnesium alloys are more difficult to control and consume more energy in the SLM process due to their higher thermal conductivity and lower laser absorption.

Due to the low density of magnesium alloys, there may be particle agglomeration in the interior due to van der Waals forces, thereby reducing the flowability of the powder. The paradox, on the other hand, is that the melting of finer particles requires lower input of laser energy, but the temperature in the molten pool is higher compared to larger powder sizes, which may lead to severe oxidation and pilling [65].

All in all, magnesium alloy is a kind of metal material that has not been widely used in AM. More work is needed to systematically study the characteristics of different kinds of magnesium alloys and get better combination of parameters in order to improve the quality of magnesium alloys in AM processes.

4.3.6 Major defects

4.3.6.1 Oxide inclusion

During the depositing process of the SLM, the upper surface is prone to be oxidized (as shown in Figure 4.22) [50]. When a new layer is deposited, the oxide layer (e.g., MgO), which usually has a very high melting temperature, is usually sandwiched between the front layer and the back layer. This oxide layer is difficult to be removed without chemical reactions. Since the oxide layers have very different properties (melting temperature, thermal expansion, shrinkage, etc.) as the deposited

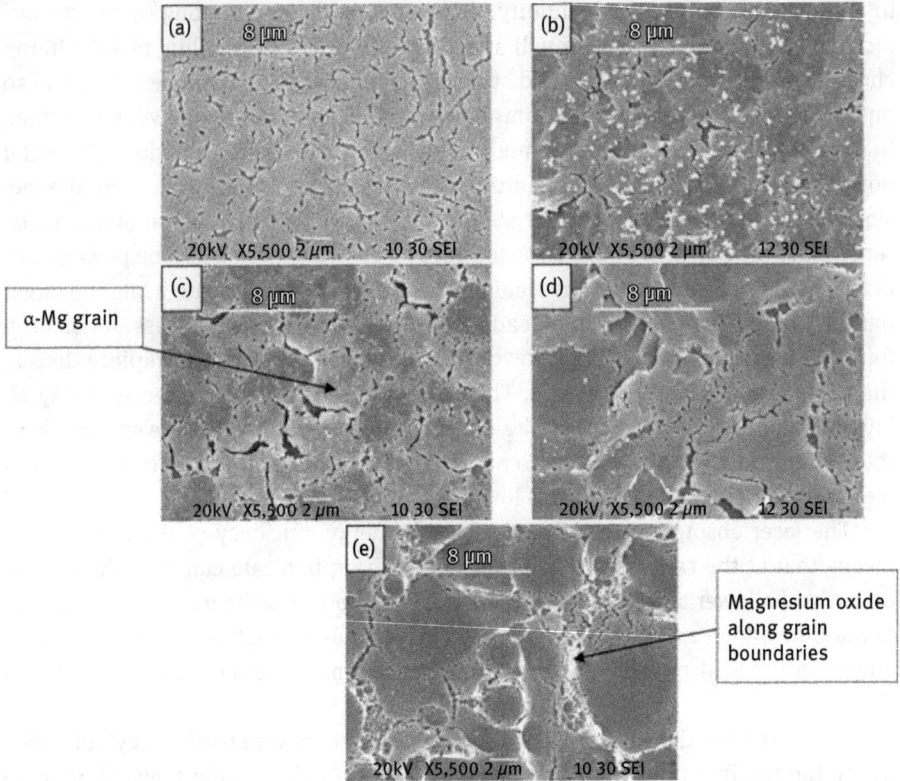

Figure 4.22: The microstructure of SLMed specimen prepared in different laser energy densities: (a) $1.27*10^9$J/m^2, (b) $2.11*10^9$J/m^2, (c) $3.92*10^{19}$J/m^2, (d) $6.33*10^9$J/m^2, and (e) $7.84*10^9$J/m^2.

layers, they tend to cause the upper and lower layers to fail to fuse effectively, which will result in delamination. At the same time, the oxide layers also generate microcracks due to their poor ductility, which greatly affect the physical properties of the workpiece.

There are usually two ways for the oxide to enter the molten pool during the processing. One is the oxide that is entrained in the raw material powder in the pretreatment, and the other is the oxide particles that are trapped in the gas on the working surface. Once these oxides enter the molten pool, an oxide film is then formed on the surface. To avoid the appearance of oxidation, one can increase the laser energy density and reduce the oxygen content in the process chamber as much as possible.

4.3.6.2 Balling

Balling is a typical phenomenon caused by loose powder bed in SLM processing. It is related to the aggregation of spherical particles, due to the low laser power, high scanning speed, and excessive layer thickness, which leads to insufficient laser energy density. During the normal laser scanning process, a cylindrical liquid trajectory is formed. However, when the surface energy of the liquid orbit is continuously reduced to equilibrium, the cylinder is decomposed into spherical blocks. Figure 4.23 [53] shows the balling phenomenon.

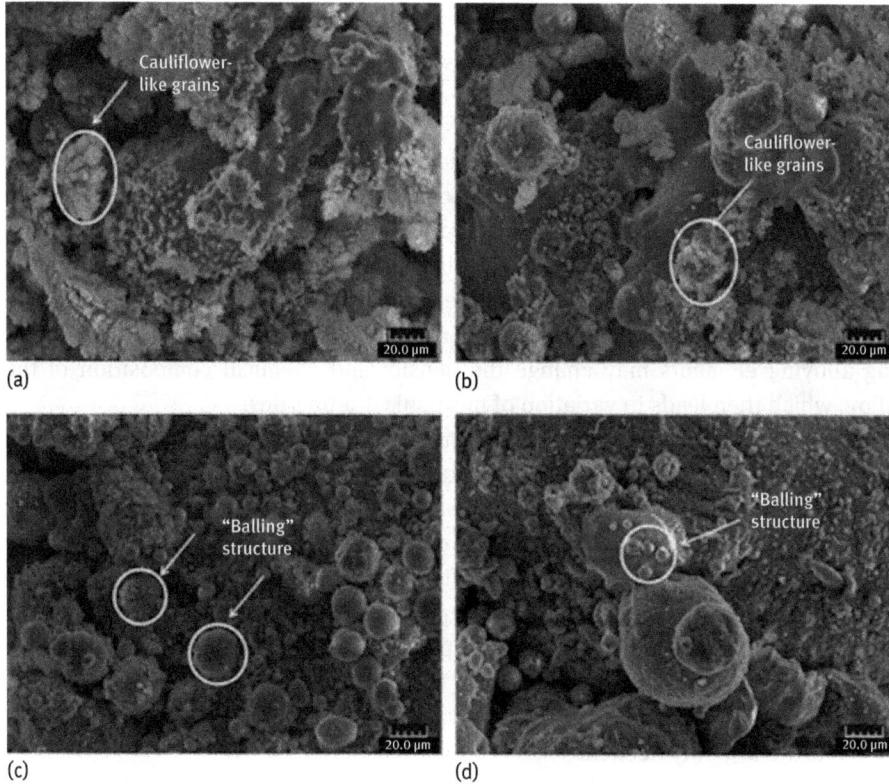

Figure 4.23: SEM images showing the characteristic microstructure of samples surface with variation of process parameters: (a) 10 W, 0.01 m/s; (b) 15 W, 0.02 m/s; (c) 20 W, 0.04 m/s; and (d) 15 W, 0.04 m/s.

To avoid this, it is necessary to improve the stability of the molten pool by reducing the aspect ratio of the molten pool or increasing the contact area. It can also be achieved by increasing the laser power or reducing the scan rate. According to Li et al. [66], the balling phenomenon is also related to the oxidation of the metal. Therefore, the oxygen content should be reduced as much as possible during the processing and the oxide film should be destroyed by repeated exposure.

4.3.6.3 Loss of alloying elements

At high laser energy densities, and due to the high vapor pressure and lower boiling point of the magnesium alloy, evaporation of some elements occurs when the laser and powder interact. In the SLM process, the temperature of the molten pool is usually much larger than the boiling point of magnesium. Therefore, evaporation of magnesium usually occurs during the manufacture of magnesium alloy by SLM. These easily evaporating elements are present in a large amount on the surface of the molten pool, and then evaporation occurs at the interface of the liquid and gas, and the evaporated metal is vaporized into the surrounding gas [67].

Some of the metal vapor entering the surrounding gas will be deposited again around it and participate in the subsequent melting, but at the same time, these volatile alloy components will also produce a black coating in the next reaction. In this way, some metals that are easily evaporated will cause an increase in the concentration of alloying elements that are difficult to be evaporated. This evaporation phenomenon also becomes intense as the energy density of the laser increases. In the final finished workpiece, the reduction in the content of these easily evaporating alloying elements may change the density and chemical composition of the alloy, which then leads to variation of mechanical properties.

At the same time, the evaporation of metal elements will have an impact on the stability of the molten pool. During the rapid evaporation of the metal, the recoil will cause the molten pool to spread outward, resulting in a "key hole" defect after solidification (see Figure 4.24) [68]. Such defects can affect the surface quality and mechanical properties of the workpiece after forming.

To further control the negative consequences caused by evaporation phenomenon, it is recommended that the laser adopt medium-high power and high scanning rate. Meanwhile, further investigation on the mechanism of evaporation as well as establishing the relationship between evaporation phenomenon and process parameters are urgently needed.

4.3.7 Perspectives

Even though there are only sparse investigation on the SLM of Mg alloys, the wide spread application of SLMed Mg alloys is expected due to their superior properties. It is reported that the ZK30-Cu alloy with high performance is obtained by processing the copper-containing magnesium alloy ZK30-Cu as raw material in SLM [69]. The alloy has good antibacterial ability and cytocompatibility and improved biodegradability, which reduce the degradation of magnesium alloy in cells. This work provides new ideas and manufacturing techniques for the study of new alloy designs. The niobium element was added to the magnesium alloy ZK60 and was produced by SLM technology by Shuai et al. [70]. This new alloy exhibits good

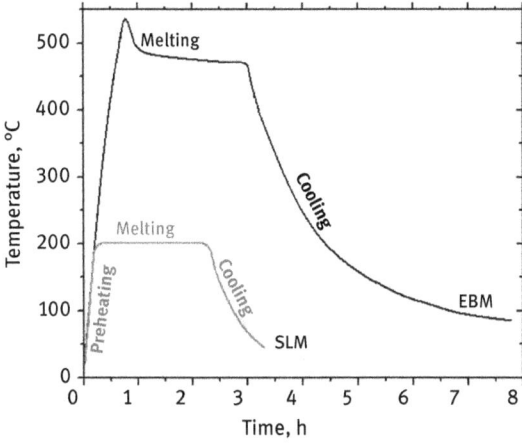

(a)

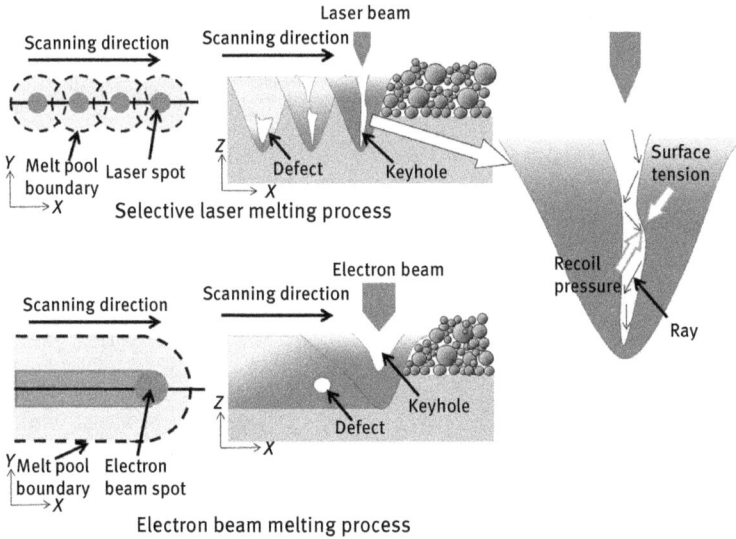

(b)

Figure 4.24: The schematic diagram of the EBM and SLM melting process.

antitumor properties, and its degradation rate is lower than that of the ordinary ZK60 due to grain refinement. Adding CaO to AZ61 alloy through SLM technology was also documented to improve the second phase characteristics of the alloy, which further improves its corrosion behavior. The formed inert (Mg, Al)2Ca phase uniformly encapsulates the Mg grains, effectively protecting them from corrosion. At the same time, AZ61-9CaO also shows good cytocompatibility and is expected to be used in orthopedic implants.

It can be seen from the above-mentioned examples that many current applications of SLM in magnesium alloys are used to develop new alloys toward the field of medical implants by adding different elemental components to make magnesium alloy components to have diverse desired functions. At the same time, in this way, it is also possible to influence the physical properties of the magnesium alloy test piece, such as improving its corrosion resistance and weakening its degradability. Such research also shows the bright future of magnesium alloy AM technology in biomedicine.

In addition, the SLM technology still have many defects and problems to be solved, which also point to the directions for the future development of SLM:

- The higher utilization of the material: One of the main reasons for the high price of SLM is the low material utilization of the powder.
- A better surface quality: We still need some postprocessing after SLM, such as polishing and CNC machining. If this is solved, AM will be more widely used.
- Suitable for more materials: SLM has material limitations when applied, a series of parameters can only be used for one material that cause the difficulties in manufacturing multiple materials.
- Some mechanisms that are currently unclear need to be further investigated such as the laser–powder interaction and the quantitative control of residual stress.

4.4 Other AM methods of Mg alloys

4.4.1 Ultrasonic additive manufacturing

4.4.1.1 Introduction

Ultrasonic additive manufacturing (USW) is a solid-phase welding process, which can weld different materials, such as different metals, metals and glasses, and nanoalloys that cannot be welded together by conventional means. During the USW process, the workpiece is clamped between the anvil and the sonotrode, the sonotrode is perpendicular to the anvil, and the vibration is transmitted to the loading shaft, as shown in Figure 4.25 [71]. This technology was first studied in the 1950s and comes to mature nowadays. In recent years, this technology has been applied in a hybrid AM technology named UAM. In the UAM technology, components are formed by layer-by-layer accumulation of raw materials. The process includes: (i) using a cylindrical sonotrode to roll the machined surface to weld the flaky metal to the pedestal, the direction of the sonotrode should be perpendicular to the direction in which it vibrates and (ii) using the CNC system to remove excess material and create complex components with the same surface roughness [72]. In the process of UAM, the main variables are sonotrode oscillation frequency (V), sonotrode oscillation amplitude (λ), normal load (F), and weld time (t).

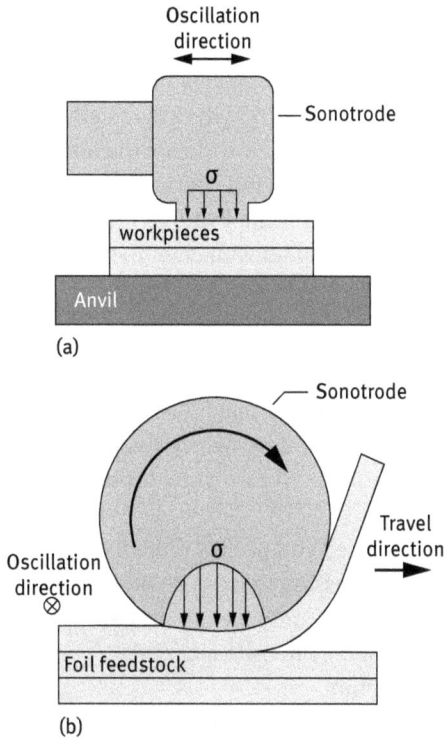

Figure 4.25: Process of ultrasonic spot welding and ultrasonic additive manufacturing.

4.4.1.2 Advantages

- Instead of using metal powders or wires, the feeding materials are ordinary commercial metal strips with a certain thickness, which are widely available at much cheaper cost.
- The ultrasonic consolidation process is a solid-state connection forming process; the temperature during the process is therefore very low, generally 25%–50% of the melting point of the metal. For this reason, this technology is also appropriate for processing Mg alloys, even though there is currently no available report in the literature. At lower temperatures, the residual internal stress inside the material is low, the structural stability is good, and no stress relief annealing is required after forming.
- The energy consumed is only about 5% of the traditional forming process; it does not produce any waste pollution such as welding slag, sewage, harmful gases, and so on, and so it is an energy-saving and environmentally friendly rapid forming and manufacturing method.

- This technology, when combines with the CNC system, can easily realize 3D complex shape parts with deep grooves, internal honeycomb structures, and so on, which cannot be fabricated by traditional metal processing methods.
- Ultrasonic consolidation not only can obtain nearly 100% physical metallurgical interface bonding rate, but also can recrystallize in the local area of the interface, locally grow nanoclusters, so that the material structure performance is improved. In addition, the oxide film on the surface of the consolidation process can be broken by ultrasonic waves without surface pretreatment of the material.

4.4.1.3 Application examples

- This technology enables rapid fabrication of layered materials. Compared to other processing methods, UAM technology can achieve nearly 100% interface bonding with good interface bonding strength, and it also provides high deposition rate with high energy efficiency.
- UAM enables the manufacture of fiber-reinforced composites that embed fibers for enhanced composite properties during processing to improve performance.
- UAM can embed optical fibers, multifunctional components, and so on in a metal matrix to manufacture metal-based smart composite materials.
- Manufacturing metal honeycomb splint structure.
- Manufacturing of metal laminated parts: As UAM technology can produce complex and precise laminated structures of internal cavities, the application prospects in the field of metal parts manufacturing have become increasingly prominent in recent years.

4.4.2 Friction stir additive manufacturing process

4.4.2.1 Introduction

Friction stirring additives were originally manufactured at Boeing and Airbus; the two companies claim that this technology can solve two major problems in AM: (i) achieving high throughput leading to faster production rates and (ii) less material wastes.

Friction stir AM is an improvement of friction stir welding (FSW), and so its basic principle is similar to FSW. Insert a custom-designed nonconsumable rotary tool (with pin and shoulder) into the overlapping surface of the plate or plate to be joined, then move along the connecting line as shown in Figure 4.26 [72]. The heat required for the welding material is provided by the frictional contact between the shoulder and the workpiece, and the strong plastic deformation is achieved by the movement of the pin.

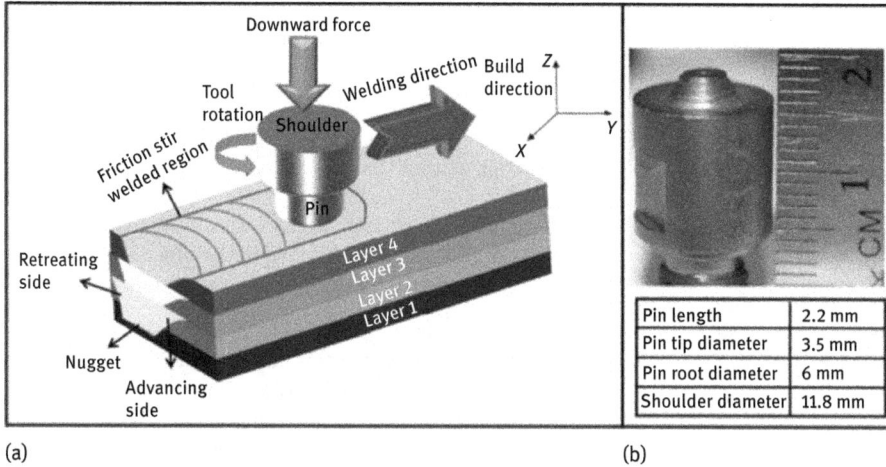

Pin length	2.2 mm
Pin tip diameter	3.5 mm
Pin root diameter	6 mm
Shoulder diameter	11.8 mm

(a)   (b)

Figure 4.26: (a) Schematic illustration of the friction stir additive manufacturing (b) the pin used in the process.

4.4.2.2 Advantages

Compared to UAM and laser AM, the advantages of FSAM are listed in Table 4.11 [73–79].

4.4.2.3 Disadvantages

i.   Tool wear occurs over time.
ii.   Significant amount of residual stresses can accumulate within the build.
iii. Scale of production and the build speed is limited by the machine dimensions and traverse speed of the tool.
iv.   Clamping the material is an issue.

4.4.2.4 Magnesium alloy application examples

In the literature, researchers from the University of North Texas applied FSWA on the WE43 magnesium alloy. They made high-performance magnesium alloy parts with higher strength and good ductility. The interested reader is referred to Palanivel et al. [72] for more details.

Table 4.11: The comparison of UAM and friction stir additive manufacturing.

Ultrasonic additive manufacturing (UAM)	– Build volume is limited by material thickness (stiffness issue) – Only materials that can be rolled into foils can be built leading to limited material selection – Preparation of foils is costly and time consuming – Affected by linear weld density – Difficult to bond materials with high work hardening rate – Bonding is achieved by plastic deformation around the asperities – Exhibit marginal fraction of the bulk material properties
Friction stir additive manufacturing	– Ability to fabricate large components – Faying surface contamination is not as vital as in UAM – Ability to bond a variety of materials that are rendered difficult by fusion welding and UAM (wide range of material selection) – Weld formation is dictated by process forces and tool geometry – High degree of reproducibility – Excellent metallurgical property in joint area – Ability to produce variable microstructures for specific applications (process flexibility) – Judicious use of materials leading to low buy to fly ratio
Fusion welding process	– Shielding gas and surface cleaning is required – Shrinkage problems coupled with high residual stresses – Poor metallurgical and mechanical properties due to directional and undesirable microstructures (low structural performance) – Environmental concerns and energy inefficient – Loss of alloying elements

4.4.3 Cold spray additive manufacturing

4.4.3.1 Introduction

Cold spray is a new type of solid-state coating preparation method. The powder is carried by high-pressure gas and accelerated by a shrink-expansion nozzle to form a supersonic fluid. The sprayed particles collide with the matrix in the solid state to produce plastic deformation and form a deposition coating. The accelerating gas is typically preheated to a temperature below the melting point of the spray material to increase particle viscosity and deformability.

4.4.3.2 Advantages

There is no traditional melting-solidification process in the cold spraying process, thus avoiding defects such as oxidation, decomposition, phase transformation, and grain growth of metal [80]. It has obvious advantages in the preparation of new coating materials, such as nano-, amorphous, and other temperature-sensitive material coatings. There is no residual stress inside the coating, which makes it appropriate for preparing thick coatings. It can prepare most metals and their alloys, including composite coatings such as metal–metals, metal–ceramics, metal–intermetallic compounds, amorphous, nanostructured coatings, and even cement.

4.4.3.3 Magnesium alloy application examples

One of the potential applications of cold spray technology in processing magnesium alloys is to repair failure parts. It is well known that magnesium alloy parts are prone to corrosion and there is no suitable technology to repair them. At present, magnesium alloy parts must be replaced after corrosion, resulting in a large amount of unnecessary waste. But cold spray technology is expected to be able to solve this problem.

In the United States, Villafuerte [81] and others used cold spray technology to repair the corrosion area of aluminum–magnesium alloy parts of aircraft engine parts. The repair process is shown in Figure 4.27 [82]. First, the corrosion surface is initially sprayed, and then machined to achieve the required surface quality and precision of the part. It is believed that the same process can be used on magnesium alloys.

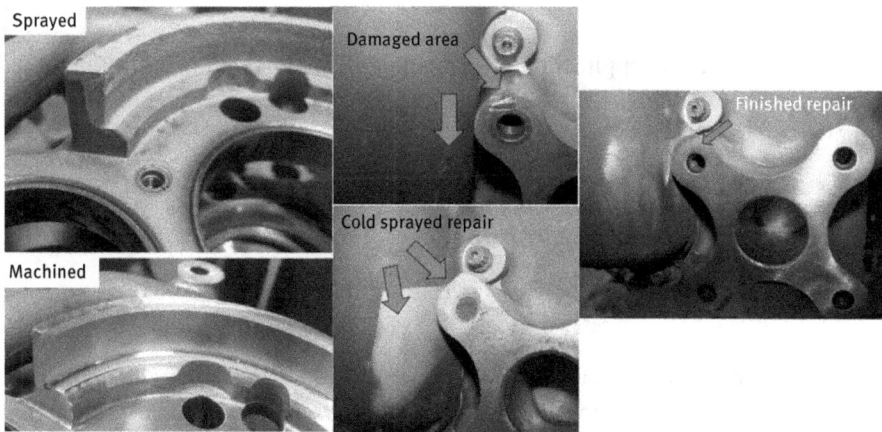

Figure 4.27: The entire process of repairing with cold spray technology.

Wei et al. [83] used shot peening-assisted cold spray technology (as shown in Figure 4.28) to prepare a corrosion-resistant nickel coating with a thickness of 150 μm for AZ31B magnesium alloy. The prepared test piece was subjected to the immersion test for 1,000 h, as well as a 1,000 h NaCl salt spray test. It has been proved that this layer of fully dense nickel coating provides excellent corrosion resistance to magnesium alloy specimens. They named the technology "in situ shot-peening-assisted cold spray process."

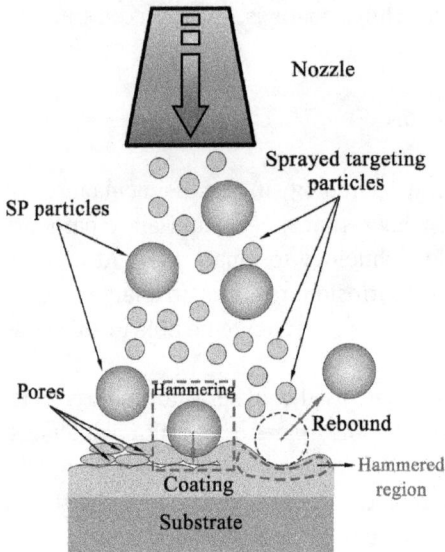

Figure 4.28: Schematic illustration of in situ shoot-peening-assisted cold spray [83].

It can be seen that the application of cold spray technology on magnesium alloys is now focused on using this technology to protect and repair magnesium alloys that are susceptible to corrosion.

4.4.4 Less common AM methods

4.4.4.1 Powder extruding deposition

Farag [84] from South Korea used PED (powder extruding deposition) technology to make porous magnesium phosphate (MgP) scaffolds with high drug absorption and release efficiency. The whole process (see Figure 4.29 [84]) can be divided into two steps: (i) first, manufacture 3D multicontrol bracket green body using PED to control the shape of the entire bracket and the position of the hole. When this step is completed, the bracket is not hardened yet; (ii) second, the stent green body is put into the binder solution for hardening. The entire process is carried out at room temperature and can therefore be used to process bioactive molecules and drugs that are susceptible to thermal denaturation.

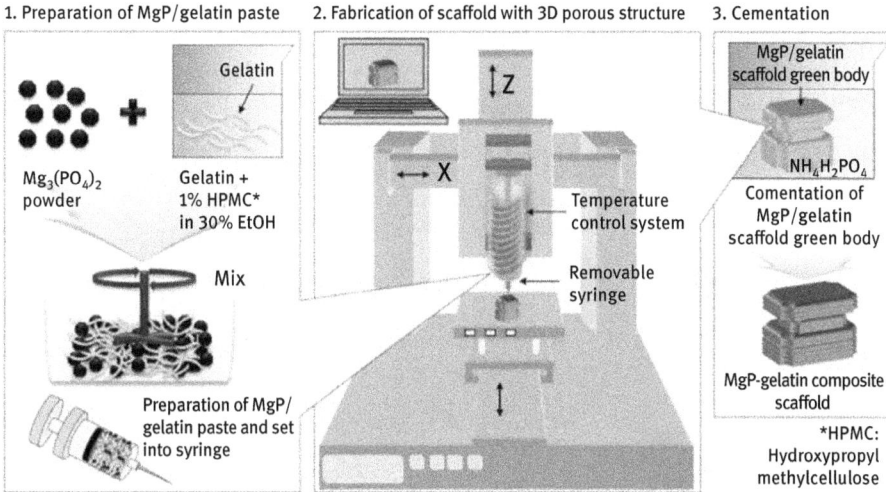

Figure 4.29: The fabrication process of PED.

4.4.4.2 Inkjet-based additive manufacturing

Researchers from Singapore [85] mentioned an improvement to the binder jet 3D printing, which they called *inkjet-based 3D printing technology*. Binder jet 3D printing is similar to the PFD technology we mentioned earlier. The purpose of developing this technology is to reduce the steps in its postprocessing without affecting the elemental composition of the final product. They applied capillary force, a force that can combine small structures, such as the fact that water-bearings can maintain a certain shape because of the existence of these forces, during AM. They used an ink based on a magnesium alloy that contained no binder or other solutes. The capillary force between the magnesium alloy particles caused the solute to form a capillary bridge between the particles as to achieve the fixation and shaping of the 3D structure. After completing the recycling step of the construction, the test piece was placed in a tube furnace for sintering for 1 h. In this way, a more economical and environmentally friendly AM process is realized, which eliminates the need for debinding processing using the binder jet 3D printing, and its deposition speed is much faster than that of other methods.

With the fast development of AM technology, the combinations between existing AM methods or with traditional processing methods to overcome the intrinsic disadvantages of individual AM method are expected; meanwhile, other new AM methods are also being developed.

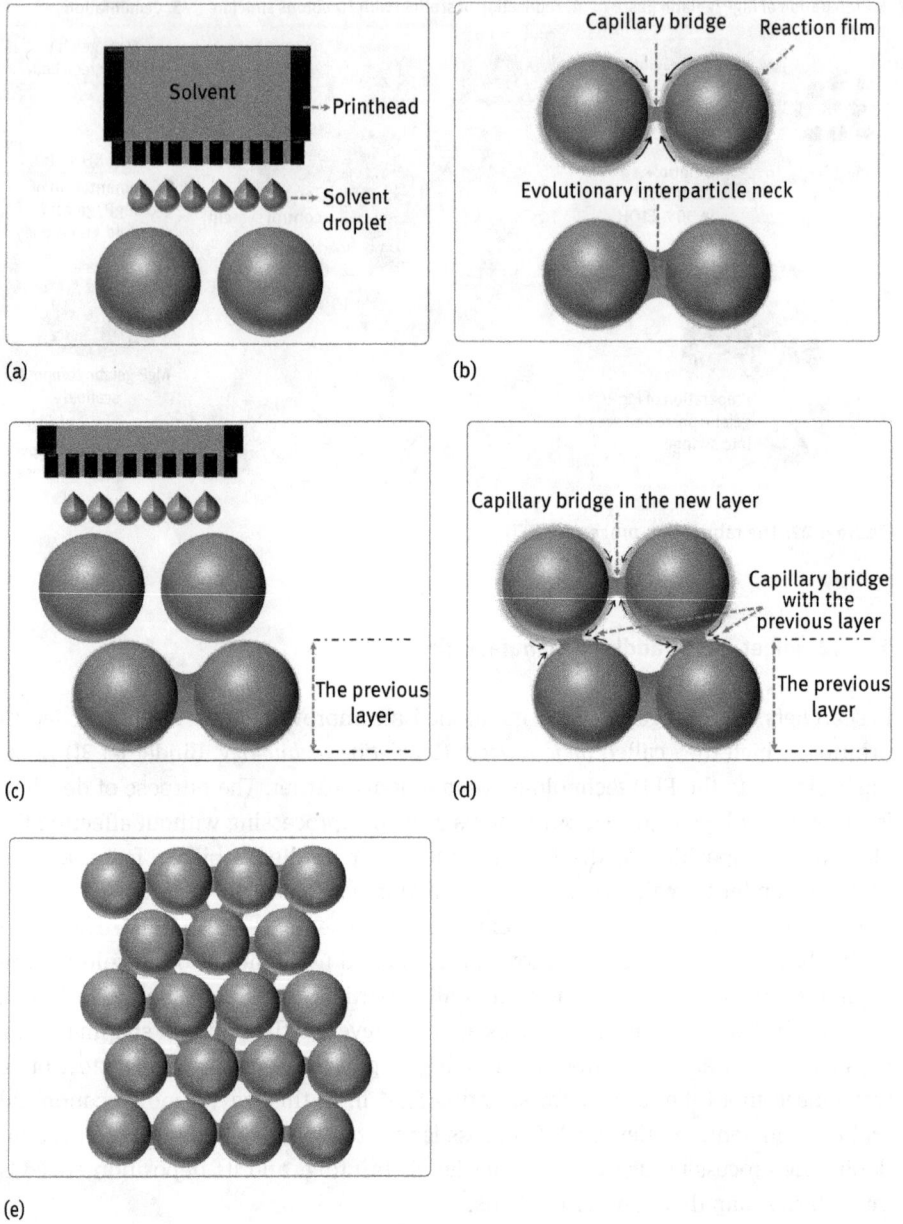

Figure 4.30: The schematic diagram of the inkjet-based additive manufacturing method [85].

Acknowledgments: This research was financially supported by the National Natural Science Foundation of China (NSFC grant number 51805415) and the Fundamental Research Funds for the Central Universities [xzd012019033].

References

[1] DebRoya, T., Weia, HL., Zubacka, JS. Additive manufacturing of metallic components – Process, structure and properties. Progress in Materials Science, 2018, 92, 112–224.

[2] Pekguleryuz, MO., Kainer, KU., Kaya, AA. Fundamentals of Magnesium Alloy Metallurgy. Woodhead Publishing, Sawston, Cambridge, 2013

[3] Kulekci, MK. Magnesium and its alloys applications in automotive industry. The international Journal of Advanced Manufacturing Technology, 2008, 39(9–10), 851–865.

[4] Li, N., Zheng, YF. Novel Magnesium Alloys Developed for Biomedical Application: A Review. Journal of Materials Science & Technology, 2013, 29(6), 489–502.

[5] Sezer, N., Evis, Z., Kayhan, SM., Tahmasebifar, A., Koc, M. Review of magnesium-based biomaterials and their application. Journal of Magnesium and Alloys, 2018, 6, 23–43.

[6] Furuya, H., Kogiso, N., Matunaga, S., Senda, K. Applications of magnesium alloys for aerospace structure system. Materials Science Forum, 2000, 305–351, 341–348.

[7] Standard terminology for additive manufacturing technologies. ASTM Int 2013; F2792–12a.

[8] Sames, WJ., List, FA., Pannala, S., Dehoff, RR., Babu, SS. The metallurgy and processing science of metal additive manufacturing. International Materials Reviews, 2016, 61(5), 315–60.

[9] Korner, C. Additive manufacturing of metallic components by selective electron beam melting – a review. International Material Reviews, 2016, 61(5), 361–77.

[10] Herzog, D., Seyda, V., Wycisk, E., Emmelmann, C. Additive manufacturing of metals. Acta Materialia, 2016, 117, 371–92.

[11] Gu, DD., Meiners, W., Wissenbach, K., Poprawe, R. Laser additive manufacturing of metallic components: materials, processes and mechanisms. International Material Reviews, 2012, 57(3), 133–64.

[12] Collins, PC., Brice, DA., Samimi, P., Ghamarian, I., Fraser, HL. Microstructural control of additively manufactured metallic materials. Annual Review of Materials Research, 2016, 46, 63–91.

[13] Lewandowski, JJ., Seifi, M. Metal additive manufacturing: a review of mechanical properties. Annual Review Materials Research, 2016, 46, 151–86.

[14] Olakanmi, EO., Cochrane, RF., Dalgarno, KW. A review on selective laser sintering/melting (SLS/SLM) of aluminium alloy powders: processing, microstructure, and properties. Progress in Materials Science, 2015, 74, 401–77.

[15] Tapia, G., Elwany, A. A review on process monitoring and control in metal-based additive manufacturing. Journal Manufacturing Science and Engineering, 2014, 136(6), 060801.

[16] Herderick, E. Additive manufacturing of metals: a review. Material Science Technology, 2011, 10, 16–20.

[17] Liu, S., Liu, W., Harooni, M. et al. Real-time monitoring of laser hot-wire cladding of Inconel 625. Optics and Lasers Technology, 2014, 62(10), 124–134.

[18] Wu, B., Pan, Z., Ding, DH. A review of the wire arc additive manufacturing of metals: properties, defects and quality improvement. Journal of Manufacturing Processes, 2018, 35, 127–139.

[19] Fang, XW., Zhang, LJ., Chen, GP. et al. Correlations between Microstructure Characteristics and Mechanical Properties in 5183 Aluminium Alloy Fabricated by Wire-Arc Additive Manufacturing with Different Arc Modes. Materials, 2018, 11(11), 2075.

[20] Sharir, Y., Pelleg, J., Grill, A. Effect of arc vibration and current pulses on microstructure and mechanical properties of tig tantalum welds. Metallurgical Technologies 2013, 5, 190–196.

[21] Kou, S. Welding Metallurgy, 2nd ed. Hoboken, NJ, USA: John Wiley & Sons, Inc., 2003.

[22] Subravel, V., Padmanaban, G., Balasubramanian, V. Effect of pulse frequency on tensile properties and microstructural characteristics of gas tungsten arc welded AZ31B magnesium alloy. Transactions of the Indian Institute of Metals , 2014, 68, 353–362.

[23] Guo, J., Zhou, Y., Liu, CM., Wu, QR., Chen, XP., Lu, JP. Wire Arc Additive Manufacturing of AZ31 Magnesium Alloy: Grain Refinement by Adjusting Pulse Frequency. Materials, 2016, 9, 823, doi:10.3390/ma9100823

[24] Shi, HC., Hu, LJ., Zheng, T. Effects of Electric Current on the Forming, Microstructure and Mechanical Properties of AZ31 Alloy Prepared by Wire Arc Additive Manufacturing. Casting Technologies, 2018, 39(10), 2285–2288. (In Chinese)

[25] Takagi, H., Sasahara, H., Abe, T. Material-property evaluation of magnesium alloys fabricated using wire and-arc-based additive manufacturing. Additive Manufacturing, 2018, 24, 498–507.

[26] ASTM; B91-12 Standard Specification for Magnesium-Alloy Forgings; West Conshohocken, PA, USA: ASTM International, 2012.

[27] David A. Martinez Holguin, Seungkyu Han, Namsoo P. Kim. Magnesium Alloy 3D Printing by Wire and Arc Additive Manufacturing (WAAM). MRS Advance, 3, 2018, 49, 2959–2964.

[28] Pan, LW., Dong, HG. Recent advances in additive manufacturing based on welding process. Welding, 2016, 29(4), 27–32 (In Chinese).

[29] Szost, BA., Terzi, S., Martina, F., et al. A comparative study of additive manufacturing techniques: Residual stress and microstructural analysis of CLAD and WAAM printed Ti-6Al-4V components. Materials & Design, 2016, 89, 559–567.

[30] Wang H, Kovacevic R. Variable polarity GTAW in rapid prototyping of aluminum parts. Proceedings of the 11th Annual Solid Freeform Fabrication Symposium, 2000.

[31] Paul A, Colegrove, Harry E. Coules, Julian Fairman, et al. Microstructure and residual stress improvement in wire and arc additively manufactured parts through high-pressure rolling. Journal of Materials Processing Technology, 2013, 213, 1782–1791.

[32] Hönnige, JR., Colegrove, PA., Ahmad, B. residual stress and texture control in Ti-6Al-4V wire + arc additively manufactured intersections by stress relief and rolling. Materials & Design, 2018, 150, 193–205

[33] Hönnige, JR., Colegrove, PA., Ganguly, S. Control of residual stress and distortion in aluminum wire + arc additive manufacture with rolling. Additive Manufacturing, 2018, 22, 775–783.

[34] Li, G., Qu, SG., Xie, MX. Effect of ultrasonic surface rolling at low temperatures on surface layer microstructure and properties of HIP Ti-6Al-4V alloy. Surface and Coatings Technology, 2017, 316, 75–84.

[35] Sun, RJ., Li, LH., Zhu, Y. Microstructure, residual stress and tensile properties control of wire-arc additive manufactured 2319 aluminum alloy with laser shock peening. Journal of Alloys and Compounds, 2018, 747, 225–265.

[36] Guo, W., Sun, RJ., Song, BW. Laser shock peening of laser additive manufactured Ti6Al4V titanium alloy. Surface & Coatings Technology, 2018, 349, 503–510.

[37] Kalentics, N., Boillat, E., Peyre, P. 3D Laser Shock Peening – A new method for the 3D control of residual stresses in Selective Laser Melting. Materials & Design, 2017, 130, 350–356.

[38] Kruth, JP., Mercelis, P., van Vaerenbergh, J., Froyen, L., Rombouts, M. Binding mechanisms in selective laser sintering and selective laser melting. Rapid Prototyping Journal. 2005, 11, 26–36.

[39] Manakari, V., Parande, G., Gupta, M. Selective laser melting of magnesium and magnesium alloy powders: A review. Metals, 2017, 71, 2.

[40] Laser Melting (LM) Technoparkstrasse 1, 8005 Zurich, Switzerland, November 1, 2017 www. additively.com/en/learn-about/laser-melting;

[41] Staiger, MP. Magnesium and its alloys as orthopedic biomaterials: a review. Biomaterials, 2006, 27(9),1728–1734.

[42] Chen, Y. Recent advances on the development of magnesium alloys for biodegradable implants. Acta biomaterialia, 2014, 10(11), 4561–4573.

[43] Yook, SW. Fabrication of porous titanium scaffolds with high compressive strength using camphene-based freeze casting. Materials Letters, 2009, 63(17), 1502–1504.

[44] Kalita, SJ. Development of controlled porosity polymer-ceramic composite scaffolds via fused deposition modeling. Materials Science and Engineering: C, 2003, 23(5), 611–620.

[45] Wycisk, E. Effects of defects in laser additive manufactured Ti-6Al-4V on fatigue properties. Physics Procedia, 2014, 56, 371–378.

[46] Zhang, BC., Liao, hl., Coddet, C. Effect of processing parameters on properties of selective laser melting Mg-9%Al powder mixture. Material, 2012, 34, 753–8.

[47] Yang, L., Mertens, R., Ferrucci, M., Yan, C., Shi, Y., Yang, S. Continuous graded Gyroid cellular structures fabricated by selective laser melting: Design, manufacturing and mechanical properties. Materials and Design, 2019, 162, 394–404.

[48] Wei, K., Wang, Z., Zeng, X. Element loss of AZ91D magnesium alloy during selective laser melting process. Jinshu Xuebao/Acta Metallurgica Sinica, 2016 52(2), 184–190.

[49] Ng, CC., Savalani, MM., Man, HC. Fabrication of magnesium using selective laser melting technique. Rapid Prototyping Journal, 2011, 17, 479–490

[50] Ng, CC., Savalani, MM., Lau, ML., Man, HC. Microstructure and mechanical properties of selective laser melted magnesium. Applied Surface Science, 2011, 257, 7447–7454.

[51] Savalani, MM., Pizarro, JM., Campbell, RI., Gibson, I. Effect of preheat and layer thickness on selective laser melting (SLM) of magnesium. Rapid Prototyping Journal, 2016, 22, 115–122.

[52] Hu, D., Wang, Y., Zhang, D. Experimental investigation on selective laser melting of bulk net-shape pure magnesium. Material Manufacturing Process, 2015, 30, 1298–1304.

[53] Zhang, B., Liao, H., Coddet, C. Effects of processing parameters on properties of selective laser melting Mg-9%Al powder mixture. Material Design, 2012, 34, 753–758.

[54] Wei, K., Gao, M., Wang, Z., Zeng, X. Effect of energy input on formability, microstructure and mechanical properties of selective laser melted AZ91D magnesium alloy. Material Science Engineering A, 2014, 611, 212–222.

[55] Wei, K., Wang, Z., Zeng, X. Influence of element vaporization on formability, composition, microstructure, and mechanical performance of the selective laser melted Mg-Zn-Zr components. Material Letter, 2015, 156, 187–190.

[56] Shuai, C., Yang, Y., Wu, P. Laser rapid solidification improves corrosion behavior of Mg-Zn-Zr alloy. Journal of Alloys and Compounds, 2017, 691, 961–969.

[57] Kruth, JP., Badrossamay, M., Yasa, E., Deckers, J., Thijs, L., van Humbeeck, J. Part and material properties in selective laser melting of metals. In Proceedings of the 16th International Symposium on Electromachining, Shanghai, China, 19–23 April 2010.

[58] Kruth, JP., Levy, G., Klocke, F., Childs, T. Consolidation phenomena in laser and powder-bed based layered manufacturing. CIRP Annals Manufacturing Technology, 2007, 56, 730–759.

[59] Yap, C., Chua, C., Dong, Z. Review of selective laser melting: Materials and applications. Applied Physics Reviews, 2015, 2, 041101.

[60] Agarwala, M., Bourell, D., Beaman, J., Marcus, H., Barlow, J. Direct selective laser sintering of metals. Rapid Prototyping Journal 1995, 1, 26–36.

[61] Das, S. Physical aspects of process control in selective laser sintering of metals. Advanced Engineering Materials, 2003, 5, 701–711.

[62] Olakanmi, E., Cochrane, R., Dalgarno, K. A review on selective laser sintering/melting (SLS/SLM) of aluminum alloy powders: Processing, microstructure, and properties. Progress In Materials Science, 2015, 74, 401–477.

[63] Attar, H., Prashanth, KG., Zhang, LC., Calin, M., Okulov, IV., Scudino, S., Yang, C., Eckert, J. Effect of powder particle shape on the properties of in situ Ti-TiB composite materials produced by selective laser melting. Journal Of Material Science & Technology, 2015, 31, 1001–1005.

[64] Liu, B., Wildman, R., Tuck, C., Ashcroft, I., Hague, R. Investigation the effect of particle size distribution on processing parameters optimization in selective laser melting process. In International Solid Freeform Fabrication Symposium: An Additive Manufacturing Conference; University of Texas: Austin, TX, USA, 2011, 227–238.

[65] Hu, D., Wang, Y., Zhang, D., Hao, L., Jiang, J., Li, Z., Chen, Y. Experimental investigation on selective laser melting of bulk net-shape pure magnesium. Materials And Manufacturing Processes, 2015, 30, 1298–1304.

[66] Li, R., Liu, J., Shi, Y., Wang, L., Jiang, W. Balling behavior of stainless steel and nickel powder during selective laser melting process. International Journal of Advanced Manufacturing Technology, 2012, 59, 1025–1035.

[67] Collur. M., Paul, A., DebRoy, T. Mechanism of alloying element vaporization during laser welding. Metallurgical Transactions B, 1987, 18, 733–740.

[68] Liu, Y., Li, S., Wang, H. Microstructure, defects and mechanical behavior of beta-type titanium porous structures manufactured by electron beam melting and selective laser melting. Acta Materialia, 2016, 113, 56–67.

[69] Xu, R., Zhao, MC., Zhao, YC. Improved biodegradation resistance by grain refinement of novel antibacterial ZK30-Cu alloys produced via selective laser melting. Materials Letters, 2019, 237, 253–257

[70] Shuai, C., Liu, L., Yang, Y. Lanthanum-containing magnesium alloy with antitumor function based on increased reactive oxygen species, Applied Sciences (Switzerland), 2018, 8(11), 2109.

[71] Ward, AA., Zhang, Y., Cordero, ZC. Junction growth in ultrasonic spot welding and ultrasonic additive manufacturing, Acta Materialia, 2018, 158, 393–406.

[72] Palanivel, S., Nelaturu, P., Glass, B., Mishra, RS. Friction stir additive manufacturing for high structural performance through microstructural control in an Mg based WE43 alloy, Materials and Design, 2015, 65, 934–952.

[73] Brandl, E., Schoberth, A., Leyens, C. Morphology, microstructure, and hardness of titanium (Ti–6Al–4V) blocks deposited by wire-feed additive layer manufacturing (ALM). Material Science Engineering A, 2012, 532, 295–307.

[74] Baufeld, B., Van der Biest O. Mechanical properties of Ti–6Al–4V specimens produced by shaped metal deposition. Science and Technology of Advanced Materials, 2009, 10, 1–10.

[75] Kelly, SM., Kampe, SM. Microstructural evolution in laser-deposited multilayer Ti–6Al–4V builds: part I. Microstructural characterization. Metall Trans, 2004, 35A, 1861–7.

[76] Kong, CY., Soar, RC., Dickens, PM. Characterization of aluminum alloy 6061 for the ultrasonic consolidation process. Material Science and Engineering A, 2003, 363, 99–106.

[77] Schick, DE., Hahnlen, RM., Dehoff, R., Collins, P., Babu, SS., Dapino, MJ. Microstructural characterization of bonding interfaces in aluminum 3003 blocks fabricated by ultrasonic additive manufacturing. Welding Journal, 2010, 89, 105–15

[78] Dehoff, RR., Babu, SS. Characterization of interfacial microstructures in 3003 aluminum alloy blocks fabricated by ultrasonic additive manufacturing. Acta Material, 2010, 58, 4305–15.

[79] American Airlines "Fuel Smart" program literature, Information can be accessed from http://hub.aa.com/en/nr/media-kit/operations/fuelsmart [accessed Sept. 2014]

[80] Papyrin A. Cold spray technology. Advanced Materials & Processes, 2001,159(9): 49–51

[81] Villafuerte, J. Current and future applications of cold spray technology [J]. Metal Finishing, 2010, 108(1), 37–39.

[82] http://www.coldsprayteam.com/W6%20Honeywell%20Greving%20061316.pdf

[83] Wei, YK., Li, YJ., Zhang, Y., Luo, XT., Li, CJ. Corrosion resistant nickel coating with strong adhesion on AZ31B magnesium alloy prepared by an in-situ shot-peening-assisted cold spray.

[84] Farag, MM. Effect of gelatin addition on fabrication of magnesium phosphate-based scaffolds prepared by additive manufacturing system, Materials Letters, 132, 111–115, 2014.

[85] Salehi, M., Maleksaeedi. S., Nai. SML., Meenashisundaram. GK., Goh, MH., Gupta, M. A paradigm shift towards compositionally zero-sum binderless 3D printing of magnesium alloys via capillary-mediated bridging. Acta Materialia, 2019, 165, 294–306.

Theodora Kontodina, Dimitrios Tzetzis,
J. Paulo Davim, Panagiotis Kyratsis

5 Additive manufacturing for patient-specific medical use

Abstract: Additive manufacturing (AM) technology is currently being promoted as the spark of a new industrial revolution. The integration of AM technologies is gaining momentum into numerous emerging markets, including the medical industry. Predominantly, AM enables the fabrication of physical parts by using initial data from medical images. Such physical models have the promising potential to affect the preoperative planning, education, and surgical simulation process, leading to various benefits regarding the surgical outcome. The focus of this chapter is to investigate the integration of AM technology in the fabrication of patient-customized medical model from scanned anatomical images. Specifically, the basic scope is to demonstrate the process of physically reproducing the T4 vertebra (thoracic spine) from a human body and measuring the dimensional accuracy of the deduced model. Following conversion and optimization of Digital Imaging and Communication in Medicine (DICOM) files to Standard Triangle Language files, the anatomical model was produced via an AM procedure. The manufactured model was scanned via 3D laser scanner in order to digitally be able to compare it with the model used (derived from the DICOM files). The flow line procedure is described and the results are discussed in terms of assessment of the dimensional declinations between the AM produced and the scanned model.

Keywords: Additive Manufacturing, 3D-printing, 3D laser scanning, Computer Aided Design, Medical Models, Digital Imaging and Communication in Medicine

5.1 Introduction

Additive manufacturing (AM) technology is implemented in a number of applications concerning various scientific fields [1]. In fact, the AM term is used to refer to a great diversity of technologies that are mainly utilized to fabricate physical models, prototypes, or functional components directly from three-dimensional (3D) computer-aided design (CAD) data [2]. The AM process manufactures these physical objects by depositing successive layers of material on the top of each other [2, 3]. Especially, the most prominent technologies encompassed by the term AM are 3D printing (3DP), rapid prototyping (RP), and rapid manufacturing (RM). In the mid-1980s, the initial use of AM is mainly referred to the fabrication of conceptual and functional prototypes, which were used as inspection tools, aiming to reduce the

https://doi.org/10.1515/9783110549775-005

production development steps of new parts and devices [4–6]. Since their invention AM technologies have been radically developed, providing a range of new processes, materials, and applications [7]. AM technologies have penetrated into various fields, such as automotive, aerospace, architectural, and fashion, and were more recently introduced in medical industry [8].

The increasing demand for quick fabrication of physical parts like customized implants or medical models in case of preorganizing an actual surgery has resulted in the growth of these unconventional methods of manufacturing with regard to the medical field [9]. Formerly, the analysis of the anatomy of a patient was based on two-dimensional (2D) data obtained by radiography and photography [10]. Nowadays, due to technological improvements, there is an arsenal of methods for acquiring 3D images of better resolution, as being the identical replica of the anatomy of a patient. In particular, raw data acquired from the computed tomography (CT) and the magnetic resonance imaging (MRI) are commonly reconstructed and stored as a Digital Imaging and Communication in Medicine (DICOM) file [11, 12]. These images are supposed to be utilized as the main source for medical software's 3D CAD, in order to produce AM medical models.

Furthermore, the production of patient-customized physical models has the promising potential to affect the preoperative planning, education, and surgical simulation process, leading to numerous advantages [13, 14]. More precisely, these benefits refer to the preplanning of a surgery, which subsequently can lead to the reduction of the operating time, the predictability in the surgical outcome, the decreased level of risk to patient, the faster patient's recovery, the improved accuracy, no geometrical restrictions, and ultimately better aesthetic and functional results [15].

The aim of this chapter is to acquire a profound understanding of the integration of AM technologies in the fabrication of patient-customized medical model from scanned anatomical images (DICOM), obtained with the help of CT or MRI scanning techniques. In particular, the main objective of this study is to present the fabrication process of a case scenario, namely the T4 vertebra of the thoracic spine of a patient, by demonstrating the three fundamental stages of this process: the design, the manufacture, and the evaluation stage. In detail, regarding the first stage, it is essential to indicate the most appropriate open-source software program for the conversion of DICOM files to Standard Triangle Language (STL) files for the specific case study. For this purpose, the required medical data was converted to the desired file format via ITK-SNAP software. Subsequently, the Meshmixer software program was utilized for the optimization of the STL file, in order to proceed to the fabrication of the medical model via AM technology. With reference to the second stage, fused deposition modeling (FDM) technology was used, by utilizing the Cura 1.0.3. software and the BCN3D Sigma R17 printer, for the creation of the anatomical model. Finally, the AM model was scanned and edited via Next Engine 3D scanner and ScanStudio HD software, respectively. The scanned model was saved in STL file format, in order to measure and evaluate the results, regarding the dimensional declinations between the AM and the scanned model. These measurements were conducted with the help of the Artec Studio 11 Professional software.

5.2 Review of additive manufacturing applications in the medical field

AM technology comprises a number of techniques, utilized for the fabrication of a physical part with the help of 3D CAD. The majority of these techniques, regardless of applying different methods to add the material, are based on the same principle of AM layer by layer [16–20]. Specifically, the most fundamental AP technologies are the following: stereolithography, selective laser sintering (SLS), direct metal laser sintering, selective laser melting, FDM, 3DP or multijet modeling, and electron beam melting, classified in accordance with the initial condition state of the material (i.e., solid, liquid, or powder).

AM technologies are gaining momentum in various scientific fields, including their penetration into the medical sector [21]. In fact, the increasing demand for customized patient care has contributed to the intense use of AM technologies, providing the possibility of quick fabrication of physical parts, like customized implants or medical models [22]. These medical models (mentioned as stereomodels [23] or biomodels [24]), obtained from individual patient data, are a 3D representation of the patient's anatomy, illustrating not only the bone structure, but additionally vascular structures, soft tissues, implants, foreign bodies, and so on. In earlier times, the fabrication of anatomical models was quite demanding and laborious process, due to the complexity of anatomical structures. Prior to AM development, medical models were produced mainly by the use of conventional processes, such as pressing, forging, machining, and casting, which proved to be time-consuming and cost a large amount of money.

Nowadays, AM plays a significant role in the production of patient-customized physical models. In the coming years, these medical models, optimized in design, have the promising potential to affect the preoperative planning, education, and surgical simulation process, providing various benefits regarding the surgical outcome. In fact, advances in biomodeling have given rise to totally additional treatment approaches and opportunities for cases regarding the preoperative planning, diagnosis of diseases, surgical simulation, and medical device prototyping, aiming to restore and reconstruct the patient anatomy to its initial state, after it has endured a physical trauma, disease, or genetic defect.

AM applications in medical sector can be classified into the following major categories:
- *Biomedical modeling*
- *Fabrication of customized implants*

- – *Fabrication of porous implants (scaffolds) and tissue engineering*
- – *Design and development of devices and instrumentation used in medical sector*
- – *Surgical planning*
- – *Medical education and training*
- – *Forensics*
- – *Drug delivery and microscale medical devices*

Biomedical modeling refers to the fabrication of physical models representing the patient's anatomy or biological structures, mainly used for preoperative planning or testing of a surgery [22]. In particular, successful implant surgery requires the fabrication of accurate medical models, manufactured by biocompatible materials. The main categories of the classified biomaterials are metals, ceramics, and polymers.

The major application of AM technologies in medical sector is the design and fabrication of customized implants and fixtures, which present the exact patient anatomy and comfortably fit the patient. These medical models can be used for prosthetic operations, including maxillofacial, dental, craniofacial, orthopedic, or spinal surgery, for rehabilitation, or even for plastic surgery, in order to increase both functionality and aesthetic appearance.

AM technology has increased the ability to fabricate medical models of complex geometry with high accuracy. Consequently, AM techniques such as FDM, SLS, and 3DP are suitable for the fabrication of controlled porous structures with special geometrical features, by utilizing biocompatible materials in the field of scaffolding and tissue engineering. Actually, scaffolds are customized permeable implants used as a vessel for supporting tissue regeneration or restoration, by providing support and guidance to defective bone or growing tissue which was damaged [1, 4, 22, 25].

Furthermore, AM technology is used not only for the presurgical phase, but also for the actual surgery. Specifically, AM technology is utilized for the design and development of medical equipment, devices, and instrumentation, including hearing aid, dental devices, and surgical tools that are used as guides during the operation process [26].

As mentioned previously, the most fundamental application of AM technology in the medical field is the fabrication of biomodels that can be utilized as an aiding tool for preplanning of a surgery and rehearsal, leading to numerous benefits. More specifically, these benefits refer to the reduction of the operating time, the predictability in the surgical outcome, the decreased level of risk to patient, the faster patient's recovery, the improved accuracy, no geometrical restrictions, and ultimately better aesthetic and functional results [27, 28].

Additionally, AM technology can be used for purposes of medical education and training. The created medical models can be utilized from medical students or young doctors in order to understand the internal or external human anatomy structure. Moreover, another application of AM technology is for investigation of criminal profile. In many cases, the models can be used for the creation of crime scenes, in order to facilitate the process of solving cases regarding the forensic field. Finally, another application of AM technologies is the fabrication of customized microsystems and therapeutic devices used to control drug delivery for certain cases. Specifically, these devices may involve networks of fluidic and electronic components that operate in an specific manner. The basic groups of these devices involve the following: biocapsules and microparticles for controlled and site-specific drug release, microneedles for transdermal and intravenous delivery, and implantable microsystems [22].

In general, AM technologies have great potential and flexibility in solving complex surgical problems. Nevertheless, they have not been adopted to a large extent in medical and healthcare sectors, due to a number of technological issues and deficiencies. Actually, the main five criteria influencing this perspective include [1, 3, 21, 22, 29]:

- *Speed.* One of the main competitive advantages of AM technology is the speed of production, though in case of batch production of medical models the AM technologies are not suitable, due to the fact that the medical data preparation can be more time consuming than the AM building process. Consequently, medical models can be fabricated for preorganized surgeries and on the contrary cannot be utilized for emergency operations.
- *Cost.* AM technologies are majorly utilized for low-cost production applications. However, the fabrication of customized medical models requires to be of improved quality, effectiveness, and efficiency, leading to the increase in the cost.
- *Accuracy.* The AM processes provide models with high accuracy. Nevertheless, the issue of accuracy is characterized as insufficient for numerous medical applications, due to poor or inaccurate medical data, obtained from 3D-imaging software tools.
- *Materials.* The availability of appropriate biocompatible materials for medical applications via AM technology is limited. Additionally, the AM machines providing the most suitable material properties for medical purposes are very expensive, leading to the general increase in the cost.
- *Ease of use.* The fabrication of medical models requires highly skilled and trained personnel in order to achieve accurate and good quality models, implying additional investment for training.

The process used for the fabrication of medical models via AM technologies is divided into two fundamental stages, namely the design stage and the manufacture stage. Specifically, the design stage comprises the following steps: acquisition of medical data (or medical imaging) and image processing via CAD. And finally, the manufacture stage is composed of the required steps for the creation and the fabrication of the model via AM technologies. Each of these steps (including both the design and the manufacturing process) can introduce geometric deviations, leading to distortions in the final AM medical model [11, 15]. Nevertheless, the main percentage of these inaccuracies in AM medical models is introduced during the process of medical imaging or during image processing, rather than during the manufacturing stage (e.g., during 3DP process), which is considered as a more accurate procedure. Consequently, further investigation regarding the accuracy of the AM in medical applications is indisputably required.

Undoubtedly, one of the most fundamental issues, regarding the AM applications in the medical sector, is the accurate acquisition of 3D anatomical models. As far as it concerns the medical imaging techniques, they are used for the visualization of internal structures of the patient's anatomy, including organs, bones, or vessels; CT, MRI, and ultrasonography are the most commonly used techniques.

The growth of medical imaging techniques in the past decade has enabled the visualization of internal structures in the human body, by generating 2D cross-sectional images of high fidelity and accuracy [7, 18]. In particular, these images are stored in DICOM data format, utilized as the main source for medical software's 3D CAD, in order to produce AM medical models [9, 12, 18]. Especially, DICOM is a worldwide information standard, under the aegis of the National Electrical Manufacturers Association, designed to ensure the storage and the transmission of medical images. Actually, the basic scope of the current DICOM standards is the accomplishment of the compatibility and the improvement of workflow efficiency between imaging systems and other information systems, regarding the healthcare sector [12].

The acquired medical data (DICOM files), obtained by medical imaging techniques (i.e., CT, MRI, or ultraphonography), can be subsequently processed, with the help of various medical image segmentation software programs, in order to create a 3D visualization of a patient anatomic part. The process of conversion of the DICOM files into compatible formats with AM machines includes a number of steps, and each of these steps requires different software programs [21, 30]. In brief, these steps comprise the acquisition of the DICOM files, the conversion of DICOM files to STL file format, the image optimization, and the creation of the model. According to research studies [21, 30], there is a variety of software solutions, both open source or commercially available, designed for the creation of 3D-printed models. The most commonly used software programs for the conversion of DICOM files to STL file format are the 3DSlicer, ITK-SNAP, Materialize (Mimics), Seg3D & ImageVis3D, and Osirix [21, 30, 31]. In particular, Materialize is considered as the market standard, providing Materialize Mimics for the direct creation of accurate 3D models, which is designed for medical applications [21], although the other solutions are free and open-source applications that can be easily utilized from nonspecialized users.

After acquiring the required medical data, obtained from CT, MRI, or ultrasonography scans in DICOM file format, a software program can be utilized in order to create a 3D model representing the patient anatomy [1, 4, 11, 18]. This model can be exported into STL file format for further processing and optimization, before proceeding to the manufacture stage. Subsequently, the model can be exported to CAD programs for further processing, by using the provided editing tools, such as smoothing, merging, and rescaling. Ultimately, the creation of the model according to the created STL file is used for the AM process and finally for the fabrication of the customized anatomical model.

5.3 Case study of T4 vertebra of thoracic spine

The main purpose of this section is the presentation of the complete process of the digital fabrication of patient-specific 3D-printed medical model. Hence, the identification of a specific case scenario is crucial for the purpose of understanding the fabrication procedure. Particularly, in the current project, the T4 vertebra of the thoracic spine of a patient is selected to be examined.

5.3.1 Working methodology

This section comprises the research methodology followed by providing the steps and the analysis of the model used to approach the work. This attempt to define the basic steps needed for the development of this work will contribute to establish a basis of the structure, leading to a profound understanding of the results obtained. Specifically, the main objective is to demonstrate the process from the acquisition of the medical data to the final AM anatomical model, in order to evaluate the final outcome, concerning the dimensional declinations between the printed and the scanned model.

More precisely, the systematic approach of this process is classified into three main sections: the *Design*, the *Manufacture*, and the *Evaluation* stage. The flowchart of the methodology followed in the current study is presented in Figure 5.1, illustrating the most fundamental steps. At this point, it is worth mentioning that the methodology illustrated in the current section comprises the basic steps of the three stages of the fabrication process. Each of these stages is analyzed and described in more detail in the corresponding section of the examined case study.

In more detail, with regard to the first section, the design stage is described in a step-by-step approach, comprising four main steps required for the acquisition of the STL file. In particular, these steps of the design study are as follows: "Acquisition of DICOM files" (via CT), "Conversion of DICOM files to STL file format" (via ITK-SNAP software), "Optimization of the STL file" (via Meshmixer software), and finally "Acquisition of the final STL file ready for the AM machine."

In the second section, the manufacture stage of the final AM medical model is presented. This stage is composed of the two following steps: the "Creation of the AM model" (via Cura 1.0.3. software) and the "Manufacture of the medical model via AM technology" (i.e., FDM technology – BCN3D Sigma R17 printer).

The third section focuses on the evaluation stage, which is composed of three principal steps. Actually, these steps consist of the above procedures: "Scanning of

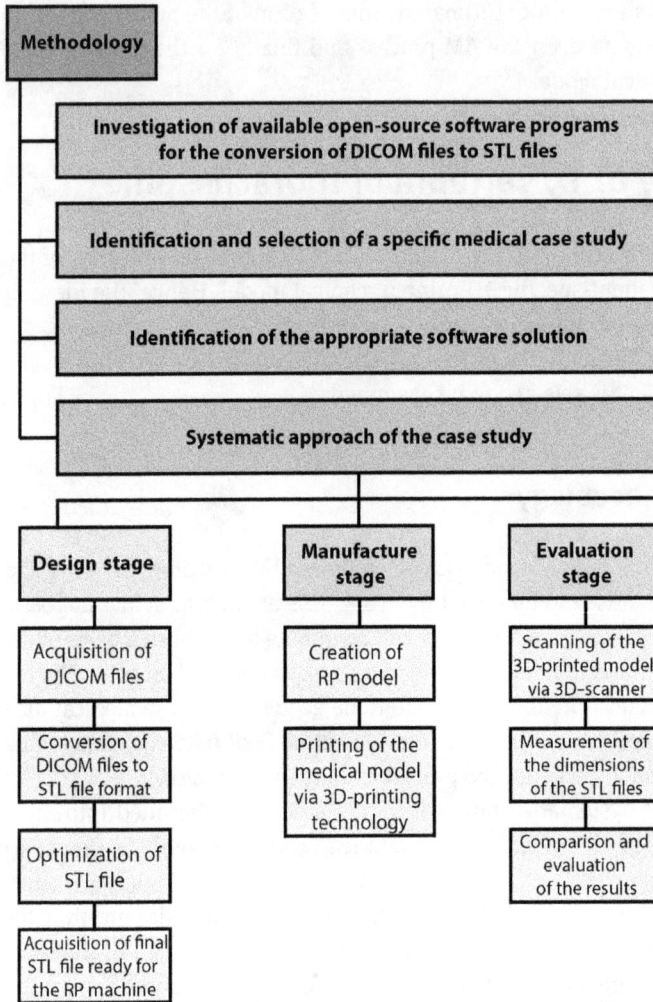

Figure 5.1: Flowchart of the methodology of the study.

the 3D-printed model via 3D-scanner" (via Next Engine 3D scanner and ScanStudio HD software), "Measurement of the dimensions of the STL files" (via Artec Studio 11 Professional software), and ultimately "Comparison and Evaluation of the Results."

5.3.2 Systematic approach

As mentioned previously, the systematic approach of the process is divided into three fundamental sections: the "Design," the "Manufacture," and the "Evaluation" stage.

5.3.3 Design stage

Initially, the most fundamental step in the fabrication process of a medical part is considered to be the *Design Stage*. This stage is composed of four main steps, which is required for the acquisition of the STL file of the model. Specifically, these steps are as follows: *Acquisition of DICOM files, Conversion of DICOM files to STL file format, Optimization of the STL file*, and finally *Acquisition of the final STL file ready for the AM machine* as demonstrated in the flowchart (Figure 5.2). The process of this stage is described in a step-by-step approach in the following sections.

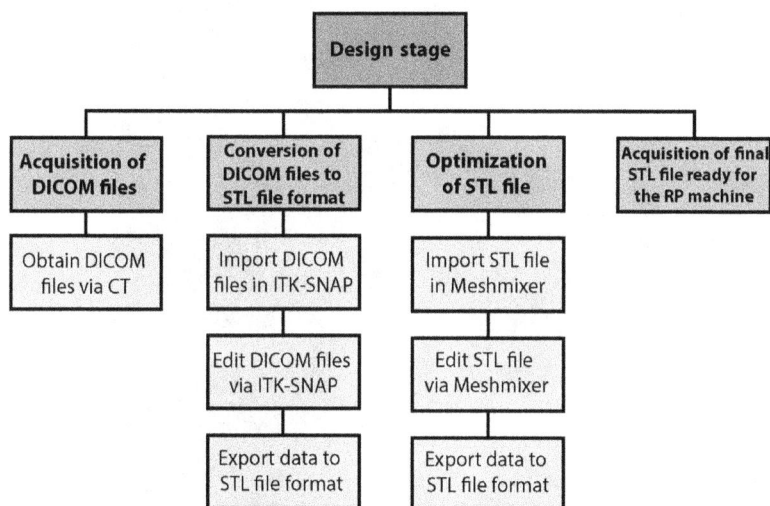

Figure 5.2: Flowchart of the design stage of the case study.

5.3.3.1 Step 1: acquisition of DICOM files

In order to launch a new project it is essential to obtain the appropriate medical files, which provide the information about the anatomical part of the patient. In the current days, the technological improvements resulted in the 3D image acquisition of the identical replica of the anatomy of a patient, with the help of CT. More specifically, the data is stored in DICOM images, utilized as the basic source for medical software's 3D-CAD, in order to produce AM medical models, as described previously.

5.3.3.1.1 Obtain DICOM files via CT
With respect to the current case study, a series of DICOM files of a patient's thorax is used, acquired from CT scan. More specifically, the DICOM files are downloaded from ITK-SNAP's official site, from the available data archive.

Furthermore, the output of the CT slices, saved in DICOM standardized file format, can be displayed as a 2D or 3D preview in various DICOM viewer software programs. In Figure 5.3, the Radiant DICOM Viewer is selected for the display of the CT scan slices of the patient's thorax, illustrated in axial, coronal, and sagittal view. Additionally, the Radiant DICOM Viewer provides the option of 3D volume rendering for six distinct views (i.e., anterior (A), posterior (P), left (L), right (R), superior (S), and inferior (I) 3D views), as presented in Figure 5.4.

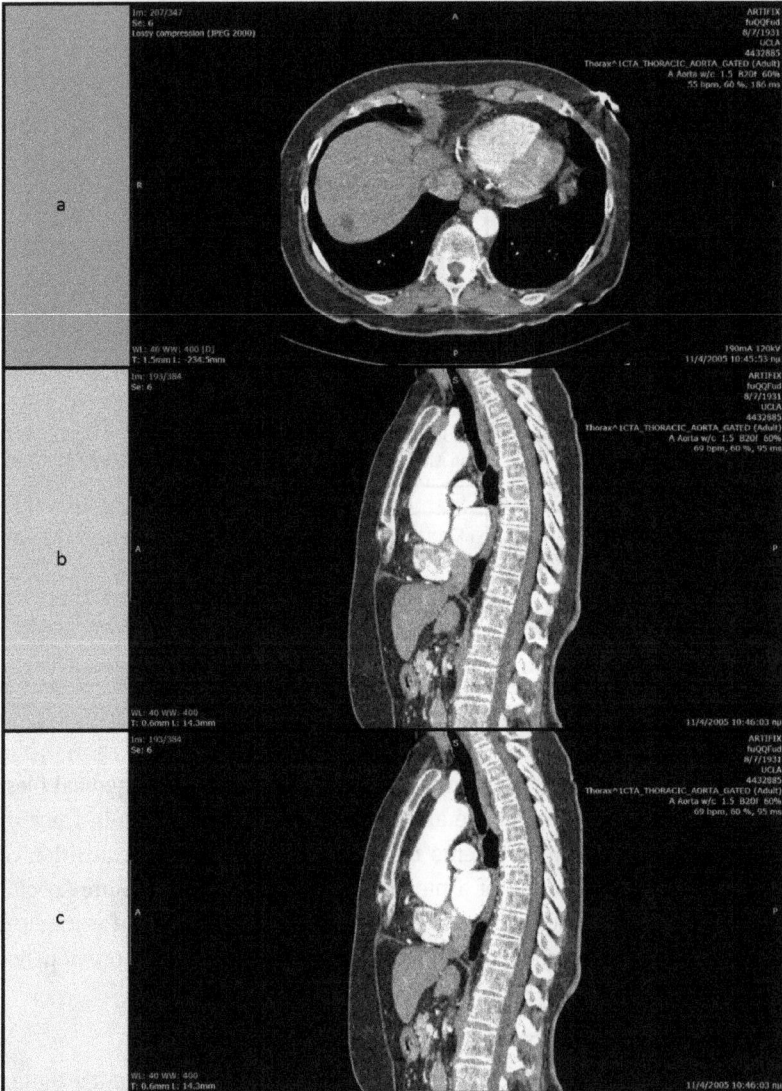

Figure 5.3: Radiant DICOM Viewer, 2D CT scan of the patient's thorax: (a) Axial view, (b) coronal view, and (c) sagittal view.

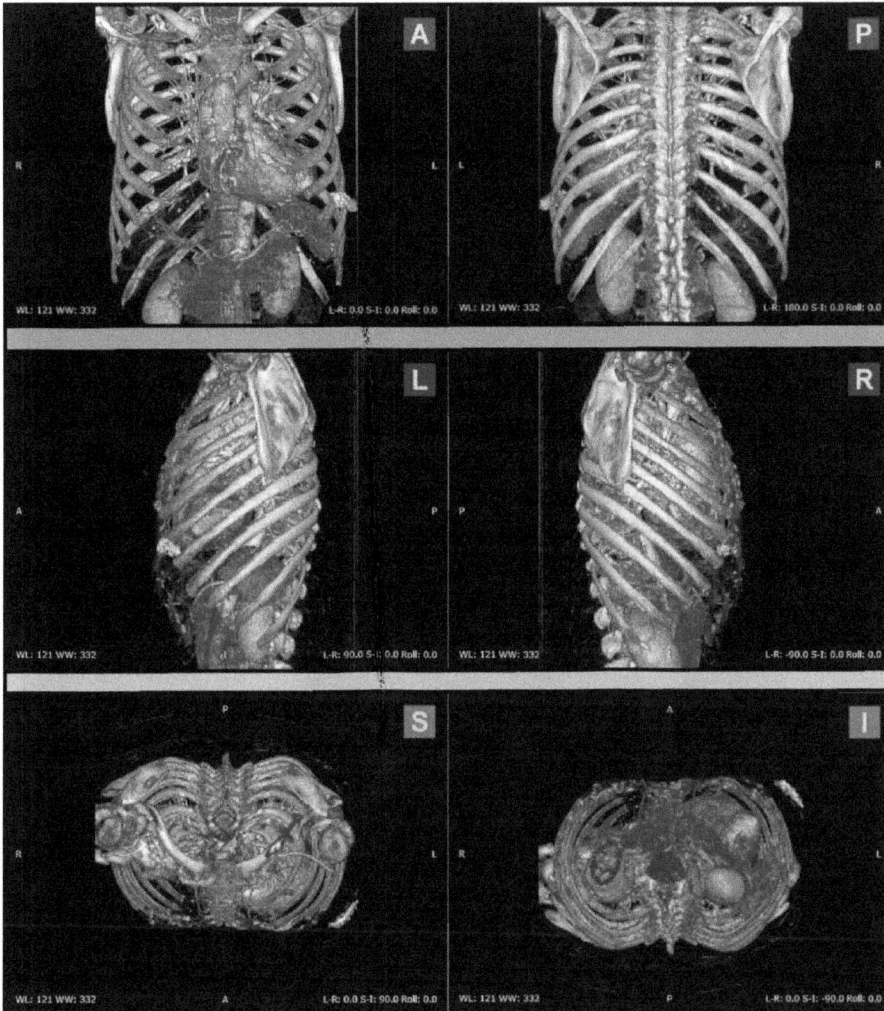

Figure 5.4: Radiant DICOM Viewer, 3D volume rendering of the patient's thorax.

5.3.3.2 Step 2: conversion of DICOM files to STL file format

After acquiring the desired DICOM files from the CT scan, the next step refers to the conversion of the DICOM files to the STL file format. The following steps summarize the conversion process.

5.3.3.2.1 Import DICOM files in ITK-SNAP

The first step of the conversion process refers to importing the DICOM files in the selected software. For the current case, the ITK-SNAP software program is chosen for the manipulation of the DICOM files.

5.3.3.2.2 Edit DICOM files via ITK-SNAP

After importing the appropriate DICOM files, it is suggested that the user proceeds to the segmentation process, in order to distinguish the *region of interest* (ROI) in the data set from the surroundings, by facilitating the creation of the desired geometric model [18]. In the current project, the ROI is selected to be the thoracic spine of the patient, as illustrated more clearly in Figure 5.5.

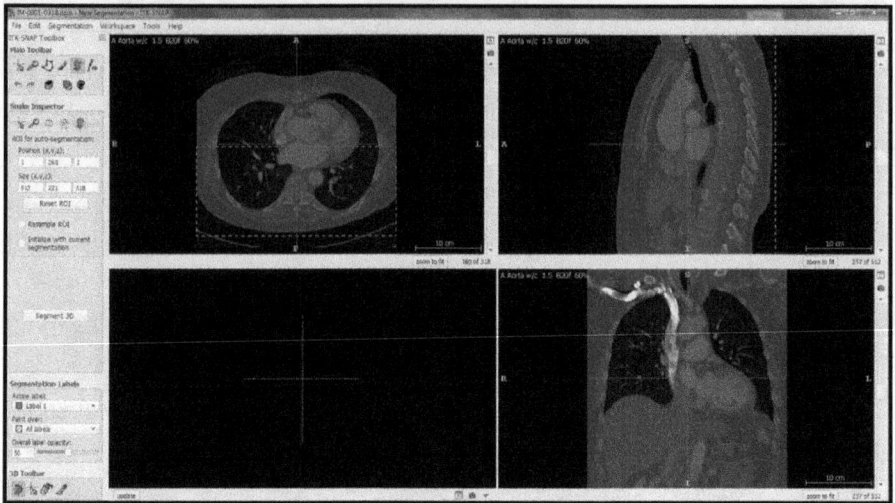

Figure 5.5: ITK-SNAP, Main Toolbar, Active Contour Segmentation Mode, Selecting the Region of Interest (ROI).

Actually, the *Automatic Segmentation* is divided into three different steps (*Presegmentation, Initialization*, and *Evolution*), which is illustrated in the right side of the screen. Initially, the first step is related to preprocessing the image by adjusting the speed function. For the current case, the speed function is selected to be generated by *Thresholding* mode (Figure 5.6). After adjusting the required settings in the *Presegmentation* step, the user is able to proceed to the second step of *Initialization* procedure. In Figure 5.6, the process is selected to be demonstrated on the second option of view, depicted with blue color, where the bones appear with white color.

Second, the user proceeds to the *Initialization* step (Figure 5.7). In this step, the software prompts the user to place bubbles in the image, in order to initialize the contour. Hence, the user should adjust the *Bubble radius*, in order to be adapted to the desired ROI. In fact, the selected bubbles for the patient's thoracic spine are illustrated in Figure 5.8, for the three orthogonal views (i.e., axial, sagittal, and coronal).

Finally, the user can start the contour evolution (Figure 5.8). When the result of the created region is satisfactory, the user is able to proceed to the ultimate step for the creation of the 3D model.

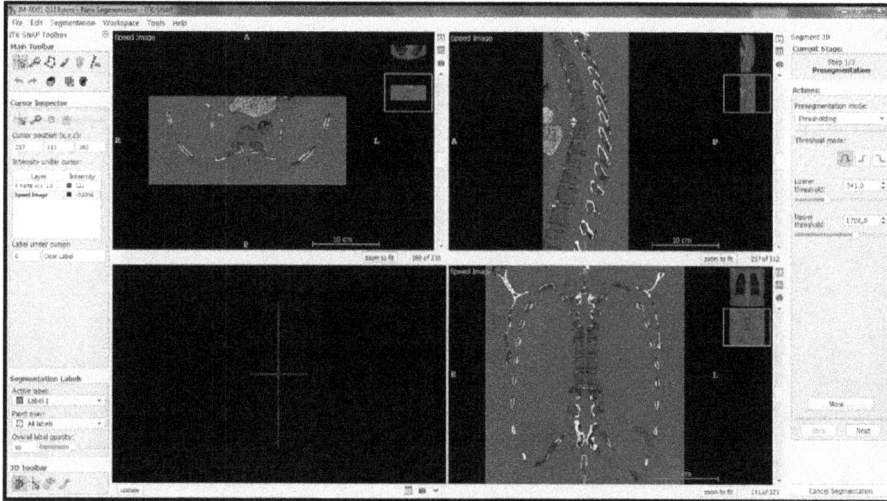

Figure 5.6: ITK-SNAP, Main Toolbar, Active Contour Segmentation Mode, Segment 3D, Step 1: Pre-segmentation.

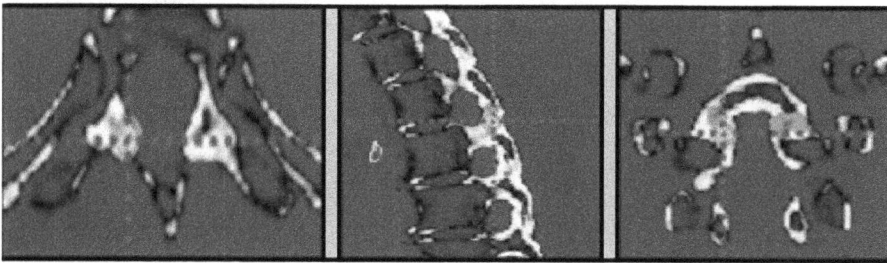

Figure 5.7: ITK-SNAP, Segment 3D, Step 2: Initialization, Adding Bubbles at Cursor.

Automatically, the software reveals a 3D image of the segmented model of the ROI. The result of the current project is presented in Figure 5.9. Additionally, the user can navigate through this 3D image by zooming or panning around in order to examine the final result. In Figure 5.10, the created 3D model of the three vertebras (T3, T4, T5) of the thoracic spine is illustrated when zoomed in.

5.3.3.2.3 Export data to STL file format

After completing the editing procedure, the user should export the created files into external formats, by selecting the name, the destination, and the type of format of the file.

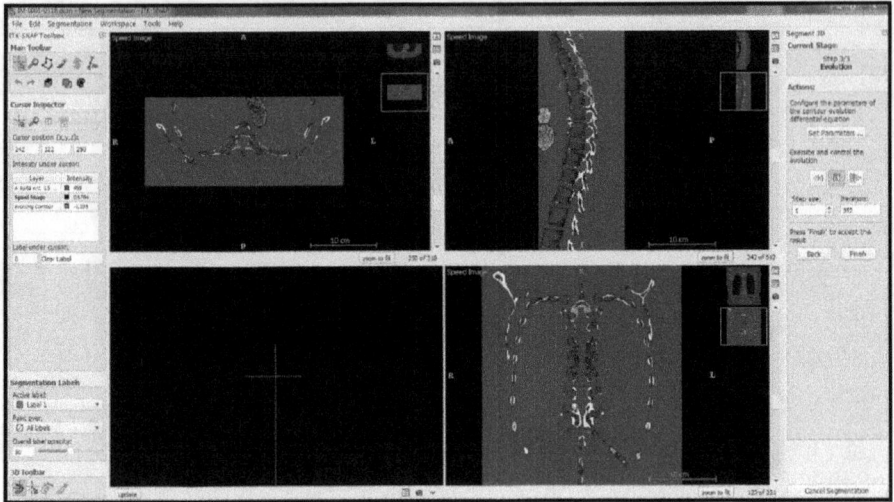

Figure 5.8: ITK-SNAP, Main Toolbar, Active Contour Segmentation Mode, Segment 3D, Step 3: Evolution, Process of evolution.

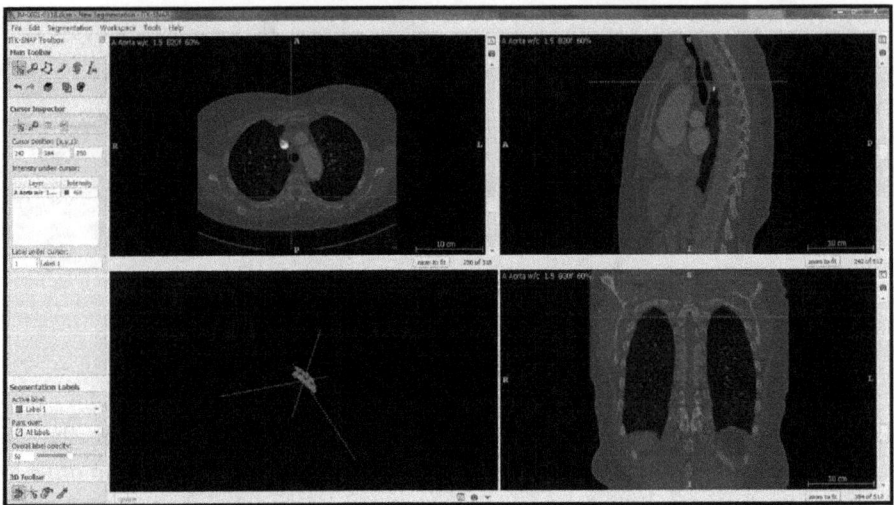

Figure 5.9: ITK-SNAP, Main Toolbar, Active Contour Segmentation Mode, Segment 3D, Step 3: Evolution, Final result, Updated 3d view.

5.3.3.3 Step 3: optimization of STL file

After exporting the 3D model from the ITK-SNAP software in STL format, it is necessary to optimize this file, before proceeding to the manufacture stage. The optimization

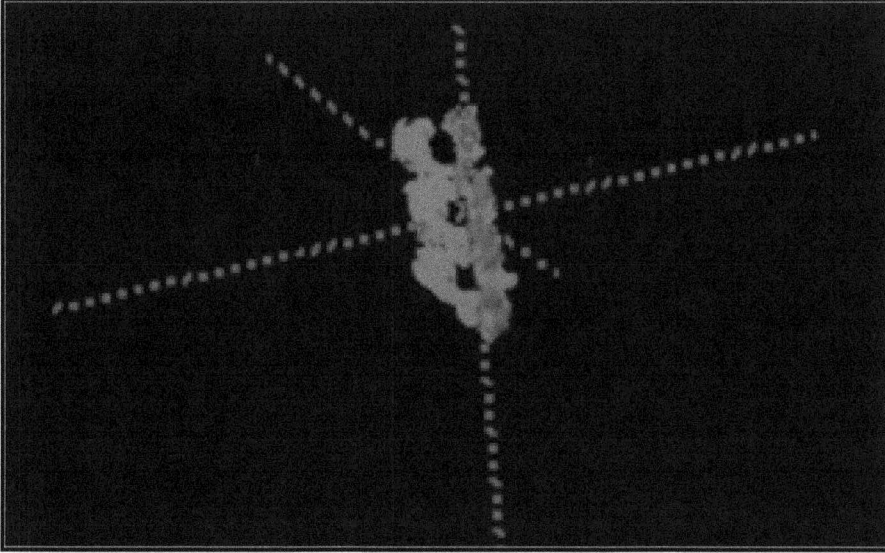

Figure 5.10: ITK-SNAP, Final result, Updated 3d view of the three vertebras (T3, T4, T5) of the thoracic spine.

process is presented in the following three sections with the assistance of the Meshmixer software.

5.3.3.3.1 Import STL file in Meshmixer

In order to import the desired STL file, the user should select the proper STL file from the directory. The imported STL file of the three vertebras (T3, T4, T5) of the patient's thoracic spine is presented in Figure 5.11.

5.3.3.3.2 Edit STL file via Meshmixer

For the current case, the T4 vertebra of the thoracic spine is selected to be examined. Consequently, by utilizing the appropriate tools of Meshmixer software, the T4 vertebra is isolated from the array of the thoracic vertebras. The selected ROI is depicted in Figure 5.12.

The result of the T4 vertebra of the patient's thoracic spine, illustrated in Figure 5.12, contains many imperfections. In order to optimize the created model, the user has at his/her disposition a variety of modifying tools. Hence, after conducting the smoothing operations properly, the final result of the 3D model of the T4 vertebra is depicted in Figure 5.13. The first figure, on the left, presents the initial model before performing the smoothing operations. The second figure, in the middle, demonstrates the model after adjusting the smoothing boundaries. The third figure, on the right, presents the final model, modified in an appropriate way, by

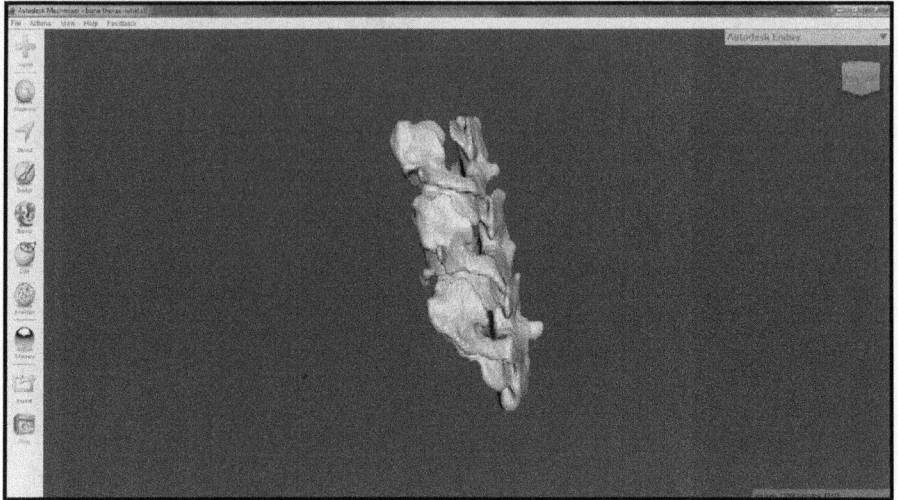

Figure 5.11: Meshmixer, Imported STL file, vertebras (T3, T4, T5) of the thoracic spine.

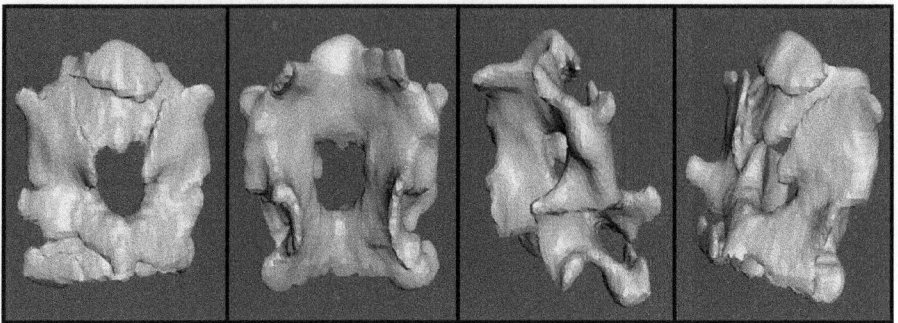

Figure 5.12: Meshmixer, T4 vertebra of the thoracic spine.

correcting and erasing the edgy points in the superior part of the volume, in order to obtain a high-quality result.

5.3.3.3.3 Export data to STL file format

After completing the editing procedure, the user should export the created file into an STL file format.

5.3.3.4 Step 4: acquisition of final STL file ready for AM machine

At last, the design stage is completed and the final model exported into STL file format is ready for the manufacture stage, utilizing the AM technology. The final result

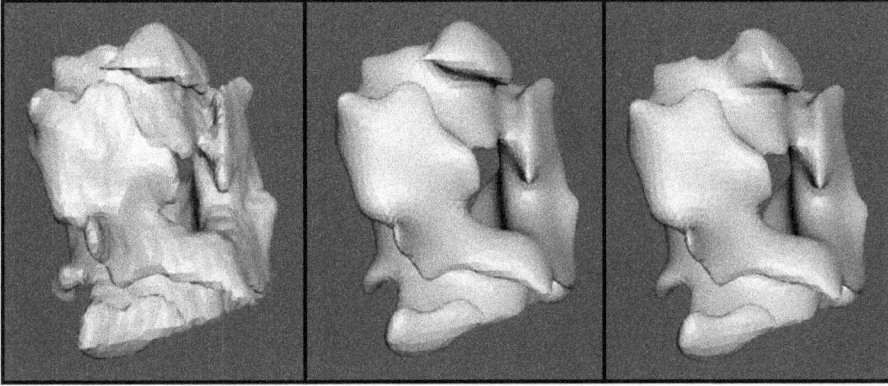

Figure 5.13: Meshmixer, Evolution of smoothing operation, T4 vertebra of the thoracic spine.

of the T4 vertebra of the patient's thoracic spine is illustrated in Figure 5.14, depicted in four different views. The first figure (from the left) presents anterior plane of model, while the second figure presents the posterior side of the model. The third and the fourth figures demonstrate the left and the right planes of the model, correspondingly.

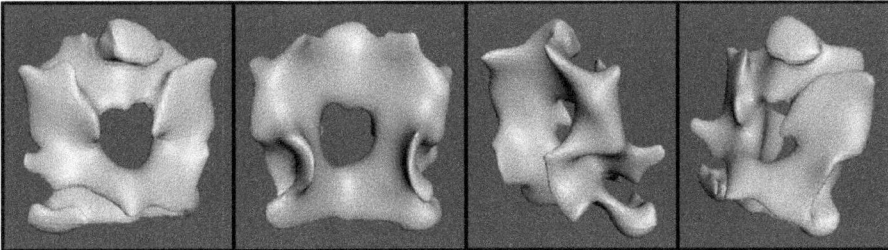

Figure 5.14: Final STL file of the created 3D model of the T4 vertebra of the patient's thoracic spine (anterior, posterior, left, and right sides of the model).

5.3.4 Manufacturing stage

Subsequently, the second stage of the fabrication process of the medical part is the *Manufacturing Stage*. In fact, this stage is composed of two main steps, which are required for the production of the medical model, the *Creation of the RP model*, and the *Printing of medical model via Fused Deposition Modeling (FDM)*, as demonstrated in Figure 5.15. More precisely, for the fabrication of the T4 vertebra, the Cura 1.0.3 software and the BCN3D Sigma R17 printer were utilized for the creation of the AM model. As a matter of fact, the process of the manufacturing stage is described in a step-by-step approach in the following sections.

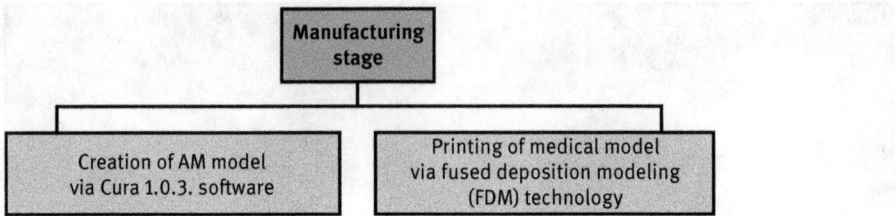

Figure 5.15: Flowchart of the manufacture stage of the case study.

5.3.4.1 Step 1: creation of AM model

Regarding the first step of the manufacturing stage, the Cura 1.0.3 software is utilized for the creation of the AM model. The process of the AM model creation of the T4 vertebra is illustrated in Figure 5.16.

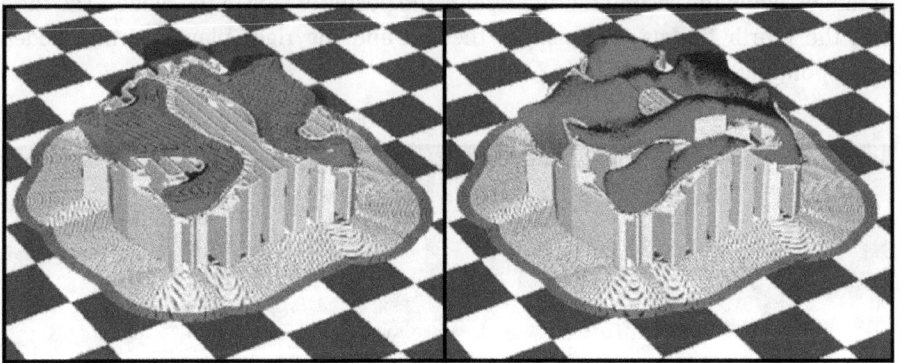

Figure 5.16: Cura 1.0.3 for creation of the AM model of the T4 vertebra.

Moreover, in order to create the AM model of the T4 vertebra, the basic properties of the AM process for BCN3D Sigma R17 printer should be adjusted concisely. In more detail, for the current case the material is selected to be PLA, the layer thickness is adjusted to 0.05 mm (50 μm), the temperature was modulated to 210 °C, the heat bed was set to 50 °C, the wall was regulated to 1.6 mm, the infill 100%, and ultimately the speed was regulated to 50 mm/s.

5.3.4.2 Step 2: printing of medical model via 3DP technology

After setting the appropriate parameters by utilizing the Cura 1.0.3 software, the model is ready to be printed via AM technology. Consequently, the AM process of the T4 vertebra is illustrated in Figure 5.17. The medical model is printed layer by layer, based on the FDM method.

Figure 5.17: The AM process of the T4 vertebra via BCN3D Sigma R17 printer.

Conclusively, the final result of the medical model, after removing the supporting material manually, is presented in Figure 5.18. The figure on the left depicts the anterior side of the obtained 3D-anatomical model, while the figure on the right demonstrates the posterior side of the model.

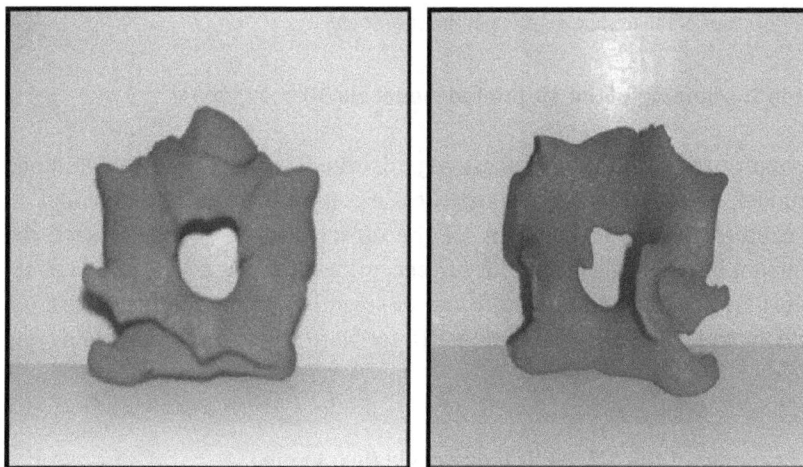

Figure 5.18: The final AM model of the T4 vertebra, anterior and posterior sides.

5.3.5 Evaluation stage

At last, the third stage of the fabrication process of the medical model is the *Evaluation Stage*. Specifically, this stage is composed of three basic steps, which are required for evaluation of the medical model, the *Scanning of the AM model via 3D-scanner* (via Next Engine 3D-scanner and ScanStudio HD software), *Measurement of the dimensions of the STL files* (via Artec Studio 11 Professional software), and ultimately *Comparison and Evaluation of the results*, as illustrated in Figure 5.19. Actually, the process of evaluation stage is presented in a step-by-step approach in the following sections.

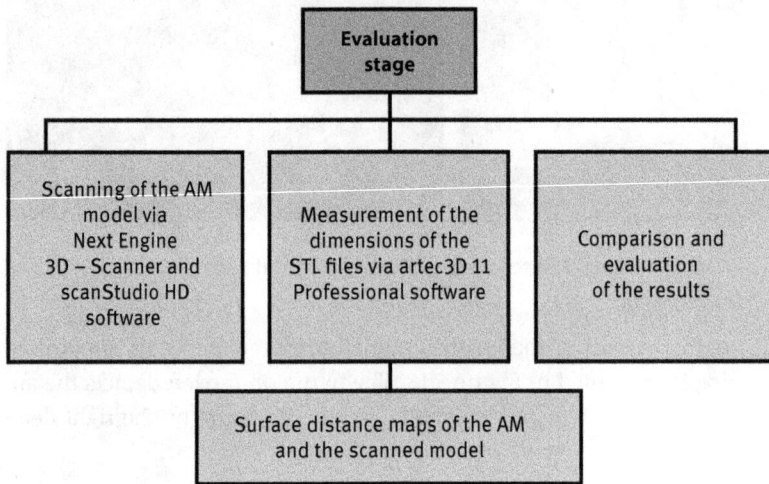

Figure 5.19: Flowchart of the evaluation stage of the case study.

5.3.5.1 Step 1: scanning of the 3D-printed model via 3D scanner

The first step of the evaluation process regards the 3D scanning of the obtained medical model. As mentioned previously, it is essential to scan the AM model, in order to capture the entire geometry of the object and subsequently save the scanned model into STL file format for further processing. Therefore, in this chapter, the Next Engine 3D scanner is utilized for scanning process of the 3D-printed anatomical model, and the ScanStudio HD software is used for processing the scanned data.

5.3.5.1.1 Scanning of the AM medical model via Next Engine
In fact, a concise presentation of the steps needed to be undertaken for scanning the AM medical model is introduced as follows. With respect to the scanning

machine, the Next Engine 3D scanner is a Non-Contact Scanner that uses laser for capturing the geometry of an object and convert it to a digital format. The Next Engine is a desktop 3D scanner that measures 50,000 points per second with multi-laser precision, at 0.005 inches accuracy, providing users' unprecedented ease of use to quickly create highly detailed, full color, and digital models.

Briefly, in order to achieve the most satisfactory results, concerning the geometrical features of the object, 21 distinct steps are conducted. Initially, the selected object is scanned, by setting the appropriate parameters in the ScanStudio HD software. Second, it is crucial to understand the peculiarities and the requirements of the model, and translate them into specifications, in order to fully capture the geometry of the object. These settings regard the positioning, the number of divisions, the points/in^2, the target, and the range.

In particular, the object was positioned vertically and horizontally to the Auto-positioner, using additionally the provided tools, in order to achieve a stable result. Additionally, it is essential to align the object on the scanning platform, by using the viewing window, so that the object is visible within the scanning field. Other editing tools (trim, align, fuse, and buff) are necessary, for optimizing the captured geometry of the object. Particularly, it is mandatory to remove the noise and the overlapping surfaces from two different captures, in order to align them in a common coordinate system and get the entire geometry of the model. The alignment of the different captures (performed four times) is achieved by specifying some points in the shaded view of the object. Moreover, the fusion process is important for covering the missing geometry and gaps. In addition, the refinement of the final model is achieved by using the buffing tool. The final model obtained from the scanning process is illustrated in Figures 5.20 and 5.21 in shaded and mesh views, respectively.

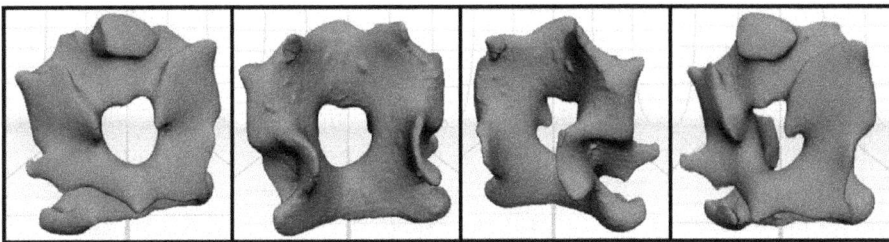

Figure 5.20: ScanStudio HD, final result of the model of T4 vertebra, shaded view.

5.3.5.1.2 Exporting data to STL file format

After completing the scanning procedure, the user should export the created file into STL file format.

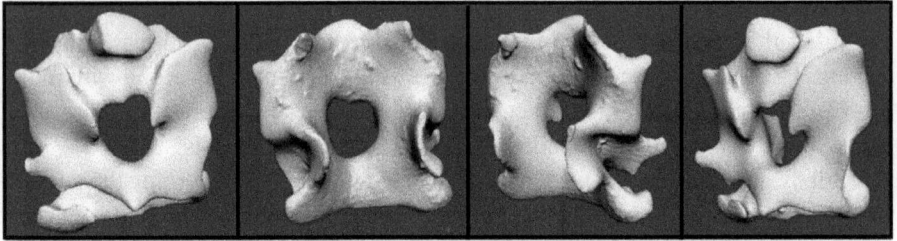

Figure 5.21: ScanStudio HD, final result of the model of T4 vertebra, mesh view.

5.3.5.2 Step 2: measurement of the dimensions of the STL files

After acquiring the desired STL files of the anatomical model of the T4 vertebra, it is fundamental to proceed to the next step of the evaluation stage. This step comprises the measurement and the evaluation of the dimensional declinations between the printed and the scanned STL models. Specifically, these measurements are conducted with the help of the Artec Studio 11 Professional software. In order to compare the two models and assess their form deviation, the Artec Studio 11 Professional provides the *Surface distance maps*, measuring tool. For the current project, this tool is used in order to compare the printed model with the scanned one. Subsequently, the user should define the *Search distance value*, considered as a maximal range in millimeters that calculated the distances between surfaces.

The created surface distance map is a render of particular regions of surfaces of the models, appearing with different colors. Related values of distances and the distribution of them can be found in the graduated scale with histogram, located next to the model. Actually, the map color can change from *blue* (corresponding to negative distance) to *red* (corresponding to positive distance). The *green* color indicates that the distance between surfaces of this region is zero. The *gray* color means that the highlighted surfaces are out of the specified search distance. The *orange* and the *bright blue* colors refer to the distances that are slightly below and above the limiting values of the scale, correspondingly. At last, the graduated scale varies within limits of the positive to the negative value of the selected *Error scale*. The value of the *Error scale* can be adjusted, although its maximal value cannot surpass the *Search distance value*. In particular, six distinct cases are examined by adjusting the *Error scale* parameter with values of 0.001, 0.045, 0.066, 0.141, 0.209, and 0.470.

5.3.5.3 Step 3: comparison and evaluation of the results

The results of the calculation of the surface distance maps are presented in Figure 5.22.

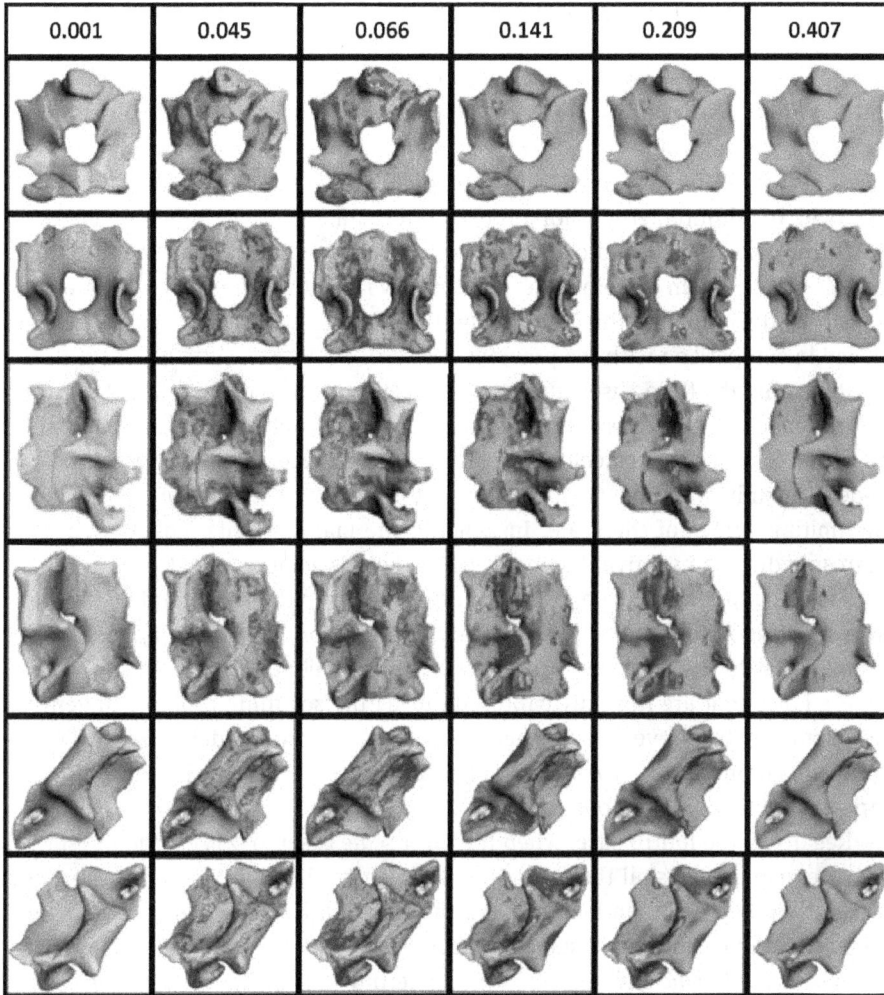

0.001	0.045	0.066	0.141	0.209	0.407

Figure 5.22: Surface distance maps at different views. Error scale from left to right image: 0.001, 0.045, 0.066, 0.141, 0.209, 0.470.

With reference to the first case, when the *Error scale* is set to 0.001 value the models present mainly *orange* and the *bright blue* colors, referring to the distances that are slightly below and above the limiting values of the scale, correspondingly. Furthermore, it can be noticed that the models present areas of *gray* color, meaning that these surfaces are out of the specified search distance. Actually, these areas are holes that are not fully captured during the scanning process. In total, the pictures revealed that the models present dimensional deviation for this error scale.

With regard to the second case, when the *Error scale* is set to 0.045 value the surface distance map changes, presenting all the color ranges. Actually, the anterior part of the model is majorly *green* color, indicating zero distance. The posterior part of the model is mainly *bright blue* color, referring to the distances that are slightly above the limiting values of the scale. In general, the map presents mostly *orange* and the *bright blue* regions (referring to the distances that are slightly below and above the limiting values of the scale, correspondingly), *red* in some areas (corresponding to positive distance), and sporadically *blue* (corresponding to negative distance). In addition, *gray* color appears in the same areas with the previous case (i.e., error scale: 0.001), indicating the holes that are not fully captured during the scanning process. In general, the pictures revealed that the models present dimensional deviation for this error scale, though the results are considerably improved in comparison with the previous case.

Regarding the third case, when the *Error scale* is set to 0.066 value the surface distance map changes, presenting all the color ranges. Actually, the anterior part of the model is majorly *green* color, indicating zero distance. The posterior part of the model is mainly *bright blue* color, referring to the distances that are slightly above the limiting values of the scale. In general, the map presents mostly *blue* areas (corresponding to negative distance), while there is some *orange* (referring to the distances that are slightly below the limiting values of the scale) and sparsely *red* areas (corresponding to positive distance). In addition, *gray* color appears in the same areas with the previous two cases (i.e., error scale: 0.001 and 0.045), indicating the holes that are not fully captured during the scanning process. In general, the overall picture revealed by these results is that the models present dimensional deviation for this error scale. The results are considerably improved in comparison with the two previous cases.

Regarding the fourth case, when the *Error scale* is set to 0.141 value the surface distance map presents all the color ranges. Actually, the anterior part of the model is majorly *green* color (indicating zero distance) with some sparse *blue* regions (corresponding to negative distance). Furthermore, it is also noticed the significant decrease of *bright blue* and regions *orange* (referring to the distances that are slightly above and below the limiting values of the scale, correspondingly). Additionally, the *red* regions (corresponding to positive distance) are decreased and presented rarely. In addition, *gray* color appears in the same areas with the three previous cases, indicating the holes that are not fully captured during the scanning process. In general, the overall picture revealed by these results is that the models present dimensional deviation for this error scale. Nevertheless, the results are considerably improved in comparison with the three previous cases.

Regarding the fifth case, when the *error scale* is set to 0.209 value, the surface distance map changes, presenting all the color ranges. In fact, the anterior part of the model is majorly *"green"* (indicating zero distance) and with some areas of *"blue"* color (corresponding to negative distance). Furthermore, it is worth mentioning that the *bright blue, orange* (referring to the distances that are slightly above and below the limiting values of the scale, correspondingly), and *red* regions (corresponding to positive distance) are significantly decreased. In addition, *gray* color appears in the same areas with the previous four cases, indicating the holes that are not fully captured during the scanning process. In general, the overall picture revealed by these results is that the models present slightly dimensional deviation for this error scale. In fact, the results are considerably upgraded in comparison with the previous four cases.

Regarding the sixth case, when the *error scale* is set to 0.470 value, the entire part is majorly *"green"* (indicating zero distance) and with some sparse areas of *"blue"* color (corresponding to negative distance). Furthermore, it is worth mentioning that the *bright blue, orange* (referring to the distances that are slightly above and below the limiting values of the scale, correspondingly), and *red* regions (corresponding to positive distance) are negligible. In addition, *gray* color appears in the same areas with the previous five cases indicating the holes that are not fully captured during the scanning process. In general, the overall picture revealed by these results is that the models present slightly dimensional deviation for this error scale, presenting satisfactory results with reference to the accuracy. Actually, the results are considerably improved in comparison with the previous five cases.

It can be noticed, according to the results of the surface distance deviation, that the printed and the scanned models do present dimensional deviation, especially when the error scale is from 0.001 to 0.066. When the value of error scale is from 0.141 to 0.470, the models present smaller dimensional declination. The regions depicted with gray color in Figure 5.22 indicate holes of the printed model that are not fully captured during the scanning process. In general, the overall picture revealed by these results is that the models present dimensional deviation, though the results are satisfactory with reference to the pilot approach of the case study.

Figures 5.23 and 5.24 show an illustration of the AM (depicted with blue color) and of the scanned (depicted with pink color) model of the T4 vertebra.

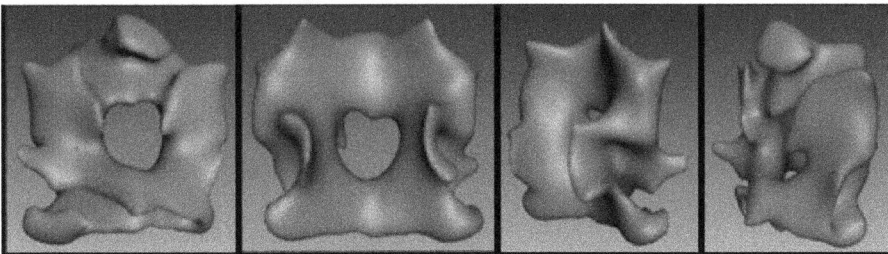

Figure 5.23: Printed model of the T4 vertebra.

Despite the fact that the current study is conducted in accordance with the methods proposed by the related literature, the issue of accuracy should be further examined. The conversion of DICOM files into the final 3D model is identified as the major source of inaccuracy, leading to distortions and geometric deviations in the

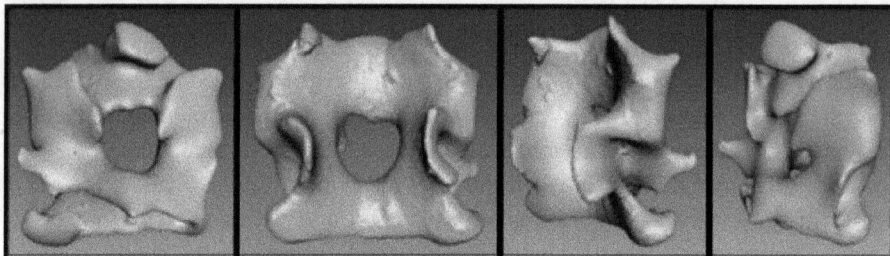

Figure 5.24: Scanned model of the T4 vertebra.

final AM medical model, though the largest percentage of these inaccuracies is introduced during the process of medical imaging or during image processing, rather than during the AM process. The semiautomatic segmentation, provided by ITK-SNAP software application, provides satisfactory results, although the improvement of process parameters needs to be further investigated before medical models are used in any real medical application requiring clinical practice.

Furthermore, the process involved a significant number of software programs for conducting distinct steps. Nevertheless, the majority of these programs are open source and not require specialized manipulation knowledge. Especially as regards the medical imaging processing, the ascertainment by a professional medical opinion is crucial for the more adequate and precise outcome.

In addition, this study is conducted in the framework of a pilot demonstration of the fabrication of patient-specific AM medical models, utilizing FDM method, though other AM techniques can be utilized for the manufacturing process. Similarly, there is a great variety of 3D-scanning methods that present more accurate results. Nevertheless, the overall picture revealed rather satisfactory results.

5.4 Conclusion

This chapter aimed at acquiring a profound understanding of the integration of AM in the fabrication of patient-customized medical model from scanned anatomical images (i.e., DICOM). In particular, the basic scope of this study is at demonstrating the process of fabricating a T4 vertebra of patient's thoracic spine, in a step-by-step approach by illustrating the three fundamental stages: the design, the manufacturing, and the evaluation stage. The first stage extended into four fundamental steps required for the acquisition of the desired STL file. These steps involve the acquisition of the DICOM files (for the current project via CT), the conversion of the DICOM files to STL file format, the optimization of the STL file, and finally the acquisition of the final STL file.

In the second section, the manufacturing stage of the final AM medical model is presented, demonstrating the required steps of the process. Hence, FDM technology is used for the creation and the medical model. After completing the AM procedure, the model required more processing, in order to remove manually the supporting material.

The third section focused on the evaluation stage where the AM model is scanned via a laser scanner and processed via a point cloud software, in order to optimize the final result. Afterward, the model is saved in an STL file format, in order to measure and evaluate the results, regarding the dimensional declinations between the printed and the scanned model. For the comparison and the evaluation of dimensional declinations between the printed and the scanned models, six distinct cases are examined by adjusting the *error scale* parameter to the following values: 0.001, 0.045, 0.066, 0.141, 0.209, and 0.470. In total, according to these results, it can be noticed that the printed and the scanned models do present dimensional deviation, especially when the error scale is from 0.001 to 0.066. When the value of error scale is from 0.141 to 0.470, the models present lower dimensional declination. The regions depicted with gray color indicate holes of the printed model that are not fully captured during the scanning process, due to the complexity of the structure. In general, the overall picture revealed by these results is that the models present dimensional deviations, although the results are satisfactory with reference to the pilot approach of the case study.

This study is an interpretative research, which mainly focused on the direction of profound understanding, regarding the integration of AM in medical sector. The obtained results indicated the capabilities of the AM application in terms of fabrication of customized medical models. Moreover, the process of creating 3D models for medical purposes is a collaborative issue that integrates the expertise from various scientific fields. Consequently, the investigation of a possible collaboration of AM researchers with medical professions is indisputably required, in order to enhance the capabilities of AM technologies in medical purposes. In the current years, despite the fact that AM technologies offer a great potential regarding the field of medical applications, they have not been adopted yet in a large extent, mainly due to the high cost and fabrication time involved. Nevertheless, it is a common fact that AM technologies are more flexible, efficient, and less time consuming in comparison with conventional manufacturing methods for designing and manufacturing customized anatomical models.

References

[1] Javaid, M., Haleem, A. Additive manufacturing applications in medical cases: A literature-based review. Alexandria Journal of Medicine, 2017, https://doi.org/10.1016/j.ajme.2017.09.003.
[2] Bibb, R., Thompson, D., Winder, J. Computed tomography characterization of additive manufacturing materials. Medical Engineering & Physics, 2011, 33, 590–596.

[3]   Gibson, I., Rosen, D., Stucker, B. Additive Manufacturing Technologies: 3D Printing, Rapid Prototyping and Direct Digital Manufacturing, New York, London: Springer, 2015.

[4]   Brennan, J. Production of Anatomical Models from CT scan Data, Master Dissertation, Dublin Institute of Technology, School of Manufacturing and Design Engineering, Leicester, United Kingdom: De Montfort University, 2010.

[5]   Weller, C., Kleer, R., Piller, F. T. Economic implications of 3D printing: Market structure models in light of additive manufacturing revisited. International Journal of Production Economics, 2015, 164, 43–56.

[6]   Santos E.C., Shiomi M., Osakada K., Laoui T. Rapid manufacturing of metal components by laser forming. International Journal of Machine Tools and Manufacture, 2006, 46(12–13), 1459–1468.

[7]   Wang, K., Ho, C.-C., Zhang, C., Wang, B. A review on the 3D printing of functional structures for medical phantoms and regenerated tissue and organ applications, Engineering, 2017, 3, 653–662.

[8]   Chua, C. K., Leong, K. F. 3D Printing and Additive Manufacturing: Principles and Applications, 5th edition, Singapore: World Scientific, 2017.

[9]   Kumar, S. M., Manmadhachary, A., Kumar, R. Y., Alwala, A. Manufacturing of Patient specific AM medical models for Complex Surgeries. Journal of Materials Today: Proceedings, 2017, 4, 1134–1139.

[10]   Petzold, R., Zeilhofer, H. F., Kalender, W. A. Rapid Prototyping technology in medicine – basics and applications. Computerized Medical Imaging and Graphics, 1999, 23, 277–284.

[11]   Eijnatten, M., Dijk, R., Dobbe, J., Streekstra, G., Koivisto, J., Wolff, J. CT image segmentation methods for bone used in medical additive manufacturing. Medical Engineering and Physics, 2018, 51, 6–16.

[12]   Lim, J., Zein, R. The Digital Imaging and Communications in Medicine (DICOM): Description, Structure and Applications, In: Kamrani, A., Nasr, E. A. (eds) Rapid Prototyping: Theory and practice, NY, USA: Springer, 2006, pp. 63–86.

[13]   Gibson, I., Cheung, L. K., Chow, S. P., Cheung, W. L., Beh, S. L., Savalani, M., Lee, S. H. The use of rapid prototyping to assist medical applications. Rapid Prototyping Journal, 2006, 12 (1), 53–58.

[14]   Salmi, M., Paloheimo, K.S., Tuomi, J., Wolff, J., Mäkitie, A. Accuracy of medical models made by additive manufacturing (rapid manufacturing). Journal of Cranio-Maxillo-Facial Surgery, 2013, 41, 603–609.

[15]   Huotilainen, E., Jaanimets, R., Valásek, J., Marcián, P., Salmi, M., Tuomi, J., Mäkitie, A., Wolff, J. Inaccuracies in additive manufactured medical skull models caused by the DICOM to STL conversion process. Journal of Cranio-Maxillo-Facial Surgery, 2014, 42, 259–265.

[16]   Farooqi, K. M., Mahmood, F. Innovations in preoperative planning: Insights into another dimension using 3D printing for cardiac disease. Journal of Cardiothoracic and Vascular Anesthesia, 2017, 32(4), 1937–1945.

[17]   Owusu-Dompreh, F. Application of Rapid Manufacturing Technologies to Integrated Product Development in Clinics and Medical Manufacturing Industries, Master Dissertation, Youngstown State University, School of Industrial and Systems Engineering, 2013.

[18]   Hnatkova, E., Kratky, P. and Dvorak, Z. Production of Anatomical Models via Rapid Prototyping. International Journal of Circuits Systems and Signal Processing, 2014, 8, 479–486.

[19]   Jardini, A. L., Larosa, M. A. Filho, R. M., Carvalho Zavaglia, C. A., Bernardes, L. F., Lambert, C. S., Calderoni, D. R., Kharmandayan, P. Cranial reconstruction: 3D biomodel and custom-built implant created using additive manufacturing. Journal of Cranio-Maxillo-Facial Surgery, 2014, 42, 1877–1884.

[20]   Marro, A., Bandukwala, T., Mak W. Three-Dimensional Printing and Medical Imaging: A Review of the Methods and Applications. Current Problems in Diagnostic Radiology, 2016, 45, 2–9.

[21] Pucci, J. U., Christophe, B. R., Sisti, J. A., Connolly, E. S. Jr. Three-dimensional printing: technologies, applications, and limitations in Neurosurgery. Biotechnology Advances, 2017, 35, 521–529.

[22] Giannatsis, J., Dedoussis, V. Additive fabrication technologies applied to medicine and health care: a review. International Journal of Advanced Manufacturing Technology, 2009, 40, 116–127.

[23] Cheung, L.K., Wong, M.C.M., Wong, L.L.S. Refinement of facial reconstructive surgery by stereomodel planning. Annals of the Royal Australian College of Dental Surgeons, 2002, 16, 129–132.

[24] D'Urso, P.S., Barker, T.M., Earwaker, W.J., Bruce, L.J., Atkinson, R.L., Langian, M.W., Arvier, J.F., Effeney, D.J. Stereolithographic biomodelling in craniomaxillofacial surgery: a prospective trial. Journal of Cranio-Maxillo-Facial Surgery, 1999, 27(1), 30–37.

[25] Moiduddin, K., Al-Ahmari, A., Al Kindi, M., Abouel Nasr, E. S., Mohammad, A., Ramalingam, S. Customized porous implants by additive manufacturing for zygomatic reconstruction. Biocybernetic and Biomedical Engineering, 2016, 36, 719–730.

[26] Chen, X., Possel, J. K., Wacongne, C., van Ham, A. F., Klink, P. C., Roelfsema, P. R. 3D printing and modelling of customized implants and surgical guides for non-human primates. Journal of Neuroscience Methods, 2017, 286, 38–55.

[27] Kurenov, S. N., Ionita, C., Sammons, D., Demmy, T. L. Three-dimensional printing to facilitate anatomic study, device development, simulation, and planning in thoracic surgery. The Journal of Thoracic and Cardiovascular Surgery, 2015, 49(4), 973–979.

[28] Lin, H.H., Lonic, D., Lo, L.J. 3D printing in orthognathic surgery – A literature review. Journal of the Formosan Medical Association, 2018, 117(7), 547–558.

[29] Singh, S., Ramakrishna, S. Biomedical applications of additive manufacturing: present and future. Current Opinion in Biomedical Engineering, 2017, 2, 105–115.

[30] Hodgdon, T., Danrad, R., Patel, M. J., Smith, S. E., Richardson, M. L., Ballard, D. H., Ali, S., Trace, A. P., DeBenedectis, C. M., Zygmont, M. E., Lenchik, L., Decker, S. J. Logistics of three-dimensional printing: Primer for Radiologists. Academic Radiology, 2018, 25(1), 40–51.

[31] Cai, T., Rybicki, F. J., Giannopoulos, A. A., Schultz, K., Kumamaru, K. K., Liacouras, P., Demehri, S., Shu Small, K. M., Mitsouras, D. The residual STL volume as a metric to evaluate accuracy and reproducibility of anatomic models for 3D printing: application in the validation of 3D-printintable models of maxillofacial bone from reduced radiation dose CT images. 3D Printing in Medicine, 2015, 1, 1–9.

Samad Nadimi Bavil Oliaei, Behzad Nasseri

6 Stereolithography and its applications

Abstract: Additive manufacturing (AM) using cost-effective, accurate, and fast processes is one of the major challenges of today's manufacturing community. Stereolithography (SL or SLA) is a promising technique of AM that is believed to satisfy these requirements. In this process, photopolymerization is used to obtain a 3D model of the desired parts directly from their computer-aided design models. The process works by focusing an ultraviolet (UV) laser on a reservoir of photosensitive polymer resin to solidify it layer by layer, resulting in the desired 3D shape. In this chapter, photopolymerization process and how photopolymers response when they are exposed to UV light sources are discussed along with the application of SLA process in different industries such as manufacturing of industrial parts, including military, medical, and biomedical applications. The physicomechanical properties of fabricated polymeric parts will be explained, including viscosity, tensile strength, elastic modulus, flexibility, and toughness. Some case studies regarding the application of this method for polymeric composite material fabrication, preoperation phantom models, scaffolds preparation used in tissue engineering, and drug-loaded models are discussed.

Keywords: Stereolithography, Mechanical Properties, Photosensitive resin, Medical Applications, Tissue engineering

6.1 Introduction

Rapid prototyping (RP) also known as solid freeform fabrication or layer manufacturing technique is a promising technology which is believed to revolutionize manufacturing. The main reason for its popularity is its capability of producing geometrically very intricate parts with internal and external features, which are impossible or require tremendous efforts to be produced by other manufacturing technologies. This popularity has made RP a very active research and development field for both academia and industry. RP techniques can be used for different purposes such as visualization, inspection, rapid design iteration, product optimization, and building functional test models [1]. All RP techniques almost follow a standard procedure; they take a 3D model of the object created using one of the computer-aided design (CAD) software, then cut it into many thin and 2D slices and subsequently print these slices one at a time on top of each other to create a desired 3D object. The need for high-resolution and high-speed 3D printing techniques has resulted in the development of various techniques of additive manufacturing (AM) capable of processing metallic materials such as direct laser metal sintering and selective laser sintering; and methods of processing polymeric materials including fused deposition

https://doi.org/10.1515/9783110549775-006

modeling (FDM) and stereolithography (SL or SLA). These technologies are used either to improve the functionality of the existing systems or as a new fabrication technology [2, 3]. In this chapter, SLA is discussed in detail. The aim is to describe the principle of the process along with its applications.

6.2 Principles of stereolithography process

SLA or SL, which is also known as photosolidification or optical printing [4], is the first and the most mature of RP technologies [5]. It is a 3D printing process, where a reactive liquid photopolymer is cured or solidified by using an irradiation light source [6]. This irradiation is responsible for the required energy for curing, where large numbers of small molecules are bonding together to form highly cross-linked macromolecules. Like any other 3D printing process, SL builds 3D physical proto-types using a 3D CAD model. The CAD model is sliced based on the resolution of the machine and required accuracy and the sliced information being recorded and is fed into the SL machine. Depending on the part geometry, some structural sup-ports may also be needed in order to ensure the stability of the prototype during fabrication process. Once the SL machine being prepared and calibrated, the build platform is moved to the initial position and curing process is started by curing layer by layer of the resin provided by the movement of the built platform. The plat-form will be moved according to a predefined value, until complete part of geome-try has been achieved. Steps of a typical SL process are shown in Figure 6.1.

Generally, any SL machine is made up of some components like a reservoir for liquid polymer, a laser source, build platform, stationary and dynamic mirrors, and retractor blade.

SLA is known for its capability to produce extreme details and the layer thick-ness in this process can be 3–4 times thinner than human hair. When compared to FDM, SLA is shown to produce parts with nice and smooth surfaces, while ridges can be seen and felt on the FDM parts. In SL process, chemical bonds are being de-veloped, which adds strength to the part, while in FDM interlayer and intralayer adhesion controls the bonding. For this reason, when FDM parts are loaded, parts quickly snap along the layer lines upon failure.

SL exhibits high fabrication accuracy and a low degree of defects when compared to FDM [7]. Part accuracy, processing time, and strength of the print are a function of several parameters, including scanning speed, laser intensity, exposure time, the characteristics of the photosensitive resin, and polymer to photoinitiator ratio. In this process, there is also the possibility to select among different photoreactive resins. The material can be selected depending on the required impact resistance, rigidity, and resistance to the temperature, moisture and humidity, colored or colorless de-pending on the desired application.

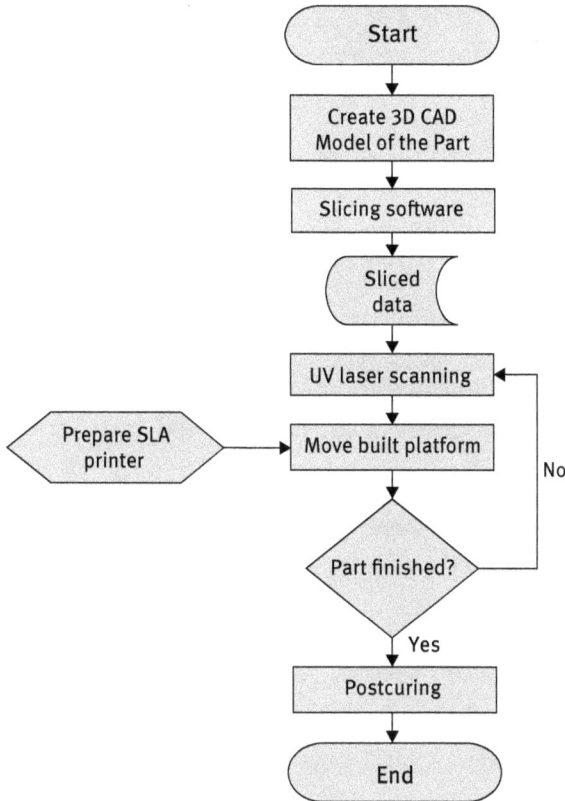

Figure 6.1: Flowchart of an SL process.

In any SLA process, several physics interact with each other as depicted in Figure 6.2. Understanding how these physics interact and their effects on process outputs are essential to have a successful SLA printing. For instance, mathematical models could be quite helpful in determining the depth of cure and over cure, having determined the laser irradiance and other laser beam characteristics by knowing laser physics, while the knowledge of chemistry and materials science would be helpful to tailor the photosensitive polymer based on the required part specifications. The precision control system can provide an increased resolution, reproducibility, and uniformity of the recoated layers. The most important and key physics involved in SL are discussed in this chapter.

6.2.1 Photopolymerization and photopolymers

Photopolymerization (PPM) is the process of hardening of a polymer through light activation, where polymer changes from liquid phase to a solid phase. In other words, the

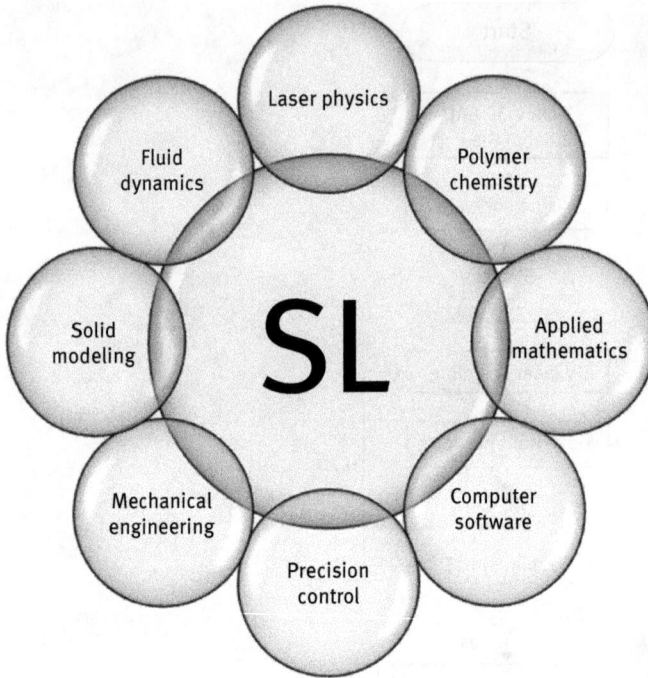

Figure 6.2: Different physics involved in SLA process.

solidification of a polymer is the same phenomenon happening in cross-linking process of the thermoset resins. During PPM by applying the initiation energy on the monomer, the free radical segments of polymer are released and the reaction is continued through propagation. Different polymerization varieties can be considered in the case of reaction initiator type as illustrated in Figure 6.3, as follows:

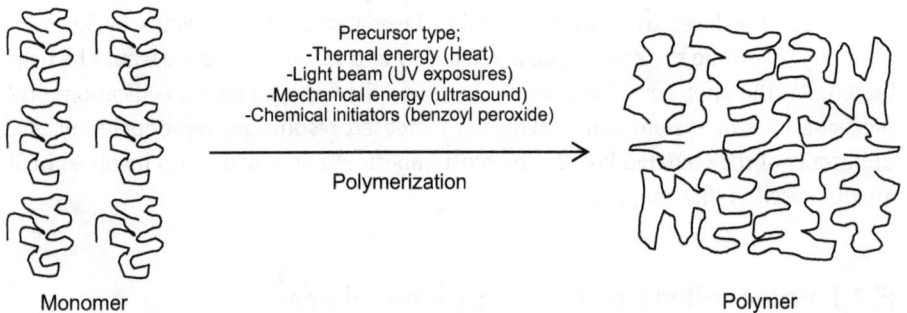

Monomer   Polymer

Precursor type;
-Thermal energy (Heat)
-Light beam (UV exposures)
-Mechanical energy (ultrasound)
-Chemical initiators (benzoyl peroxide)

Polymerization

Figure 6.3: The schematic illustration of initiator varieties in case of reaction kinetics.

- The chemicals as initiator such as benzoyl peroxide
- The optic external energy such as ultraviolet (UV) exposures or any other high-power light beam
- The mechanical external energy such as ultrasound waves

A classification for polymerization in terms of activation type for the initialization of polymerization process is shown in Figure 6.4.

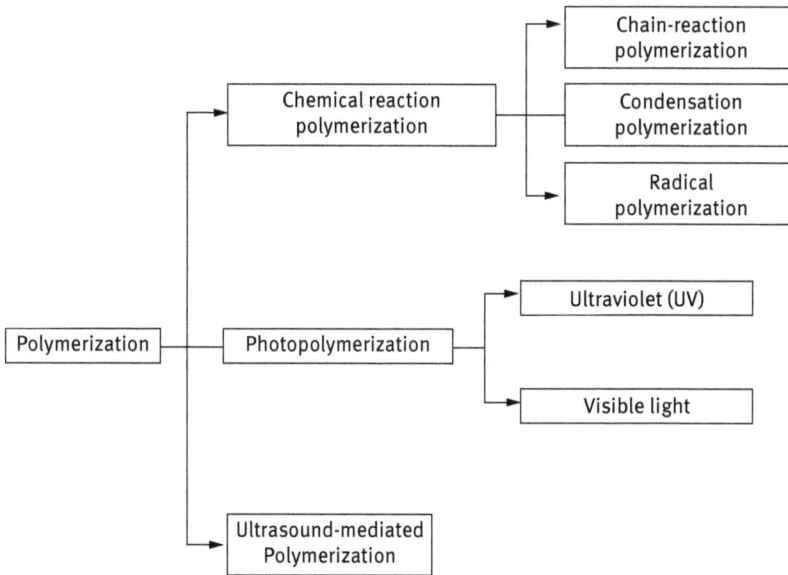

Figure 6.4: A schematic representation of polymerization modalities.

As shown in Figure 6.4, some types of polymerization reactions are illustrated regarding the mechanism of initiating precursor. The chemical techniques are considered as conventional techniques in polymerization. The UV or light energy polymerization is known as photomediated polymerization, which recently developed protocols implemented in different industries. The mechanical energy induction (ultrasound energy) is a nonconventional specific polymerization mechanism with lower application fields. Generally, the polymerization is initiated through multistage reactions; the activation, initiation, propagation, and termination, which are schematically illustrated in Figure 6.5. In the activation stage of PPM reaction, the light energy exposures (such as UV and visible light laser) initiate the production of free monomer radicals. In the initiation stage, the free monomer radicals will invade the monomer segments and produce much more monomer radicals. In this stage, the monomer, free monomer radicals, and radical/monomer combinations can

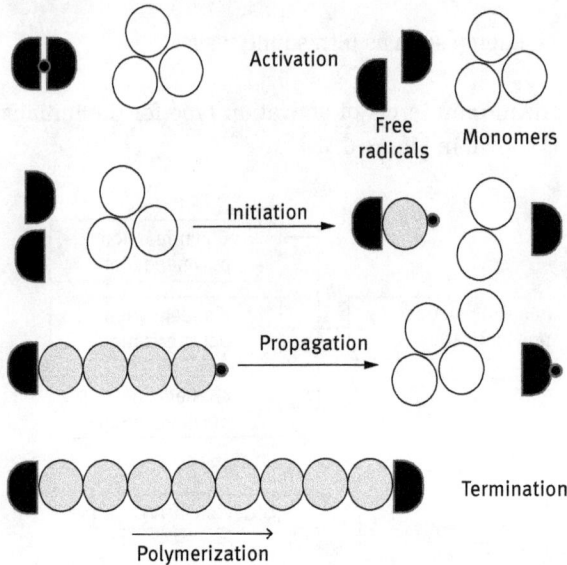

Figure 6.5: Stepwise progression in polymerization.

be seen in he reaction medium. In the propagation stage, the radical attached mono-
mers produce the free monomer radicals. Through the final stage of reaction known
as termination, the attraction of radical segments creates longer chain polymers. The
molecular weight of polymer depends on the number of radical monomer that can be
attached together. All of the aforementioned stages influence the polymer chain
length, which in turn affects the molecular weight of the polymer chain.

6.2.2 Photosensitive polymers

The selection of the type of polymeric resin plays a crucial role in SLA process [1].
The specific physical and mechanical properties in fabricated product can be
achieved by proper selection of polymer and applied fabrication protocol [8]. The
characteristics such as surface quality, and physical and mechanical properties
largely depend on these parameters. In SLA, the applied laser wavelength, exposure time, and laser power are characteristics of energy source. Also specific properties of polymer such as viscosity (related to the molecular weight) [9, 10], glass
transition temperature (T_g) [11], polymerization kinetics [12], conversion rate of reaction [13], and photocuring time are significant parameters that affect the process

outputs. These factors control the photo-induced polymerization, which determines the final product quality.

In the past decades, achievements have resulted in significant interests for SLA in various fields of industry such as prototypes for turbine blades, automotive parts, and military weapons. Other specific applications are in the art such as sculpturing and modeling and in regenerative medicine such as scaffold preparation [14].

The typical compatible type of resins with SLA applications are ethyl methacrylate (EMA) [15] and methyl methacrylate (MMA) [16]. The polymerization of MMA and EMA monomer is initiated by UV light exposures (as activation energy) via propagation and finally termination stage to achieve the thermoplastic polymers. In applications such as fabrication of dental models, surgical guides used in medical devices, investment casting, injection, and blow molding, SLA needs to satisfy certain requirements such as mechanical strength, surface quality, and sufficient manufacturing speed. The rapid manufacturing of jewelry casting and sole molds is another common use of SLA. These requirements not only resulted in developments in the resin technology but also paved the way different for SLA machines that are manufactured recently. SLA machines such as NeXt^{TM}, SLA^{TM}, and DLP movinglight^{TM} are some of the new generation of SLA machines available in the market. These high-resolution 3D printers use different types of thermoplastic resins such as epoxy-based, alkaline, and polyester resins with widespread applications in the above-mentioned industries. The epoxy resins with good physico-mechanical properties provide smooth surfaces that are important in the highly accurate medical applications. The companies such as CIBA GEIGY, HUNTSMAN, and PRODWAY technologies are the suppliers of thermoplastic liquid resins used in SLA. The PLASTCure ABS 2800 (for automotive parts modeling), PLASTCure Rigid 10,500, and PLASTCure model 320 (for dental modeling) are samples for SLA developed by PRODWAYS TECHNOLOGIES Company.

6.2.3 The polymer characteristics in SLA

In SLA, the final product specifications and its characteristics are highly influenced by the properties of the resin used for its fabrication. Thermal and mechanical properties and surface quality of the fabricated parts are affected by the nature of the resin. In turn, the nature and photoreactivity behavior of a resin are a function of its viscosity and molecular weight. Additionally, polymerization kinetic (cross-link conversion rate) of thermoplastic resin is a significant factor in SLA process. The other factors related to the laser induction properties such as dimensional issues, light-absorbing potency of the resin in SLA, and penetrability depth of laser into the resin are crucial factors that influence the properties of the fabricated part.

There are two main groups of polymers used in SLA: the first group is the synthetic polymers and the second group is the natural photosensitive polymers (biopolymers). The synthetic polymers include epoxy resins, alkaline and polyester classified resins with wide range of application domain in the industrial uses. The natural polymers include poly(lactic acid) (PLA), polyethylene glycol (high molecular), and polycaprolactone.

The limited availability of photosensitive polymers with acceptable mechanical properties and dimensional stability is the main challenge of SLA process users. The high viscosity of resins derived from hard rigid particles in the resin suspension limits their application in SLA [17]. Furthermore, the existence of rigid particles in the resin results in heterogeneity, which in turn affects the physical and mechanical properties of SLA fabricated parts. The highest viscosity indexes that can be applied in the SLA are around 5 Pa s, where resins with viscosity indexes higher than 5 Pa s are not suitable for SLA process [18]. This is mainly because of inefficient recoating of the resin. The molecular weight that is defined by polymeric segments intensity affects the resin viscosity [19]. The viscosity of resin polymer dictates the laser penetration in SLA technique [9]. It has been shown by Lee et al. [9] that resins having higher molecular weights result in less UV laser penetration. On the other hand, there is an inverse proportionality between the molecular weight of the resin and UV–laser penetration. In medical applications of laser-mediated SLA, the used resin needs to have optimum viscosity to permit the penetration of the laser to the deeper layers of the resin and to solidify the resin. By using photolinkable photosensitive resins, SLA will provide an opportunity to build the suitable scaffolds (cells nest) used in tissue engineering such as bone, cartilage, muscle, and nervous regeneration which is followed in reconstructive technique branch of medicine. Besides the advantages of SLA technique such as high accuracy and manufacturing speed, severe disadvantages can also be introduced, which limits the application of this technique. The laser spot size and the stepwise increment in the Z-direction are defined as drawbacks in the lamination of the previous cured resin layer [20].

6.2.4 Mechanical properties of parts manufactured by SLA

Mechanical properties of the prints are quite significant in most engineering applications. In SLA process, mechanical properties of the prints are a function of several parameters. Exposure is one of the most significant parameters, which dictates the mechanical behavior of prints. Since in SLA, there is an inhomogeneous exposure, the curing is supposed to be also inhomogeneous. According to the volumetric exposure equation which is defined as a function of depth below resin surface (z) and distance from laser scan axis (y) according to the following relationship:

$$E(y,z) = \sqrt{\frac{2}{\pi}}\left(\frac{P_L}{W_0 V_s}\right)\exp\left(\frac{-2y^2}{W_0^2}\right)\exp\left(\frac{-z}{D_p}\right)$$

where P_L is the laser power, D_p is the penetration depth, W_0 is the Gaussian half width, and V_s is the scan speed. Based on the above relationship, it is expected that mechanical properties also vary with the exposure resulting in a nonuniform mechanical property.

The laser wavelength is also quite a critical factor in determining mechanical properties of the prints, and either resin should be tailored according to a laser wavelength or for a given resin an appropriate laser wavelength needs to be determined through experimental investigations. Postcuring is also shown to highly affect the mechanical properties of the prints. In a study conducted by Zguris [21], the effect of various postcuring parameters (laser wavelength, temperature, and flux) on the ultimate tensile strength and elastic modulus of three different resins (castable, tough, and standard resins) have been investigated. The conditions in which highest mechanical properties can be achieved are determined. The results also revealed that unlike the common belief that more higher light fluxes can result in better mechanical properties, an optimum flux value of 1.25 mW/cm^2 is obtained for having better mechanical properties. The use of higher temperatures is also shown to reduce postcure time. The postcured mechanical properties are maximized when using a laser wavelength of 405 nm.

The addition of filler materials, especially for the case of biodegradable photopolymers, is shown to be quite promising in increasing the elastic modulus of the materials. For instance, Stampfl et al. [22] used hydroxyapatite as a filler material along with acrylate-modified gelatin in the fabrication of cellular scaffolds for bone replacement applications. Overture, which is defined as the amount of penetration of the current cure layer to the layer underneath the current layer in order to create a bond between successive layers and generate a solid part, is shown to be an important parameter affecting the tensile strength of the SL parts [23]. Recoating, which can be described in terms of Z-level wait can also be effective in determining the mechanical properties of the SL prints. It is the dwell time required for the resin to settle. In this way, defects such as bubbles can be highly reduced. However, long Z-level waiting times can result in long printing times. Therefore, long dwell times are undesirable and should be avoided in SL process. Barton and Fulton [23] have also tried to understand the effect of print orientation on the mechanical properties of SL samples. They used SL to print samples having different orientations and subjected the samples to a tensile test. Their results showed that elastic modulus, Poisson's ratio, and ultimate tensile strength are almost independent from sample position in the resin bath as far as layer orientation kept identical. However, when changing layer orientation, 11% variation in elastic modulus, 9% variation for Poisson's ratio, and 12% variation for UTS have been reported. SLA process is shown to reveal a reproducible result as far as materials are kept in an environmentally controlled condition. The effect of built orientation and postcuring on the mechanical properties of the commercial photosensitive epoxy (Watershed 11122) has been studied by Chantarapanich et al. [24]. Their results showed that orientation can result in significant variations in mechanical properties. A variation of 10% has been reported for the elastic modulus. It has shown that postcuring has a significant effect on mechanical properties of the prints up to 4 h of process; however, beyond 4 h no improvements in the mechanical properties has been observed. A statistical approach has been used by Quintana et al. [25] to analyze the effect of built orientation on the mechanical properties of the samples made by SL process. Their results revealed that UTS and elastic modulus are not affected by the axis and position of the samples. However, the orientation is shown to be quite significant on both UTS and elastic modulus.

6.2.5 Improving of mechanical characteristics in manufactured parts via SLA

As described in the previous section in SLA, applications exploiting more rigid polymers are the focus of attention in various fields [26]. This can be achieved mainly by reinforcing the resins. Recently, by applying nanotechnology in different areas of science and technology hybrid polymers with specific technical properties and improved mechanical properties are being synthesized [27]. This can be achieved by adding nanomaterials (dispersed phase) into the matrix network. The nanomaterials can be used in the forms of nanoclay, metallic nanoparticles, and nanofibers [28]. The mechanical properties such as flexural modulus, elasticity, and impact strength can be improved by selecting appropriate engineering materials based on the requirements of in-service functionality of the part. specific for defined service area but in some cases it is mandatory to achieve the extreme properties of used parts according to the service purposes [29]. In addition to the mechanical properties, thermal and electrical properties such as the higher thermal and electrical conductivities can be realized by appropriate formulation in the nanocomposites [29]. In the biomedical applications by adding a certain amount of reinforcement agent such as metallic nanoparticles, the designed part can show extreme improvements in the case of mechanical properties that were mentioned previously [30]. Besides, the increase in mechanical properties, other properties such as antibacterial effects of used metallic nanoparticles can be achieved. In a study of Oyar and her colleagues, different sizes of gold nanoparticles blended with filled matrix material of teeth in order to improve the flexural strength of the material.

6.3 Applications of SLA

In this section, different application of SLA process is discussed.

6.3.1 Stereolithography for polymer composite manufacturing

Generally, the resins used for 3D printing have poor mechanical properties such as elastic modulus, tensile strength, and fracture toughness. The use of reinforcements is a practical way of improving mechanical properties of resins. SLA has been used by Sano et al. [31] to produce discontinuous and continuous reinforced composites. Using glass powder and fiberglass fabrics, a remarkable increase in tensile strength and elastic modulus has been achieved. In terms of appearance, it has been shown that glass powder completely disappears at the matrix materials; however, short glass fibers do not disappear. When increasing glass powder content up to 50%, tensile strength and elastic modulus increase, while fracture strain decreases. SLA has shown to be a successful method of fabricating microcomponents out of ceramic-reinforced composite materials [32]. A suspension made up of photosensitive resin and ceramic particles can be used to manufacture complex and precise ceramic microcomponents. This suspension can be used in a same way as unreinforced resin in SLA process. During curing process, ceramic particles will be anchored in the polymer networks, which highly improve the mechanical properties of the prints. In order to have a successful printing process, in addition to SLA process parameters, the suspension needs to be controlled and examined in terms of its rheological behavior and curing characteristics [32]. The rheological behavior is shown to be improved by an appropriate addition of dispersants [33]. The application of silica, alumina, and silicon nitride ceramic powders with a volume fraction of 0.4–0.55 in an aqueous acrylamide-based suspension has been reported by Griffith and Halloran [34]. The application of Al_2O_3 suspensions in the fabrication of ceramic microcomponents such as micro-end mills has been reported in a study of Xing et al. [32]. In their study, ditrimethylolpropane tetraacrylate is used as photosensitive resin matrix with a density of 1.15 g/mL and viscosity of 350 mPa which is mixed with 1-hydroxycyclohexyl phenyl ketone photoinitiator.

6.3.2 General concepts of tissue engineering by using SLA

The progresses made in the reconstructive medicine, where the tissue engineering is a substantial part of that, largely depends on the preparation of complex, cell-seeded 3D structures such as synthetic scaffolds. In the late 1980s, a proliferation in the applications of 3D fabrication techniques in biomedical studies has been started. One of the most promising techniques in tissue repairing is applying an appropriate 3D synthetic scaffolds as cells nest for their proliferation and refabrication of loosed tissue [35, 36]. The different characteristics of manufactured 3D structure are known to affect the scaffold performance. In the cell/scaffold constructs, the mean pore size, porosity ratio, mean pore size distribution, and interconnectivity between cavities are important. Other characteristics of interests are surface topography and mechanical behavior of the synthetic scaffolds [37]. The schematic description of SLA-centered tissue engineering is depicted in Figure 6.6.

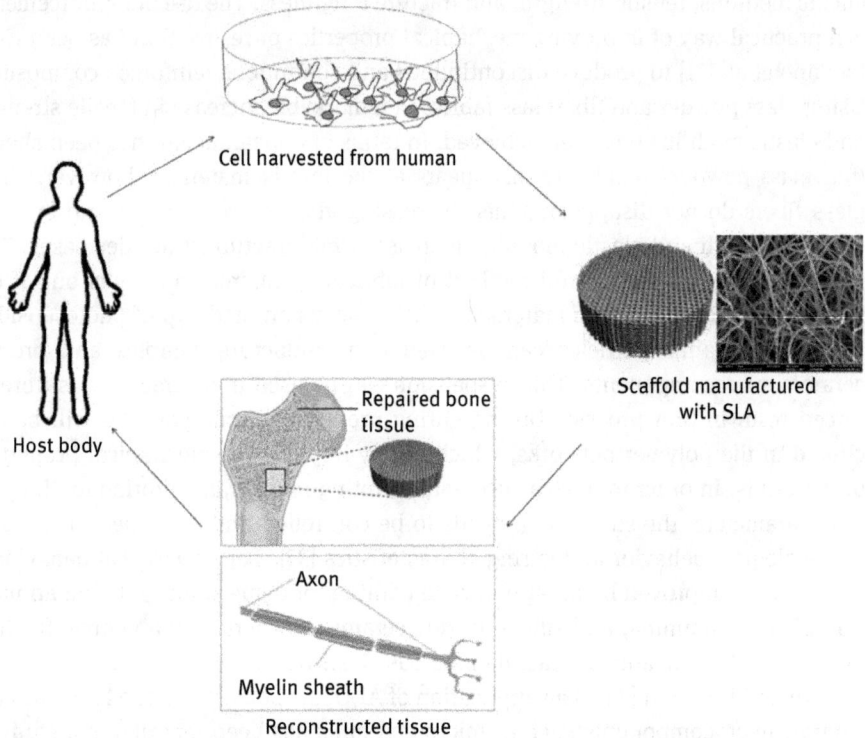

Figure 6.6: Schematic illustration of SLA application in the tissue engineering.

6.3.3 SLA in medical applications

SLA has emerged as a process with high interests in the field of pharmaceutical manufacturing between years 1996 and 2016. Along the mentioned period, the FDM and powder-based jet-based 3D manufacturing were introduced. The SLA has been known as Food and Drug Administration (FDA) approval technique in 3D-printed procedure in pharming. The first FDA-approved 3D-printed therapeutic agent was presented as Spritam (levetiracetam) tablet for epileptic seizures treatment. Also, other biomedical applications of 3D printing such as SLA prospects the new hopes for new generation of treatment and controlled released rate of reagent due to the 3D architecture. The drug-loaded SLA-fabricated platform for drug administration also can be a suitable aperture for therapeutic agent administration in the interest tissue [38]. The morphology of treating region is needed to be properly modeled to provide the highest therapeutic efficiency of manufactured prototype. The 3D gel type of polymers is suitable for mentioned purpose (as shown in Figure 6.7).

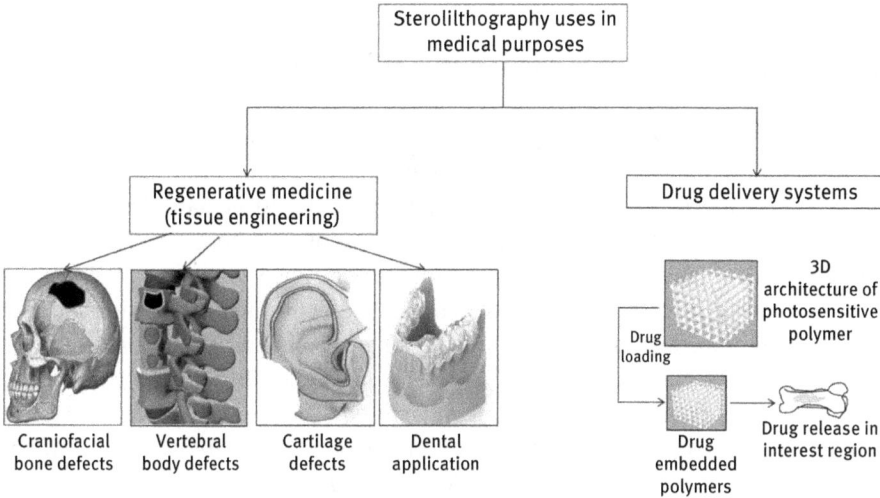

Figure 6.7: Schematic illustration for biomedical applications of SLA.

Besides the dealing of SLA in medical applications, the attention to the different classifications of AM techniques used in biomedical area as implantable biomaterials in the body are important [39]. The SLA mediated lateral medical applications are as follows:
– Medical phantoms as pre- and postoperative templates for education and training purposes
– Rapid orthosis and prosthesis preparation
– Anatomical modeling
– Surgical planning
– Prototype instruments for medical applications
– Inert implants
– Biofabrication

The other uses of SLA can be introduced in drug delivery and tissue engineering, which will be discussed in the upcoming sections. The different applications of SLA in medical achievements can be used in regenerative medicine for repairing of the defects of injured tissues [35, 40]. This can be achieved by fabricating scaffolds that can be applied as cell's nests prepared by photosensitive polymers (synthetic and natural) through SLA. Also, SLA can provide proper drug delivery carriers to region of interest with a few benefits compared to thermal methods. With photon energy-mediated SLA, the fast polymerization is carried out without overheating of the administered drug. In tissue engineering protocol with SLA, the mimicking of damaged tissues is experienced via restoration of the injured scaffold for achieving the functional characteristics of tissue.

6.3.4 Reconstruction of damaged hard tissue by SLA

Normally, the bone healing is a postnatal repair process that for defect sizes larger than 6 mm, the biological reconstruction mechanism can't act well without applying graft implants or biomaterial substitution [36]. In the novel, hard-tissue regeneration techniques of tissue engineering, the using of SLA bone scaffolds can be addressed. Generally, a rebuilding technique is performed to recover the bone defects happening in cases such as accidents, acute bone tumors resulting in bone loose, and chronic diseases. In SLA, a wide range of repaired tissue characteristics are needed to be complied [41, 42]. The mechanical specifications such as tensile strength and compression set in the manufactured 3D bone scaffolds and physical properties such as porosity, pore size, and interconnectivity are main factors that affect the efficiency of therapeutic technique. Each of the above-mentioned parameters identifies the ability of SLA as a repairing method.

Generally, in SLA technique for rebuilding of bone tissue the shape of damaged region of bone is modeled which was fabricated by applying biocompatible minerals and polymers such as hydroxyapatite and PLA blend. Investigations show that the scaffold rigidity, porosity also pore sizes impress on the signaling of osteogenic cells during proliferation of prematured cells and bone growth in post stages. Since the cells use the fabricated scaffold as a nest for keeping up the cell growth, the mean pore sizes can affect the cell attachment, proliferation, and migration [43].

The size of the pores should be compatible with the cell size. The pores should be large enough to let the cells penetrate into the pores and at the same time they should not be large to adversely affect the mechanical performance of the part. The investigations clarify that, increasing of the fabrication speed dramatically, which increases the mean pore sizes in the 3D porous scaffolds [37]. By increasing the pore sizes, the tensile and compressive strengths can be affected dramatically, which can lead to the performance deficiency of synthetic scaffold in the applied region [44]. In addition to the high therapeutic impact of scaffolds manufactured by SLA in the damages healing, the cost-effective SLA method can increase the demands of SLA between bioengineering scientists. Figure 6.8 presents the schematic description of SLA in bone tissue reconstructing.

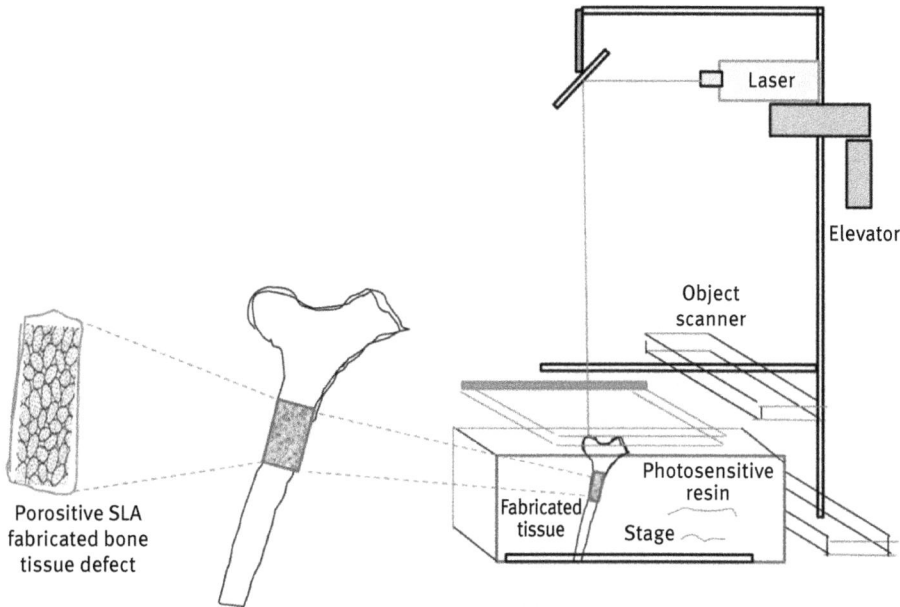

Figure 6.8: Hard tissue reconstruction via stereolithography as tissue engineering.

6.3.5 Cartilage scaffold rebuilding by SLA

As mentioned in the previous section, the hard tissue reconstruction is a remarkable area in medicine because of sever damages that can be happened in connective tissues [45] such as bones and cartilages [46]. Similar to the other tissues such as hard tissues, the damaged scaffold of tissue needs to be repaired by suitable manufacturing technique [47]. The SLA is known as a proper technique for rebuilding of injured scaffold. Either synthetic or natural polymers can be used in SLA to prepare cartilage scaffold structure [48]. Since scaffold is in a direct contact with the cells, the cytotoxicity of the scaffold materials is of vital importance and special attention should be given when selecting resin material for these applications.

As cartilages lack vascular architectures, the interconnectivity and pore size require to be optimized in a way that acts as a capillary structure to be able to provide enough blood and growth factors flowing to the cells as a result of capillary action. Since nonvascularized extracellular architecture in cartilages limits the biofactors transferring to the damaged area, an efficient technique with sufficient regeneration impact needs to be applied in injured tissue (Figure 6.9).

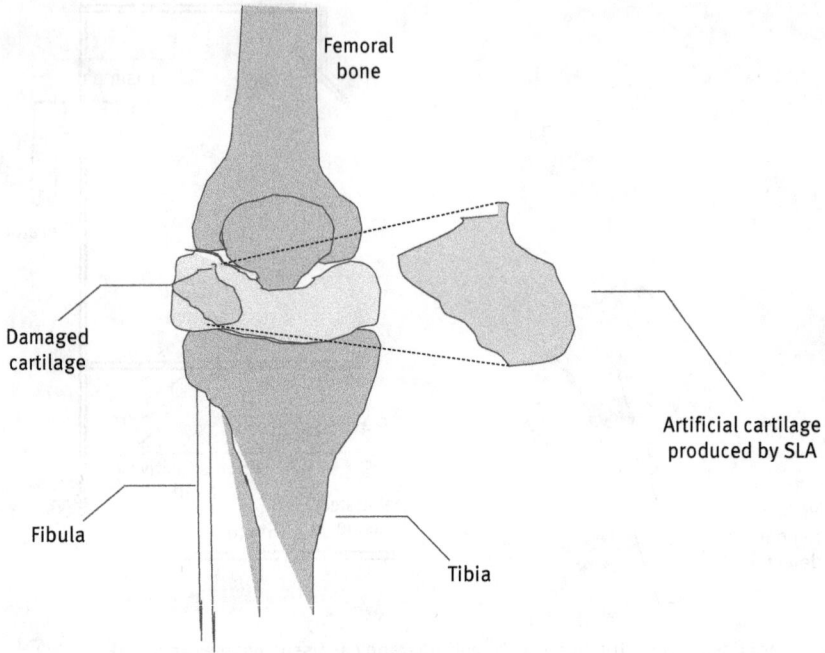

Figure 6.9: Cartilage scaffold refabrication by stereolithography.

6.3.6 Drug delivery in tissues by using SLA-fabricated implants

Implants promise high impact in the operative treatments of medicine. In conventional techniques, the use of prosthesis is considered for providing the permanent stability or fixation of damaged tissue, which was specifically implemented in the case of hard tissues. The inflammation or infection in the damaged area resulted in serve lesions, and their operations of injured tissue were generally inevitable due to the mal sterilization of operation rooms. In novel techniques, in addition to the fixation of tissues for administrating of drugs and providing its therapeutic effect by using impregnated implants are observed [49]. The mentioned techniques can be placed in the semiactive drug delivery techniques. In passive targeting of drug, the excess amounts of administered drug are excreted by hepatic metabolism and subsequent renal excretion system. The drug administration through SLA is a patient-specific technique for transferring the biological objects and therapeutic factors (drugs). In the drug-infused implants manufactured by SLA technique, the drug-embedded biocompatible polymers (with sufficient mechanical properties) can be used. In this method, the therapeutic agent is implemented in the damaged tissue to provide the fixation and healing effect simultaneously [50]. The accurate dimensional size of damaged region resulted from computerized tomography scan equipment was fed to the SLA machine for fabricating the appropriate implant, which uses the therapeutic agent (drug) combined with the polymeric material. The controlled releasing of drug in the desired tissue is the main factor in the therapy efficiency. Figure 6.10 illustrates the schematic image of drug-impregnated SLA implants in the medicine.

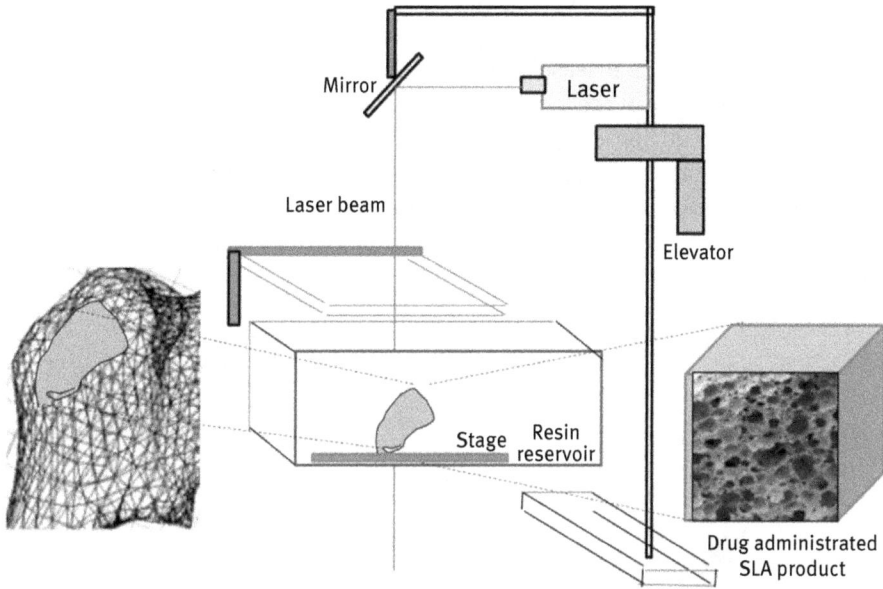

Figure 6.10: Drug mixed stereolithography reconstructed scaffold applicable for severe traumatic defects of bone.

6.3.7 Reducing the disease risks by studies on SLA-mediated models

In this section, two diseases with high risks will be presented. The models fabricated by SLA are shown to be used to plan the optimum therapeutic protocol with minimal operational mortality. Although the new generation of therapeutic methods present the satisfied operation qualities in the case of operation risks, the application of SLA can be helpful in the treatment protocols in medicine by implementing preoperation studies on abnormalities that can occur during operation. This can be developed in all clinical applications, especially for high-risk regions in human body such as tissues, organs, or systems.

In a research in the university hospital of Lubeck in Germany, the scientists used highly accurate SLA 3D-printed models of brain arteries with sufficient precision to reduce the risks of brain operations in the diseases like stroke. Dr. Kemmling and his research team used the CT scan files of patient's brain and used the high-resolution

Formslab SLA machine to fabricate the artificial tissue for this purpose [51]. The aneurismal arteries abnormality is another disease case, which can be modeled by SLA. Aneurismal aorta artery dysfunctionality is one of the high-risk diseases related to the cardiovascular diseases category. The high mortality risk artery aneurism needs to get action in the disease treatment. In some cases, before aorta rupturing, the physicians have the opportunity to involve in the curing of the disease for operation. In these cases, the 3D models can help physicians to take the best decision about the disease treatment. The SLA as a manufacturing strategy will help conclude about the fabricated model according to the scanning 3D file of interest region reducing surgery risks [52]. The schematic illustration of SLA-mediated model on aneurismal aorta artery is given in Figure 6.11. The accurate dimensions of vessels are mandatory to use more accurate manufacturing protocols in artificial arteries modeling and their production that happens in cases such as renal diseases.

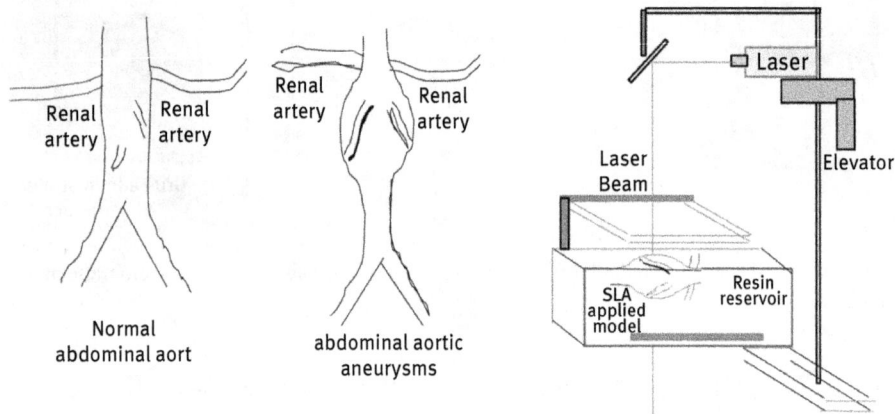

Figure 6.11: The normal and aneurismal aorta artery and disease investigation model fabrication via SLA.

The endovascular surgery can be possible by applying high-tech microscopic woven wire grafts and polymeric hoses for delivering blood. This high-precision flow transporters need to be mounted accurately inside arteries. This can be implemented exactly with a few risks due to misplacement or misalignments. By using the accurate model of interest region of artery, a suitable model can be built to provide the optimum condition of operation in the case mounting place, decreasing adjustment impossibilities, and minimization of the blood leakage in the artery/synthetic graft junction area [51, 53]. SLA is a quite powerful and cost-effective tool for such applications.

6.4 Conclusions

Studies related to SLA process presented in this chapter are shown that SLA is a capable and reliable AM technology, which can provide wide range of benefits such as higher speed, tighter manufacturing tolerances, higher resolution, and accuracy and can be economically cost-effective. In this process, parameters such as resin viscosity, laser wavelength, laser beam power, and laser/polymer induction duration affect the functionality of the final part. They need to be well calibrated to get the desired performance. In the biomedical applications of SLA, the preoperational observations can be achieved by applying the high accurate models fabricated by SLA technique. The drug-loaded artificial scaffolds usable for wound healing of the tissues defects can be counted as an important biomedical application of SLA with high therapeutic properties. The mentioned application fields of SLA will have high contribution impacts in the biomedical progresses.

References

[1] Ligon, S.C., et al., Polymers for 3D printing and customized additive manufacturing. Chemical reviews, 2017, 117(15), 10212–10290.
[2] Sadia M., Alhnan M.A., Ahmed W., Jackson M.J., 3D Printing of Pharmaceuticals. In: Jackson M., Ahmed W. (eds) Micro and Nanomanufacturing, 2018, Volume II, Springer, Cham.
[3] Kruth, J. P., Leu, M. C., & Nakagawa, T., Progress in additive manufacturing and rapid prototyping. Cirp Annals, 1998, 47(2), 525–540.
[4] Jacobs, P.F. Fundamentals of stereolithography. in 1992 International Solid Freeform Fabrication Symposium, 1992.
[5] Jacobs, P.F., Rapid prototyping & manufacturing: fundamentals of stereolithography, 1992: Society of Manufacturing Engineers.
[6] Bártolo, P. J. (Ed.). Stereolithography: materials, processes and applications, 2011, Springer.
[7] Kruth, J.-P., Leu, M.-C., and Nakagawa, T., Progress in additive manufacturing and rapid prototyping. Cirp Annals, 1998, 47(2), 525–540.
[8] Dizon, JRC., A.E., Chen, Q., Advincula, RC. Mechanical characterization of 3D-printed polymers. Additive Manufacturing, 2017, 20(-), 47.
[9] Lee, K.-W., et al., Poly (propylene fumarate) bone tissue engineering scaffold fabrication using stereolithography: effects of resin formulations and laser parameters. Biomacromolecules, 2007, 8(4), 1077–1084.
[10] de Leon, AC., Napolabel, Q.C., Palaganas, B., Palaganas, Jerome O., Manapat, Jill, Advincula, Rigoberto C., High performance polymer nanocomposites for additive manufacturing applications. Reactive and Functional Polymers, 2016, 103, 11.
[11] Stansbury, J.W. and M.J. Idacavage, 3D printing with polymers: Challenges among expanding options and opportunities. Dental Materials, 2016. 32(1), 54–64.
[12] Esposito Corcione, C., et al., Organically modified montmorillonite polymer nanocomposites for stereolithography building process. Polymers for Advanced Technologies, 2015, 26(1), 92–98.
[13] Makvandi, P., et al., Antimicrobial modified hydroxyapatite composite dental bite by stereolithography. Polymers for Advanced Technologies, 2018, 29(1), 364–371.
[14] http://www.forecast3d.com/sla.html (accessed: April 2016).

[15] Medina, F., Wicker, R., Palmer, J. A., Davis, D. W., Chavez, B. D., & Gallegos, P. L., U.S. Patent No. 8,252,223. 2012, Washington, DC: U.S. Patent and Trademark Office.

[16] Zanchetta, E., et al., Stereolithography of SiOC ceramic microcomponents. Advanced Materials, 2016, 28(2), 370–376.

[17] Taormina, G., et al., Special resins for stereolithography: In situ generation of silver nanoparticles. Polymers, 2018, 10(2), 212.

[18] Hinczewski, C., Corbel, S., and Chartier, T., Ceramic suspensions suitable for stereolithography. Journal of the European Ceramic Society, 1998, 18(6), 583–590.

[19] Melchels, F.P., Feijen, J., and Grijpma, D.W., A review on stereolithography and its applications in biomedical engineering. Biomaterials, 2010, 31(24), 6121–6130.

[20] Manapat, J.Z., et al., 3D printing of polymer nanocomposites via stereolithography. Macromolecular Materials and Engineering, 2017, 302(9), 1600553.

[21] Zguris, Z., How mechanical properties of stereolithography 3D prints are affected by UV curing. Somerville, MA: Formlabs Inc, accessed Mar, 2016, 7, 2017.

[22] Stampfl, J., et al., Biodegradable stereolithography resins with defined mechanical properties. Virtual and Rapid Manufacturing, Proceedings of VRAP, Leira, Portugal, 2007, 283–288.

[23] Dulieu-Barton, J. and Fulton, M., Mechanical properties of a typical stereolithography resin. Strain, 2000, 36(2), 81–87.

[24] Chantarapanich, N., Puttawibul, P., Sitthiseripratip, K., Sucharitpwatskul, S., & Chantaweroad, S., Study of the mechanical properties of photo-cured epoxy resin fabricated by stereolithography process. Songklanakarin J. Sci. Technol, 2013, 35(1), 91–98.

[25] Quintana, R., et al., Effects of build orientation on tensile strength for stereolithography-manufactured ASTM D-638 type I specimens. The International Journal of Advanced Manufacturing Technology, 2010, 46(1–4), 201–215.

[26] Bartolo, P.J., J.G., Metal filled resin for stereolithography metal part. CIRP Annals – Manufacturing Technology, 2008, 57, 4.

[27] Kumar, S., et al., Reinforcement of stereolithographic resins for rapid prototyping with cellulose nanocrystals. ACS applied materials & interfaces, 2012, 4(10), 5399–5407.

[28] Eng, H., et al., 3D Stereolithography of Polymer Composites Reinforced with Orientated Nanoclay. Procedia engineering, 2017, 216, 1–7.

[29] Chiu, S.-H., et al., Mechanical and thermal properties of photopolymer/CB (carbon black) nanocomposite for rapid prototyping. Rapid Prototyping Journal, 2015, 21(3), 262–269.

[30] Oyar, P., et al., Effect of green gold nanoparticles synthesized with plant on the flexural strength of heat-polymerized acrylic resin. Nigerian journal of clinical practice, 2018. 21(10), 1291.

[31] Sano, Y., Matsuzaki, R., Ueda, M., Todoroki, A., & Hirano, Y., 3D printing of discontinuous and continuous fibre composites using stereolithography. Additive Manufacturing, 2018, 24, 521–527.

[32] Xing, H., Zou, B., Lai, Q., Huang, C., Chen, Q., Fu, X., & Shi, Z., Preparation and characterization of UV curable Al2O3 suspensions applying for stereolithography 3D printing ceramic microcomponent. Powder technology, 2018, 338, 153–161.

[33] Zhang, S., N. Sha, and Z.J.J.o.t.E.C.S. Zhao, Surface modification of α-Al2O3 with dicarboxylic acids for the preparation of UV-curable ceramic suspensions. 2017, 37(4), 1607–1616.

[34] Griffith, M.L. and J.W.J.J.o.t.A.C.S. Halloran. Freeform fabrication of ceramics via stereolithography, 1996, 79(10), 2601–2608.

[35] Dhariwala, B., Hunt, E., and Boland, T., Rapid prototyping of tissue-engineering constructs, using photopolymerizable hydrogels and stereolithography. Tissue engineering, 2004, 10(9–10), 1316–1322.

[36] He, Y.-X., et al., Impaired bone healing pattern in mice with ovariectomy-induced osteoporosis: A drill-hole defect model. Bone, 2011, 48(6), 1388–1400.

[37] Kim, K., et al., The influence of stereolithographic scaffold architecture and composition on osteogenic signal expression with rat bone marrow stromal cells. Biomaterials, 2011, 32(15), 3750–3763.

[38] Kempin, W., et al., Assessment of different polymers and drug loads for fused deposition modeling of drug loaded implants. European Journal of Pharmaceutics and Biopharmaceutics, 2017, 115, 84–93.

[39] Jukka Tuomi, K.-S.P., Vehviläinen, Juho, Björkstrand, Roy, Salmi, Mika, Huotilainen, Eero, Kontio, Risto, Rouse, Stephen, Gibson, Ian, Mäkitie, Antti A., A Novel Classification and Online Platform for Planning and Documentation of Medical Applications of Additive Manufacturing. Surgical Innovation, 2014, 21(6), 7.

[40] Gauvin, R., et al., Microfabrication of complex porous tissue engineering scaffolds using 3D projection stereolithography. Biomaterials, 2012, 33(15), 3824–3834.

[41] Kebede, M. A., Asiku, K. S., Imae, T., Kawakami, M., Furukawa, H., & Wu, C. M., Stereolithographic and molding fabrications of hydroxyapatite-polymer gels applicable to bone regeneration materials. Journal of the Taiwan Institute of Chemical Engineers, 2018, 92, 91–96.

[42] Kim, K., et al., Stereolithographic bone scaffold design parameters: osteogenic differentiation and signal expression. Tissue Engineering Part B: Reviews, 2010, 16(5), 523–539.

[43] Murphy, C.M., Haugh, M.G., and O'Brien, F.J., The effect of mean pore size on cell attachment, proliferation and migration in collagen–glycosaminoglycan scaffolds for bone tissue engineering. Biomaterials, 2010, 31(3), 461–466.

[44] Moutos, F.T., Freed, L.E., and Guilak, F., A biomimetic three-dimensional woven composite scaffold for functional tissue engineering of cartilage. Nature materials, 2007, 6(2), 162.

[45] Puppi, D., et al., Polymeric materials for bone and cartilage repair. Progress in polymer Science, 2010, 35(4), 403–440.

[46] Zhang, W., et al., Cartilage repair and subchondral bone migration using 3D printing osteochondral composites: a one-year-period study in rabbit trochlea. BioMed research international, 2014.

[47] Mondschein, R.J., et al., Polymer structure-property requirements for stereolithographic 3D printing of soft tissue engineering scaffolds. Biomaterials, 2017, 140, 170–188.

[48] Aisenbrey, E.A., et al., A Stereolithography-Based 3D Printed Hybrid Scaffold for In Situ Cartilage Defect Repair. Macromolecular bioscience, 2018, 18(2), 1700267.

[49] Gulati, K., et al., Local drug delivery to the bone by drug-releasing implants: perspectives of nano-engineered titania nanotube arrays. Therapeutic delivery, 2012, 3(7), 857–873.

[50] Chia, H. N., & Wu, B. M., Recent advances in 3D printing of biomaterials. Journal of biological engineering, 2015, 9(1), 4.

[51] Kemmling, D. Reducing Risks in Brain Operations with 3D Printed Arteries 2017; Available from: https://formlabs.com/blog/reducing-risks-in-brain-operations-with-3D-printed-arteries/.

[52] Elomaa, L., & Yang, Y. P., Additive manufacturing of vascular grafts and vascularized tissue constructs. Tissue Engineering Part B: Reviews, 2017, 23(5), 436–450.

[53] Marija Vukicevic, B.M., Min, James K. and Little, Stephen H., Cardiac 3D printing and its future directions. Journal of American College of Cardiology, 2017, 10(2), 14.

Van-Du Nguyen, Ngoc-Hung Chu

7 Ultrasonic-assisted deep-hole drilling

Abstract: Drilling is one of the most common subtractive processes to make holes. In deep drilling, when the hole depth is larger than two times the diameter, the drilling torque increases exponentially. The most accepted explanation is that, as the depth increases, the increased amount of chips filled up the flutes, leading to chip clogging and thus causing the total torque increase. Several undesirable effects could be occurred due to excessive torque, such as chipping of the drill lips, thermal softening of the tool, or torsional failure of the drill. It is difficult and inefficient to take out the broken part of the tool from the workpiece. Such issues are critical problems for the smart machining processes and become a blockage in production lines. Specialized tools and techniques to improve chip evacuation in deep drilling have been developed, such as single-lip gun, tube drills, grooved tool bits, or peck drilling. Ultrasonic-assisted drilling (UAD) is a new technique providing an alternate and promising solution. This chapter presents how effective this technique can provide for deep hole drilling. Experimental data revealed that ultrasonic vibration can significantly reduce not only the cutting torque, but also the chip evacuation torque. A new and useful tool to analyze and evaluate the effectiveness of UAD is provided via a new model of the drilling torque, which consists of three components. Steps of design and realization of an UAD are also summarized with the hope of helping readers to apply in practical applications.

Keywords: Ultrasonic-assisted machining, deep drilling, chip evacuation, drilling torque, stick-slip torque

7.1 Introduction

Subtractive processes are very broadly employed in manufacturing, where material (usually metal) is removed from a crudely shaped initial part to achieve the final shape and dimensions [1]. Drilling is a common and widely used subtractive process, taking about one-third of all machining operations [2], to make holes from solid parts. A machining progression typically consists of as much as 30% drilling, following by 20% turning, 16% milling, 15% threading (15%), 6% engraving, and others (13%) [3].

Unlike other conventional machining processes such as turning and milling where the chips can freely evacuate from the cutting area, the chips in drilling are constrained by the drill flute and the drilled hole, leading to variations of chip shape, of the drilling forces and of the torque [4]. In drilling of ductile materials, long and persistent chips can be clogged easily inside the drill flutes, leading to

https://doi.org/10.1515/9783110549775-007

restrict the chip evacuation [5]. The chip evacuation practically challenges the dry drilling, an environmentally friendly machining technique, of ductile materials [6]. In deep drilling, where the aspect ratio, that is, the ratio of the hole depth, L, to the drill diameter, D, is higher than 3 [5], the drilling torque appeared to rise up exponentially [4, 7–11], although the thrust (axial) force remains comparatively constant [9, 10] or increases slightly [11]. At higher depths, the accumulated chips would fill up the constrained flutes, leading to chip-clogging and thus increase the resistant torque [4, 7–11]. The addition of chip evacuation torque results in excessive torsional stress, which resulted in a common failure of the tools – the drill breakage [7, 10, 12]. It is challenging and scathing to extract the tool fragment from the workpiece [13]. Consequently, chip evacuation has become the main constraint for the deep drilling [14], especially in smart and automatic machining processes. Specialized drilling tools and techniques have been developed for deep hole drilling, such as using gun drills, tube drills [15], groove-type chip breaker [16], peck drilling [12], or minimum quantity lubricant [17]. Nevertheless, dry drilling with twist tools is still preferable as a dominant tool for their environment-friendly, cost-effective, versatility, and flexibility [18].

Another issue has been widely considered in drilling is burrs, the phenomenon of plastical deformation of workpiece material that at the entry and the exit sides of the drilled holes. Burrs appeared after drilling would require a secondary operation, making difficult to automate and thus would become a bottleneck in a production line. Consequently, limiting burr formation is preferable than deburring them in a subsequent finishing operation [19]. It has been said that deburring and edge finishing of precision components may require as much as 30% of the cost of the finished parts [20]. For drilling of metal sheet, burrs can appear on both the entry and the exit sides of the hole leading to critical problems such as functional problems, small injuries of assembly workers, and assembly matters, requiring an additional deburring process [21]. The investigation [21] concluded that lower thrust force would cause less workpiece material is subject to deformation and thus leads to smaller burrs. Experimental studies were also made to carry out optimum cutting condition so as to minimize burrs (see [22–24] for example).

Recently, ultrasonic-assisted machining (UAM) has been considered as a very useful and promising method compared to both conventional and advanced manufacturing processes. In UAM, a small-amplitude and ultrasonic-frequency vibration is superimposed on the relative motion between the tool and the workpiece. Major advantages of UAM can be concluded as [25] reduced cutting forces, extended tool life, reduced surface roughness and improved form accuracy, greater depth of cut, and suppression of burr formation. Numerous benefits from ultrasonic-assisted drilling (UAD) of aluminum alloys have been reported in several studies [26–31]. Compared to conventional drilling (CD), UAD can offer significant reduction of thrust force [26–28, 30], improvements in built-up edge [27, 28], burr size [29, 32], tool life [33], and hole oversize [27]. Improvements of drillability of aluminum plates due to UAD in terms of the penetration rate has also been evaluated [31]. For other difficult-to-cut materials, several studies have been implemented [7, 34–41], such as for Inconel 738-LC [34–36], carbon fiber composite [37, 38, 40], carbon/epoxy laminates [39], and cortical bone [41]. However, most of the investigations have been conducted with the aspect ratio of drilled depth (L) and drilling diameter (D) that ranged from 2 to 4. Consequently, the chip evacuation in dry and deep hole drilling of ductile material has still been an issue needed to be addressed. This chapter summarizes a major approach and the results obtained in recent investigations by the authors to deal with that issue in ultrasonic-assisted deep hole drilling. The readers may want to find out other benefits of the technique in recent publications of the authors in references [42–44].

The chapter is constructed as follows: Section 7.2 gives a description of the practical system which can be used for UAD as well as step-by-step guides to realize it. Some discussions on the drilling torque in both normal (shallow) and deep drilling processes are given in Section 7.3. The reductions of the drilling torque obtained from UAD compared to CD are analyzed in Section 7.4 by means of the new proposed model. The chapter conclusions are given in the last section.

7.2 Practical system of ultrasonic-assisted drilling

The ultrasonic assistant unit for drilling typically consists of three major parts: the transducer, the horn, and the tool bit. Figure 7.1 depicts the structure of the unit clamped on the spindle of a lathe, providing both rotation and vibration to the drill bit.

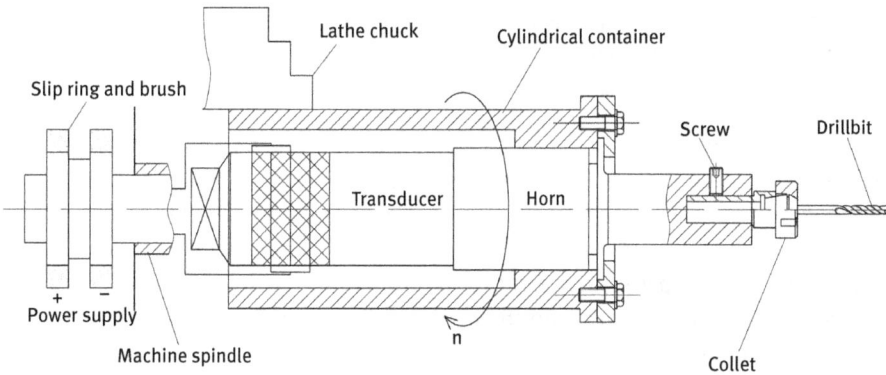

Figure 7.1: The structure of the vibratory unit [45].

In Figure 7.1, the ultrasonic horn and transducer are fixed inside the cylindrical container whose clamped in the chuck of a lathe. The transducer converts electrical energy, which is excited by the power source via a couple of rings and brushes, into proper mechanical vibration. In high-power ultrasonic applications for metal cutting, Langevin transducers are usually used [46, 47]. Langevin transducers are commercially available for ultrasonic cleaning and welding applications, with a wide range of power capacity and working frequency. For UAD applications, Langevin transducers with power capacity of 200 W or higher are suitable [48]. The working frequency of a Langevin transducer is actually its resonant frequency, provided by the vendor. Given a commercial transducer, the horn and tool attached should be designed and fabricated so as to match with the provided frequency of the transducer in order to transfer the maximum energy.

The horn, sometimes called as the sonotrode, is regularly used to magnify and concentrate vibration energy into the tool. The horn can be made in either an exponential, catenoidal, cosine, tapered, or cylindrical stepped form (see Figure 7.2).

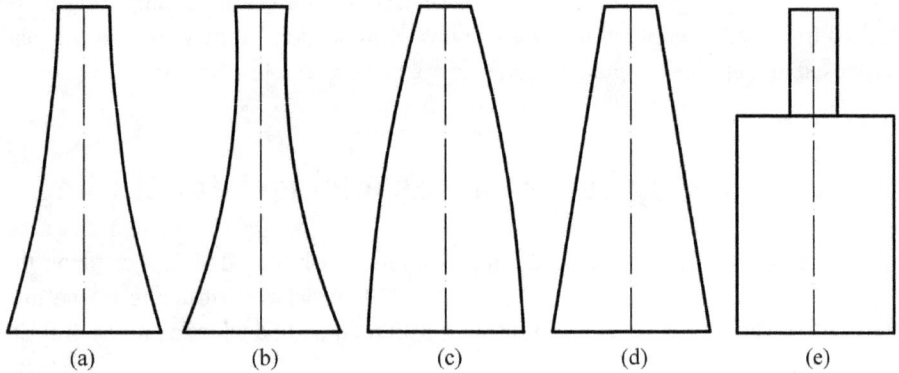

(a)   (b)   (c)   (d)   (e)

Figure 7.2: Different forms for the horn: (a) exponential, (b) catenoidal, (c) cosine, (d) tapered, and (e) cylindrical stepped shapes [49].

The cylindrical stepped horn is preferable because it is simple, easily designed, and realized. It is noted that geometric characteristics of the horn must be carefully determined and validated. Any modifications in the dimension of the tool and/or assembly structure will straightforwardly change the resonant frequency and thus weaken the working performance of the vibratory unit.

The assembly of the horn and the cutting tool can be further scrutinized in detail as shown in Figure 7.3.

In Figure 7.3, the tool 1 can be attached to the horn 4 by means of a collet 2. The collet is secured inside the horn by four screws 3. The longitudinal dimensions L_i ($i = 1/4$) can be crudely calculated by using wave transmission theory in order to achieve the maximum amplitude at the cutting lips. The ratios between D_i can be determined depending on the amplification gain.

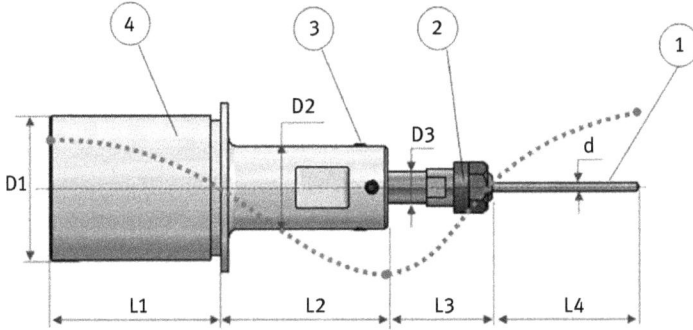

Figure 7.3: Assembly of the horn and the drill bit [45].

Dimensions of the horn can be carried out based on the half-wavelength principle, which has been well known in several literatures. A basic calculation is briefly described as below (readers can see details in [47, 45]).

Noting λ is the wavelength of the ultrasonic oscillation, c_1 is the sound velocity transferred in the horn material, w is the angular frequency; ξ_1 and ξ_2 are element movements at $x = 0$ and at $x = L$, respectively; E is the elasticity modulus and ρ is the density of the horn material; the following expressions can be obtained:

$$c = \sqrt{E/\rho} \tag{7.1}$$

$$\lambda = c/f \tag{7.2}$$

then the movement ξ_x at the position x can be expressed as [50]

$$\begin{cases} \xi_x = \xi_1 \cos\left(\frac{wx}{c}\right) \cos(wt) \text{ for } 0 \leq x \leq \lambda/4 \\ \xi_x = \xi_2 \cos\left(\frac{w(L-x)}{c}\right) \cos(wt) \text{ for } \lambda/4 \leq x \leq \lambda/2 \end{cases} \tag{7.3}$$

The total length, L, of a cylindrical half-wave stepped horn can be calculated as [51]

$$L = k_1 \frac{c}{4f} + k_2 \frac{c}{4f} \tag{7.4}$$

To simplify the calculation, the correction factors k_1 and k_2 can be set at unity. The location of the flange, used to fix the horn (the "nodal plane"), can be determined by the following expression:

$$\cos\left(\frac{wx_{\text{node}}}{c_1}\right) = 0 \Leftrightarrow L_1 = L_2 = \frac{L}{2} \tag{7.5}$$

Note: the values of L_1 and L_2 obtained at this stage are preliminary and will be further corrected.

In the stepped horn, a lower amplitude ξ_2 at the large end is gained to a larger amplitude ξ_1 at the small end of the horn. The gain factor, G, for a half-wave, double-cylinder horn is

$$G = \frac{\xi_2}{\xi_1} = \frac{S_1}{S_2} = \frac{D_1}{D_2} \tag{7.6}$$

where S_1 and S_2 are the cross areas of the horn at $x = 0$ and at $x = L$, respectively.

The small diameter D_2 is selected so as to have sufficient space for the hole containing the collet. Given the gain factor of the horn G, one can then calculate the larger diameter D_1 of the horn as

$$D_1 = D_2 \sqrt{G} \,(\text{mm}) \tag{7.7}$$

The lengths of the cantilever segments of the collet L_3 and the drill bit L_4 must be determined to maximize the vibration amplitude at the drill tip (see Figure 7.2). Substitute sound transmission velocities in the collet and the drill bit materials into eq. (7.3), one can get pilot values of L_3 and L_4.

The dimensions D_i and L_i above are obtained with assumptions that the horn and the collet with the tool attached are two cylindrical double-step solids, without hole, thread, and flange structures. Such differences between supposed and actual structures cause changes in the resonant frequency of the whole system. For power ultrasonic applications, the resonant frequency of the whole system must match the working frequency of the selected transducer. Therefore, the designed unit must be checked and refined. The finite element analysis technique is usually employed at this stage. A 3D solid and assembled model of the horn, the screws, the collet, and the tool is placed in a computer-assisted engineering (CAE) environment, such as ANSYS, Abaqus. Applying modal analysis, resonant frequency of the system can be easily carried out. It is advised that shortening the total length can increase the resonant frequency. In contrast, reducing the segment length leads to decrease in the resonant frequency. Several loops may be required until the resonant frequency of the system is the same as or very near to the working frequency of the transducer. Figure 7.4 shows an example for the horn proposed to work with a 25 kHz commercial transducer.

Once the refining stage is done, the final structure will be fabricated and assembled. Because the fabricating and assembly processes may have several errors, leading to change the resonant frequency of the system, another checking and refining stage must be implemented. One of the simplest and easiest methods is testing the electrical impedance of the whole unit. If available, a commercial impedance analyzer should be used. The impedance can also be measured by the V–I method [52]. The principle of the measurement circuit is shown in Figure 7.5. The transducer is excited by a swept frequency voltage of several volts. At each value of the excitation frequency, the corresponding output voltages V_{A1}, V_{A2} and the phase differences

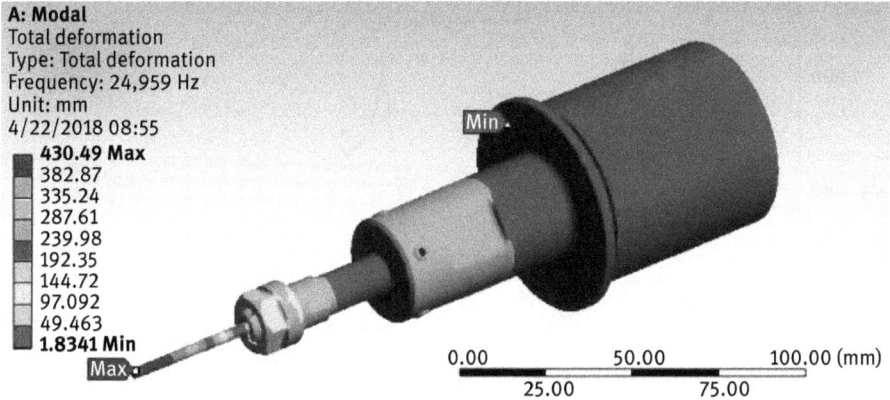

Figure 7.4: A modal analysis of the horn assembled with the tool bit [45].

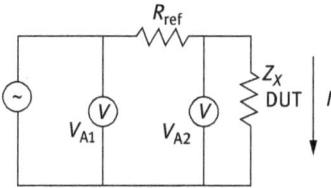

Figure 7.5: A simple circuit to measure ultrasonic impedance using V–I method [45].

between V_{A1} and V_{A2} are collected. The impedance of the ultrasonic unit is then approximately calculated as

$$Z_x = \frac{V_{A_2} R_{\text{ref}}}{\sqrt{V_{A_1}^2 - 2V_{A_1} V_{A_2} \cos\theta + V_{A_2}^2}} \tag{7.8}$$

A plot of the impedance Z_x as a function of the excitation frequencies is then printed out. Figure 7.6 depicts an example of the impedance plot versus excitation frequency of the aforementioned system.

The resonant frequency of the system can be found where the impedance line reaches its local peaks. Further tuning tasks such as changing the dimensions of the parts or the cantilever lengths of the assembly may be needed at this stage to finally obtain the proper resonant frequency of the system. A final setup of the study on UAD is shown in Figure 7.7.

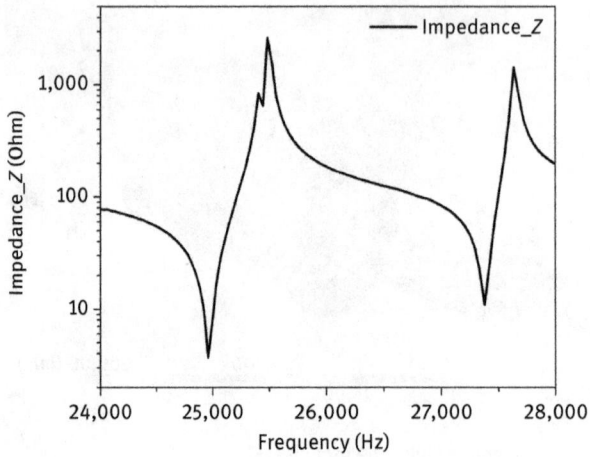

Figure 7.6: The measured impedance of the whole assembly [45].

Figure 7.7: A photograph of the experimental setup [42, 43, 53]: 1, the ultrasonic unit; 2, lathe chuck; 3, tool bit; 4, thermal sensor; 5, sample; 6, the jig; 7, bearing; 8, force sensor; 9, torque sensor.

7.3 Torque in deep hole drilling

Typically, the drilling torque, T, is basically provided in an exponential form as follows [54]:

$$T = C.f^{\alpha}.d^{\beta} \tag{7.9}$$

where C, α, β are constants for particular workpiece material and cutting conditions. As can be seen, the torque is a function of only the drilling diameter, d, and the feeding rate, f. The effect of the drilled depth on the torque is omitted. Such effects are usually considered to be small and are thus neglected in most engineering textbooks and handbooks (see, e.g., in [55–57]). The above opinion would be acceptable for typical drilling, where the depth-to-diameter ratio is smaller than 5 [9].

In deep drilling, the drilling torque reflects the addition of the chip-evacuation performance. The exponential increase of the drilling torque at large aspect ratios can be found in numerous studies (see, e.g., in [4, 7–11, 42]). In those studies, the total drilling torque was divided into two parts: the cutting torque and the chip-evacuation torque (see Figure 7.8).

The cutting torque was considered constant and independent of the drilled depth. The reason for the increase of the chip-evacuation torque at higher drilling depths is assumed to be the growing force required to move the cumulative amount of chips. Several studies also mentioned the evacuation torque component in deep drilling, in addition to the cutting torque (see [58, 14]).

Mathematical models to describe and predict the chip evacuation torque have been developed in numerous studies. In the work of Mellinger et al. [7], the chip-evacuation torque was expressed as a function of the drilled depth with several parameters as follows:

$$M = \frac{R\mu_w S_w F_c(0)}{B}(\exp((kBD/A0)z) - 1) \tag{7.10}$$

where R and D are the radius and diameter of the drill tool, respectively; μ_w is the friction coefficient between the chips and hole wall; S_w is the length of the contact arc between the chip element and the hole wall; $F_c(0)$ represents the initial chip-evacuation force at the depth, $z = 0$; B is the factor of workpiece material, the friction between the chips and the flutes; and A_0 is the cross section of the chip [7].

In order to simplify the above model, a study of Han et al. [14] proposed a model where the chip evacuation torque, $T_{ch}(z)$ is expressed in the following form:

$$T_{ch}(z) = K_{tch}(\exp(K_{ch} \cdot z) - 1)\frac{D}{2} \tag{7.11}$$

where K_{tch} and K_{ch} are coefficients, regressed from experimental data, D and z are the tool diameter and the drilled depth, respectively [14].

In the mentioned studies, the instability of the torque signals were smoothed by applying low-pass filter or the average techniques. Consequently, the significant changes of the chip evacuation torque would be ignored. The study of Chu et al. [53] revealed that the fluctuating component of the torque at deeper hole is practically significant. The author then considered the drilling torque as a combination of three components (see Figure 7.9): the cutting torque T_1, the chip evacuation torque T_2, and the stick-slip torque T_3. The cutting torque is the steady value measured

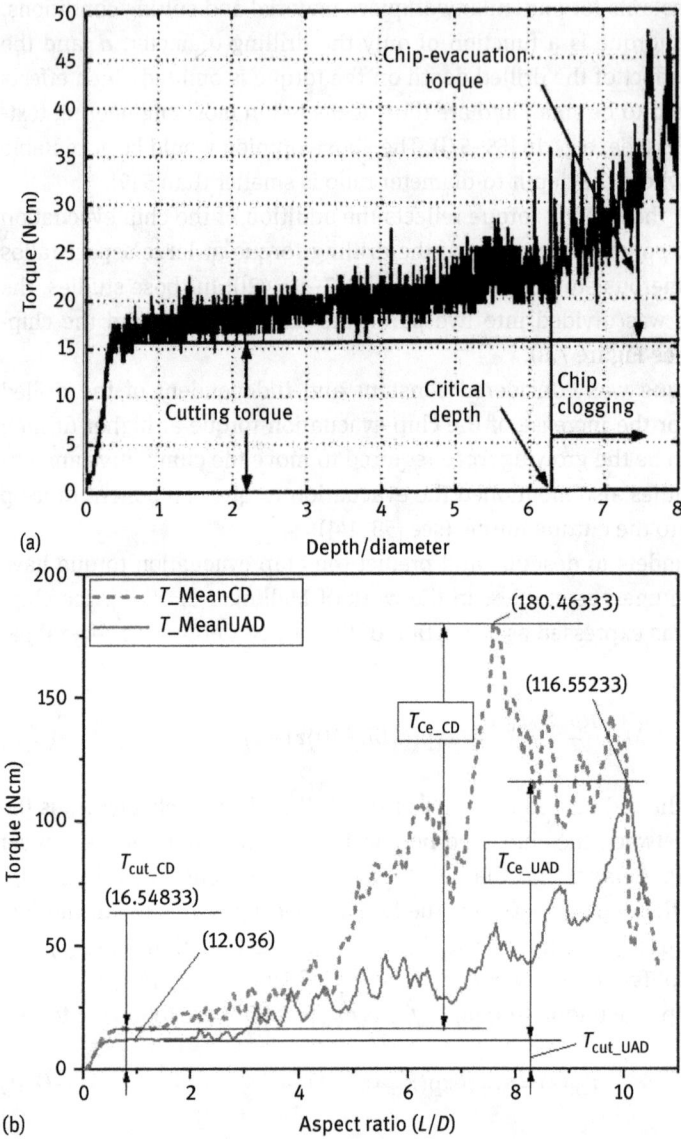

Figure 7.8: Experimental cutting torque and chip evacuation torque from (a) Mellinger's et al. [7] and (b) Nguyen et al. [42].

once the tool lips fully engaged into the workpiece. The chip-evacuation torque is the continuously increasing component of the total signnal (see Figure 7.9a). The stick-slip torque is obtained by subtracting the total torque by the cutting and chip evacuation torques, reflecting the fluctuation part of the torque. Figure 7.9 depicts those three components of a typical drilling torque.

(a)

(b)

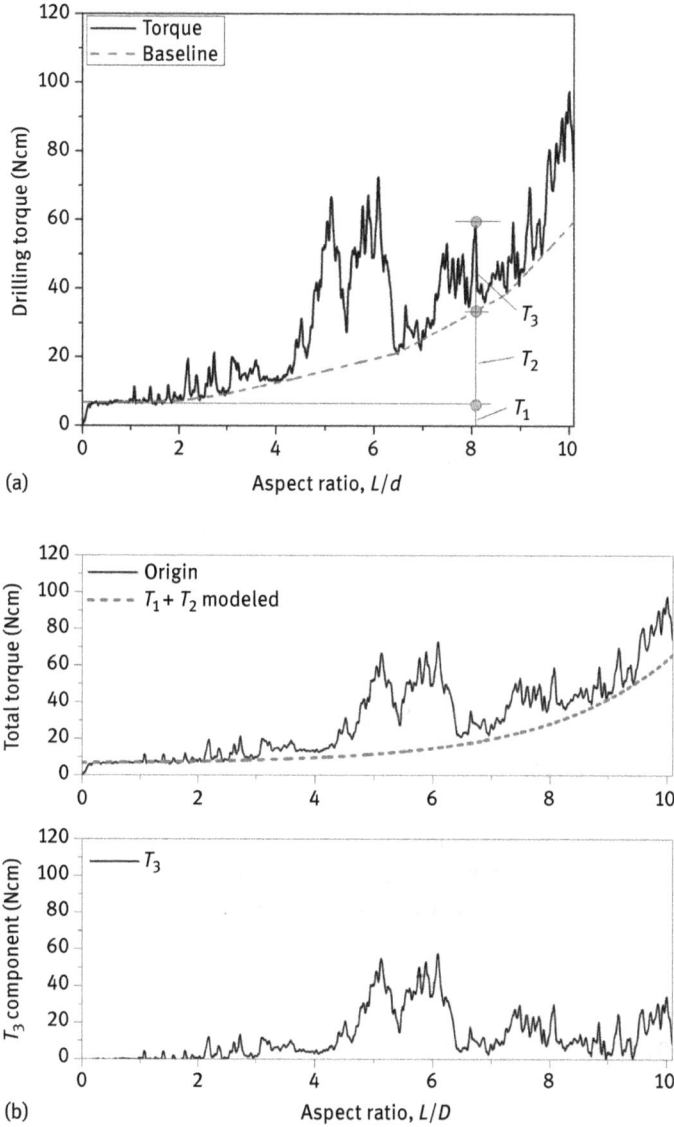

Figure 7.9: The three parts of the total torque and a base line of original torque signal (a), the modeled cutting and chip evacuation torque and the rubbing torque (b) [53].

The cutting torque, T_1, is calculated using numerous formulas of the total drilling torque provided in engineering textbooks for normal drilling with the aspect ratio less than 5. For example, the cutting torque can be expressed as [54, 59, 60]

$$T_1 = C.f^{(1-\alpha))}D^{(2-\alpha)} \tag{7.12}$$

where the constants C and the material factor α can be given in handbooks or practically determined from calibration experiments.

The chip evacuation torque T_2 is expressed as a function of the drill depth [53]:

$$T_2 = A(e^{(A \cdot z)} - 1) \tag{7.13}$$

where A is the coefficient determined by calibration tests, and z is the drilled depth. The coefficient A can be determined using the spindle speed of the drill bit n and the feeding rate f as follows:

$$A = a_0 + a_1 \ln(n) + a_2 \ln(f) + a_3 \ln(n) \ln(f) \tag{7.14}$$

where the constants a_i ($i = 0 \div 3$) are determined from experimental data by the regression technique.

The stick-slip torque component reflects the stick-slip phenomenon occurred in the relative rotation between the tool and workpiece (see details in [53]). Because the component is stochastic, a statistical analysis was applied to carry out the probability of its peak values and the standard deviation. By dividing the total torque in deep drilling into three parts, several enhancements of UAD over CD can be clearly emphasized, as shown in the next section.

7.4 Torque reduction in ultrasonic-assisted deep-hole drilling

As mentioned earlier, the most critical challenge in deep drilling of ductile material is the chip evacuation, the major factor leading to the increase in drilling torque at deeper depths. UAD would provide smaller cutting torque and, furthermore, a considerable reduced chip evacuation torque, compared to CD.

7.4.1 Reduction of the cutting torque

Looking at the expression of the cutting torque in eq. (7.12), it can be seen that with the same workpiece material (i.e., the same α) and the same tool diameter (D), the smaller the factor C, the smaller the cutting torque. Experimental data can help calibrate the values of C in UADs and in CDs. The following example gives an illustration of such approach.

Two sets of the calibration tests were implemented separately, one for CD and another for UAD. The tests were implemented by drilling on squared Al6061-T6 samples with dimensions of 10×10×30 mm³, using HSS twist drill bits with diameters of 3 and 4 mm, and under dry cutting conditions. In each set, nine experimental tests were planned as depicted in Table 7.1. All experiments were replicated

Table 7.1: Cutting parameters and their levels in experiments [53].

Parameters / Level	Speed (rpm)	Feed rate (mm/rev)
−1	1,000	0.050
0	1,250	0.065
1	1,500	0.085

twice to avoid random effects. The experimental parameters are presented in Table 7.1.

Table 7.2 shows statistical results obtained from the calibration experiments.

Table 7.2: Statistical results to calibrate factor C (source: [53]).

Drill diameter	Cutting conditions	Value of C	R^2
4 mm	CD	0.154	0.99869
	UAD	0.121	0.99863
3 mm	CD	0.150	0.97633
	UAD	0.121	0.9032

As shown in Table 7.2, the values of C in CD tests are around 0.15, and those in UAD are about 0.121. It means that the cutting torque in UAD was lower at 0.15/0.12 = 1.25 times than that in CD. It can be said that ultrasonic assistance can reduce the cutting torque at an average of 25% in the drilling tests.

7.4.2 Reduction of the chip evacuation torque

The model described in eq. (7.13) provided a convenient way to compare the chip evacuation torques between UAD and CD processes. It can be seen that reduction of the parameter A results in decreasing the chip evacuation torque.

Using the data obtained in the experimental tests described in Table 7.1, a regression process can be applied to calibrate the value of the coefficient A. The results are listed in Table 7.3.

It is shown in Table 7.3 that the data were fitted well with the torque model, with an average regression correlation value R^2 mostly higher than 0.9.

The values of the coefficient A obtained from two cutting conditions, CD and UAD, will be compared in pair. The paired t-test technique helps to predict the difference

Table 7.3: Regression results of the parameter A (source: [53]).

$n(v/p)$	f(mm/v)	CD, Φ3		UAD, Φ3		CD, Φ4		UAD, Φ4	
		A	R^2	A	R^2	A	R^2	A	R^2
1,000	0.05	0.423	0.964	0.421	0.956	0.702	0.989	0.519	0.968
1,000	0.05	0.495	0.952	0.430	0.812	0.679	0.982	0.526	0.813
1,000	0.065	0.470	0.993	0.340	0.982	0.684	0.943	0.507	0.970
1,000	0.065	0.494	0.974	0.318	0.941	0.670	0.913	0.632	0.991
1,000	0.085	0.510	0.920	0.477	0.971	0.623	0.902	0.536	0.983
1,000	0.085	0.515	0.928	0.468	0.837	0.673	0.915	0.590	0.978
1,250	0.05	0.603	0.976	0.479	0.956	0.608	0.975	0.392	0.873
1,250	0.05	0.589	0.983	0.500	0.952	0.633	0.975	0.546	0.979
1,250	0.065	0.568	0.923	0.435	0.912	0.699	1.000	0.453	0.995
1,250	0.065	0.599	0.863	0.432	0.945	0.594	0.988	0.609	0.998
1,250	0.085	0.587	0.976	0.360	0.981	0.658	0.942	0.505	0.992
1,250	0.085	0.584	0.986	0.346	0.857	0.634	0.983	0.365	0.993
1,500	0.05	0.509	0.953	0.441	0.993	0.621	0.944	0.527	0.955
1,500	0.05	0.521	0.942	0.511	0.904	0.659	1.000	0.533	0.850
1,500	0.065	0.541	0.966	0.433	0.985	0.704	0.988	0.584	0.948
1,500	0.065	0.561	0.968	0.485	0.994	0.654	0.902	0.630	0.999
1,500	0.085	0.503	0.991	0.437	0.927	0.751	0.999	0.516	0.900
1,500	0.085	0.530	0.954	0.495	0.988	0.760	0.999	0.575	0.853

between two populations of data based on limited experimental data. The results of the statistical tests are shown in Table 7.4.

Table 7.4: Paired t-test for difference of the chip-evacuation coefficient A between CDs and UADs [53].

Drill diameter (mm)	Mean of μ_difference	St. dev	SE mean	95% lower boundfor μ_difference	t-Value	p-Value
3	0.1040	0.0728	0.0140	0.0801	2.43	0.011
4	0.1190	0.0580	0.0137	0.0952	2.12	0.024

μ_difference: mean of (A_CD – A_UAD).

As mentioned in Table 7.4, with a confidence of 95%, the coefficient A in UAD population data is smaller than that in CD of at least 0.104 for the drill of 3 mm, and at least 1.119 for the drill of 4 mm diameter.

To double confirm the hypothesis, it is declared that the coefficient A of UAD is smaller than that of CD, a two-sample t-test can be made. Figure 7.10 shows the plots of such tests [53].

Figure 7.11 illustrates the chip evacuation torque calculated by eq. (7.13) using the mean values of the parameter A.

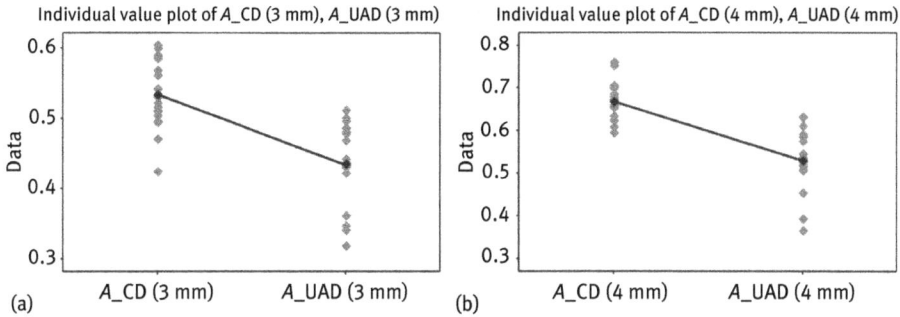

Figure 7.10: Individual plots from two-sample t-test for difference of coefficients A between CD and UAD tests with drill diameters of 3 mm (a) and mm (b) [53].

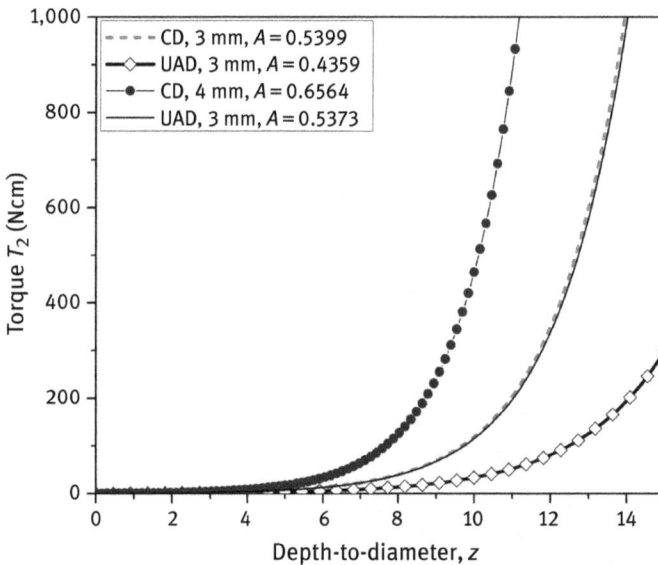

Figure 7.11: Plots of the chip evacuation torque calculated from the mean values of A.

As shown in the figure, a small difference of the coefficient A resulted in a significant difference of the torque.

7.4.3 Reduction of the stick-slip torque

The peak values of the component T_3 obtained from all investigated tests were statistically inferred. The probability distributions of the population data for T_3 are depicted in Figure 7.12 (see details in [53]).

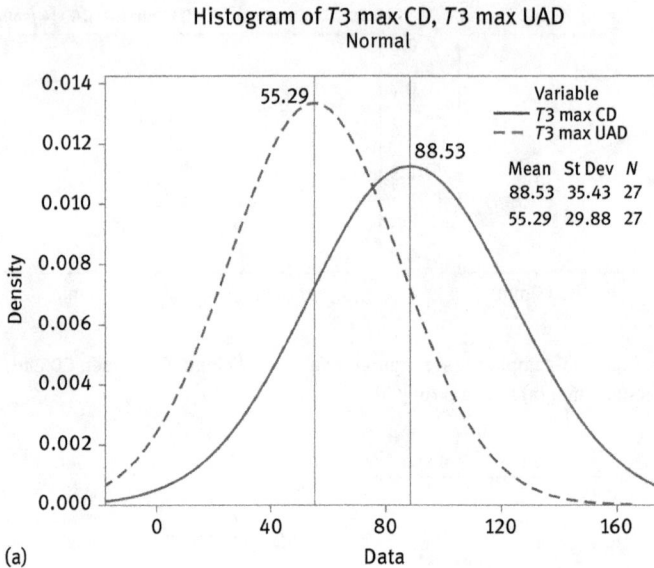

(a)

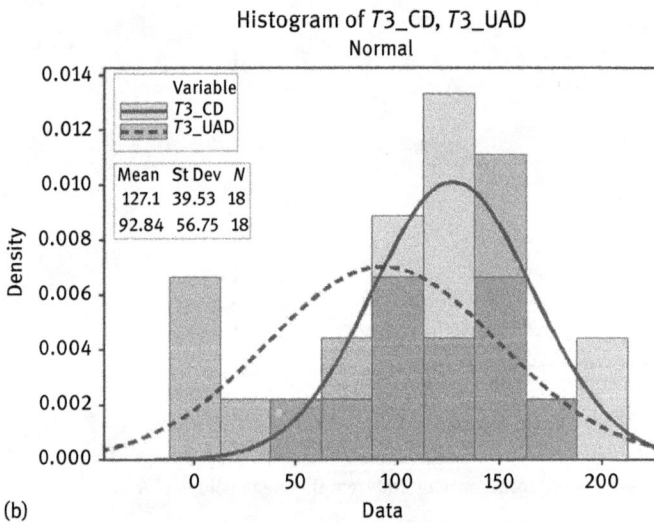

(b)

Figure 7.12: Probability of the maximum values of stick-slip torques for the tool diameter of (a) 3 mm and (b) 4 mm [53].

As shown in the figure, the mean value of stick-slip torque in UADs was significantly smaller than that in CDs. Besides, the distribution of T_3 is either narrower (in case of 3 mm diameter in Figure 7.12(a)), or tended to the lower side of the range (in case of 4 mm diameter in Figure 7.12(b)). Another worth remark is that the smallest stick-slip torque (higher than 55 Ncm as shown in Figure 7.12(b)) is much higher than the cutting torque (less than 20 Ncm, see Figure 7.8(b)). Consequently, the stick-slip torque would significantly affect on the total torque and thus should not be neglected.

7.5 Conclusion

In this chapter, a new model of the total torque was used to emphasize the effectiveness of ultrasonic-assisted deep drilling. The proposed model with three components provide the following benefits:

- The reduction of cutting torque of UAD can be easily pointed out, as high as 25% compared to that of CD.
- The effect of ultrasonic vibration in assisting the chips that evacuate easier can be simply carried out by comparing one coefficient of the torque model.
- The fluctuated component of torque in deep drilling is significant and should not be neglected, as usually treated in several previous studies. Further investigation should be made to modelize this component of the torque.

The model described in this chapter would be useful to predict the total torque required in both CD and UAD processes.

References

[1]   Hutchings, I. and Shipway, P., 9 – Applications and case studies. in Tribology: Friction and Wear of Engineering Materials (Second Edition), I. Hutchings and P. Shipway, Eds., ed: Butterworth-Heinemann, Elsevier, Netherlands, 2017, 303–352.
[2]   Liu, H. S., Lee, B. Y., and Tarng, Y. S., In-process prediction of corner wear in drilling operations. Journal of Materials Processing Technology, 2000, 101, 152–158, 2000/04/14/.
[3]   Rivero, A., Aramendi, G., Herranz, S., and López de Lacalle, L. N., An experimental investigation of the effect of coatings and cutting parameters on the dry drilling performance of aluminium alloys. The International Journal of Advanced Manufacturing Technology, 2005, 28, 1–11.
[4]   Ke, F., Ni, J., and Stephenson, D. A., Chip thickening in deep-hole drilling, International Journal of Machine Tools and Manufacture, 2006, 46, 1500–1507.
[5]   Drozda, T. J., Tool and Manufacturing Engineers Handbook: Machining, Society of Manufacturing Engineers, USA, 1983.
[6]   Roy, P., Sarangi, S. K., Ghosh, A., and Chattopadhyay, A. K., Machinability study of pure aluminium and Al–12% Si alloys against uncoated and coated carbide inserts. International Journal of Refractory Metals and Hard Materials, 2009, 27, 535–544.
[7]   Mellinger, J. C., Burak Ozdoganlar, O., DeVor, R. E., and Kapoor, S. G., Modeling Chip-Evacuation Forces and Prediction of Chip-Clogging in Drilling. Journal of Manufacturing Science and Engineering, 2002, 124, 605.

[8]   Mellinger, J. C., Ozdoganlar, O. B., DeVor, R. E., and Kapoor, S. G., Modeling Chip-Evacuation Forces in Drilling for Various Flute Geometries. Journal of Manufacturing Science and Engineering, 2003, 125, 405.

[9]   Nagao, T., Hatamura, Y., and Mitsuishi, M., In-Process Prediction and Prevention of the Breakage of Small Diameter Drills Based on Theoretical Analysis. CIRP Annals, 43, 1994, 85–88, 1994/01/01/.

[10]   Furness, R. J., Ulsoy, A. G., and Wu, C. L., Feed, Speed, and Torque Controllers for Drilling, Journal of Engineering for Industry, 1996, 118, 2.

[11]   Arzur-Bomont, A., Confente, M., Schneider, E., Bomont, O., and Lescalier, C., Machinability in drilling mechanistic approach and new observer development. International Journal of Material Forming, 2010, 3, 495–498.

[12]   Ravisubramanian, S. and Shunmugam, M. S., Investigations into peck drilling process for large aspect ratio microholes in aluminum 6061-T6. Materials and Manufacturing Processes, 2017, 33, 935–942.

[13]   Choi, Y. J., Park, M. S., and Chu, C. N., Prediction of drill failure using features extraction in time and frequency domains of feed motor current. International Journal of Machine Tools and Manufacture, 2008, 48, 29–39, 2008/01/01/.

[14]   Han, C., Zhang, D., Luo, M., and Wu, B., Chip evacuation force modelling for deep hole drilling with twist drills. The International Journal of Advanced Manufacturing Technology, 2018, 98, 3091–3103.

[15]   Aized, T. and Amjad, M., Quality improvement of deep-hole drilling process of AISI D2. The International Journal of Advanced Manufacturing Technology, 2013, 69, 2493–2503.

[16]   Sahu, S. K. B. O., DeVor, O., Kapoor, Richard E., Shiv G., "Effect of groove-type chip breakers on twist drill performance. International Journal of Machine Tools and Manufacture, 2003, 43, 617–627.

[17]   Kishawy, H. A., Dumitrescu, M., Ng, M., and Elbestawi, M. A., Effect of coolant strategy on tool performance, chip morphology and surface quality during high-speed machining of A356 aluminum alloy. International Journal of Machine Tools and Manufacture, 2005, 45, 219–227.

[18]   Khan, S. A., Nazir, A., Mughal, M. P., Saleem, M. Q., Hussain, A., and Ghulam, Z., Deep hole drilling of AISI 1045 via high-speed steel twist drills: evaluation of tool wear and hole quality. The International Journal of Advanced Manufacturing Technology, 2017, 93, 1115–1125.

[19]   Zedan, Y., Niknam, S. A., Djebara, A., and Songmene, V., Burr Size Minimization When Drilling 6061-T6 Aluminum Alloy. 2012, 1053–1059.

[20]   Gillespie, L. R. K., Deburring and Edge Finishing Handbook: Society of Manufacturing Engineers, USA, 1999.

[21]   Pilný, L., De Chiffre, L., Píška, M., and Villumsen, M. F., Hole quality and burr reduction in drilling aluminium sheets. CIRP Journal of Manufacturing Science and Technology, 2012, 5, 102–107.

[22]   Kundu, S., Das, S., and Saha, P. P., Optimization of Drilling Parameters to Minimize Burr by Providing Back-up Support on Aluminium Alloy. Procedia Engineering, 2014, 97, 230–240.

[23]   Heisel, U. and Schaal, M., Burr formation in short hole drilling with minimum quantity lubrication. Production Engineering, 2009, 3, 157–163.

[24]   Thakre, A. A. and Soni, S., Modeling of burr size in drilling of aluminum silicon carbide composites using response surface methodology, Engineering Science and Technology, an International Journal, 2016, 19, 1199–1205.

[25]   Brehl, D. E. and Dow, T. A., Review of vibration-assisted machining. Precision Engineering, 2008, 32, 153–172.

[26]   S. S. F. a. Chang, B., Gary M., Thrust force model for vibration-assisted drilling of aluminum 6061-T6. International Journal of Machine Tools and Manufacture, 2009, 49, 1070–1076.

[27] Amini, S., Paktinat, H., Barani, A., and Tehran, A. F., Vibration Drilling of Al2024-T6. Materials and Manufacturing Processes, 2013, 28, 476–480.

[28] Barani, A., Amini, S., Paktinat, H., and Fadaei Tehrani, A., Built-up edge investigation in vibration drilling of Al2024-T6. Ultrasonics, 2014, 54, 1300–10, Jul.

[29] Chang, S. S. F. and Bone, G. M., Burr height model for vibration assisted drilling of aluminum 6061-T6. Precision Engineering, 2010, 34, 369–375.

[30] Li, X. F., Dong, Z. G., Kang, R. K., Wang, Y. D., Liu, J. T., and Zhang, Y., Comparison of thrust force in ultrasonic assisted drilling and conventional drilling of aluminum alloy. Materials Science Forum, 2016, 861, 38–43.

[31] Babitsky, V. I. A., Meadows, V. K., A., Vibration excitation and energy transfer during ultrasonically assisted drilling. Journal of Sound and Vibration, 2007, 308, 805–814.

[32] Chang, S. S. F. B., Gary, M., Burr size reduction in drilling by ultrasonic assistance. Robotics and Computer-Integrated Manufacturing, 2005, 21, 442–450.

[33] Chang, S. S. F. and Bone, G. M., Thrust force model for vibration-assisted drilling of aluminum 6061-T6. International Journal of Machine Tools and Manufacture, 2009, 49, 1070–1076, 2009/11/01/.

[34] Azarhoushang, B. and Akbari, J., Ultrasonic-assisted drilling of Inconel 738-LC. International Journal of Machine Tools and Manufacture, 2007, 47, 1027–1033.

[35] Azghandi, B. V., Kadivar, M. A., and Razfar, M. R., An Experimental Study on Cutting Forces in Ultrasonic Assisted Drilling. Procedia CIRP, 2016, 46, 563–566.

[36] Baghlani, V. M., Akbari, P., Nezhad, J., Sarhan, Erfan Zal, Ahmed A. D. Hamouda, A. M. S., An optimization technique on ultrasonic and cutting parameters for drilling and deep drilling of nickel-based high-strength Inconel 738LC superalloy with deeper and higher hole quality. The International Journal of Advanced Manufacturing Technology, 2015, 82, 877–888.

[37] Dahnel, A. N. A., Helen Barnes, Stuart, The Effect of Varying Cutting Speeds on Tool Wear During Conventional and Ultrasonic Assisted Drilling (UAD) of Carbon Fibre Composite (CFC) and Titanium Alloy Stacks. Procedia CIRP, 2016, 46, 420–423.

[38] A. Gupta, H. Ascroft, and S. Barnes, "Effect of Chisel Edge in Ultrasonic Assisted Drilling of Carbon Fibre Reinforced Plastics (CFRP)," Procedia CIRP, 46, 619–622, 2016.

[39] Phadnis, V. A., Makhdum, F., Roy, A., and Silberschmidt, V. V., Drilling in carbon/epoxy composites: Experimental investigations and finite element implementation. Composites Part A: Applied Science and Manufacturing, 2013, 47, . 41–51.

[40] Sanda, A. A., Garcia Navas, Iban, Bengoetxea, Virginia, Gonzalo, Ion Oscar, Ultrasonically assisted drilling of carbon fibre reinforced plastics and Ti6Al4V. Journal of Manufacturing Processes, 2016, 22, 169–176.

[41] Wang, Y., Cao, M., Zhao, X., Zhu, G., McClean, C., Zhao, Y., et al., Experimental investigations and finite element simulation of cutting heat in vibrational and conventional drilling of cortical bone. Medical Engineering & Physics, 2014, 36, 1408–15, Nov.

[42] Nguyen, V.-D. and Chu, N.-H., A Study on the Reduction of Chip Evacuation Torque in Ultrasonic Assisted Drilling Of Small and Deep Holes. International Journal of Mechanical Engineering and Technology, 2018, 9, 899–908.

[43] Chu, N.-H. and Nguyen, V.-D., The Multi-Response Optimization of Machining Parameters in the Ultrasonic Assisted Deep-Hole Drilling Using Grey-Based Taguchi Method. International Journal of Mechanical and Production Engineering Research and Development, 2018, 8, 417–426.

[44] Chu, N.-H., Nguyen, V.-D., and Do, T.-V., Ultrasonic-Assisted Cutting: A Beneficial Application for Temperature, Torque Reduction, and Cutting Ability Improvement in Deep Drilling of Al-6061. Applied Sciences, 2018, 8(10), 1708.

[45] Chu, N.-H., Ngo, Q.-H., and Nguyen, V.-D., A Step-by-Step Design of Vibratory Apparatus for Ultrasonic-Assisted Drilling. International Journal of Advanced Engineering Research and Applications, 2018, 4, 139–148.

[46] Yan, Z., Fang, Q., Huang, J., He, B., and Lin, Z., Considerations and guides of the wattmeter method for measuring output acoustical power of Langevin-type transducer systems – II: experiment. Ultrasonics, 1997, 35, 543–546, 1997/11/01/.

[47] Short, M. and Graff, K. F., 16 – Using power ultrasonics in machine tools in Power Ultrasonics, J. A. Gallego-Juárez and K. F. Graff, Eds., ed Oxford: Woodhead Publishing, 2015, 439–507.

[48] Li, X., Meadows, A., Babitsky, V., and Parkin, R., Experimental analysis on autoresonant control of ultrasonically assisted drilling. Mechatronics, 29, 57–66, 2015.

[49] Babitsky, V. K. A. V. I., Ultrasonic Processes And Machines: Dynamics, Control and Applications, Berlin Heidelberg: Springer-Verlag, 2007.

[50] Dale Ensminger, F. B. S., Ultrasonics Data, Equations and Their Practical Uses, Taylor & Francis Group, LLC, 2009.

[51] Allaparthi, M., Khan, M. R., and Addepalli, S. N., FE Modal and Harmonic Analysis of Micro Drill with Ultrasonic Horn, in Materials Design and Applications, L. F. M. d. Silva, Ed., ed Cham: Springer International Publishing, 2017, 281–293.

[52] Pan, Q., Xiao, D., Deng, M., Ren, H., Xu, Y., and Xu, C., A voltage-current method of measuring ultrasonic transducer impedance, in IEEE conference "2013 Far East Forum on Nondestructive Evaluation/Testing: New Technology and Application", Jinan, China, 2013, 125–128.

[53] Chu, N.-H., Nguyen, D.-B., Ngo, N.-K., Nguyen, V.-D., Tran, M.-D., Vu, N.-P., et al., A New Approach to Modelling the Drilling Torque in Conventional and Ultrasonic Assisted Deep-Hole Drilling Processes. Applied Sciences, 2018, 8(12), 2600.

[54] Juneja, B. L. and Seth, N., Fundamentals of Metal Cutting and Machine Tools: New Age International Publishers, New Delhi, India, 2003.

[55] Carvill, J., 5 – Manufacturing technology, in Mechanical Engineer's Data Handbook, J. Carvill, Ed., ed: Butterworth-Heinemann, Elsevier, Netherlands, 1993, 172–217.

[56] Smith, G. T., Cutting Tool Technology Industrial Handbook, London: Springer-Verlag, 2008.

[57] Shaw, M. C., Metal Cutting Principles, Published by 198 Madison Avenue, New York, New York 10016: Oxford University Press, Inc., 2005.

[58] MacAvelia, T., Ghasempoor, A., and Janabi-Sharifi, F., Force and torque modelling of drilling simulation for orthopaedic surgery, Computer Methods in Biomechanics and Biomedical Engineering, 2014, 17, 1285–1294, 2014/09/10.

[59] Dandekar, C., Orady, E., and Mallick, P. K., Drilling Characteristics of an E-Glass Fabric-Reinforced Polypropylene Composite and an Aluminum Alloy: A Comparative Study. Journal of Manufacturing Science and Engineering, 2007, 129, 1080–1087.

[60] Oxford, C. J., On the Drilling of Metals 1-Basic Mechanics of the Process, ASME Transaction, USA, 1955, 77, 103–114.

Karmjit Singh, Ibrahim A. Sultan

8 Information and computational modeling for sustainability evaluation and improvement of manufacturing processes

Abstract: Over the last few decades, manufacturing industries are working toward sustainable manufacturing due to the high cost of raw material and depletion of natural resources. There is also a great concern for manufacturing companies to produce products by considering three aspects of sustainability, that is, economic, environmental, and social. This pressure has motivated modern manufacturing industries to, commit to, adopt the ethos and methods of sustainable manufacturing. An outcome is manifested by the current undertakings of many research projects that are designed to propose automated systems for determination of CO_2 emissions, solid waste, and energy consumption of various manufacturing plants.

An effort for developing a generic model for different manufacturing processes, graphic user interface system, is presented in this chapter. The chapter presents detailed discussions of the proposed models for two different manufacturing processes. Sustainability analysis case studies, based on the models developed, are presented for die-casting, injection molding, and turning processes. It will be shown that the developed models are effective for sustainability assessment process and have the potential of helping the manufacturing companies. Moreover, the work presented is highly beneficial in determining the sustainability assessment of a manufacturing process at the beginning of product design. An overarching motive of this chapter is to reduce emissions from energy production into atmosphere and alleviate impact on the environment.

Keywords: Manufacturing processes, Sustainability evaluation, Energy consumption, environment impact

8.1 Introduction

8.1.1 Manufacturing processes

Manufacturing is defined as the production of different products by using different types of materials, machines, tools, chemicals, and manpower. In a manufacturing process, various activities are taking place for converting the raw material to finished product. For manufacturing a product, the initial stage is product design and drafting followed by material classification and type of machinery that is used to manufacture the product [1].

https://doi.org/10.1515/9783110549775-008

Any manufacturing industry is a combination of various unit processes that include both inputs and output information related to the product. Input module includes information related to machine, material, and various types of energy that is required to operate the machinery. While the output module contains the finished product along with the several types of wastages. In the unit process, the output of one process is converted into the input of the next process. The output characteristics of the final product include the outcome of each unit process that features a sequence of production machines [2].

8.1.2 Background of manufacturing industries

It is true that manufacturing industries are affecting the environment in various ways. This is a fact that motivated industry to pay more attention to sustainable manufacturing approaches [3]. The chief objective of the work presented here is to help the manufacturing companies to make the product more sustainable. To propose more sustainable manufacturing frameworks, industry needs more detailed knowledge of three pillars of sustainable development manufacturing (i.e., environmental, economic, and social). The work presented in this chapter has been motivated by the need to propose a computational model for sustainable manufacturing evaluation and improvement to increase the product's overall sustainability and reduce environmental, social, and economic challenges.

8.1.3 Need and motivation of present work

Most methods currently employed for the determination of sustainability analysis for a product have the following limitations:
- complex mathematical relations solved manually,
- data are not readily available, and
- the whole process is iterative and time consuming [4].

Hence, there is a need for a computer-aided system that automatically assesses the sustainability of different process plans at early design stage to guide the manufacturing decision-making process. Such a system would help in choosing efficient processes to minimize waste, decrease the cost of manufacture and labor, and reduce carbon emissions [5].

Sustainable manufacturing assessment models would enable the prediction of energy consumption and carbon emission, material usage and wastage, cost and production time for each stage of the process. Industry can then utilize these predictions to plan and implement sustainable measures and approaches.

8.1.4 Review of literature and objective of present work

The study for methods and models to evaluate sustainability for manufacturing processes are conducted with the purpose of identifying the formulations for different sustainability indicators to choose the most important indicators for manufacturing sustainability. Bhanot et al. [6] propose a framework to analyze different machining processes based on environment and economic indicators. The authors consider wet and dry turning conditions for this study. This work also contributes toward optimization of environment and economic parameters by using different optimization techniques. Bourhis et al. [7] propose a new methodology to calculate electricity consumption, solid wastage, and usage of several types of coolants for addictive manufacturing processes for direct metal removal processes. This methodology is based on both analytical and practical models and provides optimization for additive manufacturing with respect to sustainable assessment. Chen et al. [8] propose a sustainability assessment tool for different manufacturing industries in relation to the key performance indicators (KPIs) and identify importance of each indicator. The tool provides decision for improvements and makes a comparison among sample companies. Ciceri et al. [9] propose a methodology to calculate product material and energy as per the given BOM (bill of materials). The method helps determine the energy consumed by manipulating present data from embodied figures, empirical and BOM. Culaba and Purvis [10] develop a computer model for the sustainability evaluation of manufacturing processes considering different approach from the use of hybrid model. This model is used for evaluating the sensitivity analysis for manufacturing processes. Dalquist and Gutowski [11] propose a framework to evaluate environmental impact for a conventional die-casting process. An approach of energy and material usage in die-casting process is presented, and improvements are demonstrated. Dalquist and Gutowski [12] present a methodology for sand-casting processes, including all stages starting from mold preparation to the final casting product. The methodology calculated the amount of different types of wastage that occur in sand-casting process in the form of liquids, gases, and solid wastes. Huanga and Badurdeena [13] propose a tool to evaluate sustainably at the process level, which contains a set of metrics for five different hierarchies. Jeswiet and Kara [14] propose a methodology that relates electrical energy with the carbon emission for different manufacturing processes. Their model determines product-related carbon emissions from the electricity usage. Kim et al. [15] propose a decision-guidance structure to improve sustainability in manufacturing. Their framework contains six different stages. For implementation of this methodology, step turning was performed on a computer numerical control (CNC) machine. Moldavska and Welo [17] develop an assessment tool on the basis of practical application of system thinking to improve sustainability in manufacturing organizations. Ocampo and Clark [18] propose a methodology that is used for early design making to make the production process more sustainable. The proposed framework is beneficial for manufacturing industries and provide a better decision at the early design stage. Thirez and Gutowski [19] proposed an approach to identify the effect of machine type on energy consumption. A system-level environmental analysis is carried out for injection molding (IM), and this led to a conclusion that the emissions associated with that process are higher than those associated with other processes. Vimal and Vinodh [20] propose a methodology, which is combination of analytical network process with life cycle assessment for calculating the impact score and single point process sustainability. In this study, comparison of different manufacturing processes is considered on the basis of sustainability performance. This methodology is beneficial for decision-making in practical conditions. The major aim of this work is to motivate manufacturing organizations to use sustainable manufacturing practices in their processes and provide modeling approach for economic sustainability.

Researchers have also considered various modeling and simulation tools and frameworks to improve manufacturing sustainability. Feng and Joung [21] proposed a tool for sustainable machining. A science-based methodology has been proposed to calculate the various indicators of sustainable manufacturing. Seow and Rahimifard [22] proposed a framework that enables designers to select the most energy-efficient materials and processes. The models presented calculate the use of energy at both plant and process. Then the amount of energy contribution for each unit manufacturing process is assessed. After that a model was proposed for embodied product energy during manufacturing. Madan et al. [23] proposed a methodology for energy calculation for subprocesses of manufacturing. Computer-based method is used to measure the sustainability. For implementation of proposed methodology, authors have considered IM as an example. Lee et al. [24] introduced computer modeling for sustainability in manufacturing industries. They provided practical implication for automobile company to verify their model. Zhang et al. [25] developed a process-oriented information gathering with the product design information.

8.1.4.1 Objectives of the work featured in this chapter

1. To study sample manufacturing operations and their unit processes with the purpose of developing a sustainability assessment methodology for these operations
2. To propose a science-based approach for sustainability assessment of the manufacturing processes with respect to KPIs.
3. To develop a framework that helps assess sustainability for product manufacturing with proposed manufacturing processes and process plans. This framework will provide comparison among various processes and process parameters and enable the manufacturing engineers to select the most sustainable process plan

8.2 Determination of sustainability indicators for manufacturing processes

The methodology to evaluate key sustainability performance indicators, in this work, is grounded in energy consumption, wastage, and the impact of a given manufacturing process on the environment and the availability of resources.

8.2.1 Sustainability indicators for die-casting process

8.2.1.1 Electrical energy

Electric energy consumption and fossil fuel consumption are determined using the well-known laws of physics. Theoretical electrical energy consumption depends on the cycle time and rated power load characteristics. In fact, theoretical energy consumption, E, is determined by the following equation:

$$E = P \times T \tag{8.1}$$

where P is the rated power of a given process and T signifies the time used to achieve that process.

The units for electrical energy consumption are kilowatt-hours (kWh). However, actual energy use for manufacturing process can be more accurately determined from practical calculation in manufacturing companies. To calculate practical energy consumption, we have to require the current, voltage, and cycle time values. The following expression is used to calculate practical electrical energy:

$$E = \sqrt{3} \times V \times I \times T \times \cos \varphi \tag{8.2}$$

where V and I are, respectively, the measured voltage and current and φ is the power phase angle.

8.2.1.2 Quantity of fuel consumption

Die casting is associated with the process of melting metals in oil-fired furnaces, where heat energy is provided by the burning of fuel (e.g., furnace oil, diesel, and other elements). To calculate the value of fuel consumption primarily depends on the heat energy required to increase the temperature of a material from the ambient temperature to the required molding temperature. During the melting process phase transformation (from the solid state to the liquid state) takes place. The total heat requirement calculated using quantitative formulae is stated as follows:

$$Q_s = MC_s(T_s - T_a) \tag{8.3}$$

$$Q_f = M(C_s (T_1 - T_s) + H_f) \tag{8.4}$$

$$Q_{sh} = MC_s (T_{sh} - T_1) \tag{8.5}$$

$$Q_t = Q_s + Q_f + Q_{sh} \tag{8.6}$$

where M is the mass of metal in the furnace; C_s is the specific heat of the metal; T_a is the ambient temperature; T_s is the temperature at the liquid state; T_1 is the temperature at the molding state; Q_s is the heat to raise temperature from room temperature to the start of melting; Q_t is the total heat required for the melting process; Q_f is the heat to raise temperature from the solidus to liquid temperature; Q_{sh} is the heat to raise temperature from superheat casting to holding furnace temperature; H_f is the latent heat of fusion; and T_{sh} is the temperature at the saturation heat state [26].

By knowing such parameters as the mass of metal processed, temperature at each process and total heat required in melting can be calculated.

For calculating the quantity of fuel required, the following expression is used:

$$V_f = \frac{Q_t}{\eta \times H_f \times \rho_f} \tag{8.7}$$

where V_f is the fuel volume flow rate required; H_f is the heat value of fuel per unit mass; ρ_f is the fuel density; and η is the efficiency of the furnace.

For the actual measurement of fuel consumption, values are taken from meters installed on furnaces.

8.2.1.3 Air emissions

Emissions possess a particular significance due to their harmful impact on the environment and humans. Of note are gasses such as CO_2, CH_4, SO_2, and NO_x. Among these stands out CO_2 (or carbon dioxide) as a major contributor to global warming conditions. Carbon dioxide value mainly depends on the use of electricity and fuel consumption.

8.2.1.4 Carbon dioxide emissions due to electricity

Carbon dioxide emissions are calculated on the basis of electricity consumption. The formulation used to calculate electrical energy for machining is discussed in the below expression, where C_w denotes carbon emissions:

$$C_w = E \times f \tag{8.8}$$

where f represents the emission factor.

In the above equation, the units used to quantify the emission factor are either mass of CO_2/energy produced.

8.2.1.5 Carbon dioxide emissions due to fuel usage

Some amount of air emissions occur due to combustion of fuel. Carbon dioxide emissions for fuel usage are calculated by using the following expression:

$$C_w = V_f \times f \tag{8.9}$$

Note that the units used to quantify f is the above equation is mass of CO_2/amount of fuel burnt.

8.2.1.6 Solid waste

Material is wasted at each stage of the manufacturing process. The wastage of material depends upon the type of material, type of fuel used and nature of process. Material loss, M_w, can be quantified by using the following equation:

$$M_w = M_m \times \gamma \tag{8.10}$$

where M_m represents the mass of molten metal (kg) and γ is the percentage of mass waste.

The value of γ used in the computer models presented in this work has been taken from cast metal coalition (CMC) [26]. For example, for induction furnaces, CMC proposes that γ should fall between 0.75% and 1.25%. The solid wastage is calculated in kg.

8.2.2 Sustainability indicators for the machining process

In a machining process, the total energy, E_t, consists of the sum of idle, basic, and machining energies. The equation used for calculating total energy is, therefore, given as follows:

$$E_t = (P_b \times T_b) + (P_i \times T_i) + (P_m \times T_m) \tag{8.11}$$

whereP_b represents the basic power; T_b the basic time;P_i the ideal power;T_i the ideal time;P_m the machining power; and T_m the machining time.

It is worth noting here that the above parameters that determine the energy consumption for a specific machining process depends on such aspects as the workpiece material, the type of process, and the machine used to affect the cutting.

The chip mass, which is the material lost by the workpiece during the machining process, can be calculated from the material density, ρ, workpiece length, l, and both the depth and width of cut as follows:

$$V_m = l \times w_c \times d_c \tag{8.12}$$

$$M_c = V_m \times \rho \times \left(1 \text{ m}^3 / 1E + 09\text{mm}^3\right) \tag{8.13}$$

where w_c is the width of cut; d_c the depth of cut; V_m the volume of material; and M_c the mass of chip.

It is worthy of noting here that, in case of turning, the volume of the removed metal is calculated as follows:

$$V_m = \pi \times \frac{D_i^2 - D_f^2}{4} \times l \tag{8.14}$$

where D_i is the initial workpiece diameter; D_f the final workpiece diameter.

8.2.3 Sustainability indicators for injection molding process

In case of IM process, the counting of cycle time, t_{cycle}, starts with the mold closing and ends as the mold opens. Total cycle time is usually given by the following equation:

$$t_{cycle} = 2t_{action} + t_{inject} + t_{cooling} \tag{8.15}$$

where the subscripts clearly signify the various phases during which time is consumed.

The action time, t_{action}, is the time needed to close or open the mold, and the cooling time, $t_{cooling}$, is the cooling time for the mold from the nozzle temperature down to removal temperature. An estimate of cooling time is of paramount importance to an estimate of the energy consumption. As per Madan et al. (2014) cooling time can be estimated as

$$t_{cooling} = (2 \text{ to } 3)h^2 \tag{8.16}$$

where h is the wall thickness of the specimen in mm.

The IM machine injects the molten plastic into a mold piece. The plastic then cools, as it is packed into the mold to take the desired shape. The entire process of an IM machine cycle can be broken down into four main energy consuming phases: plasticizing, injecting, cooling, and clamping/unclamping. The total energy consumption, E_{total}, of IM is calculated as follows:

$$E_{total} = E_{melting} + E_{process} + E_{cooling} \tag{8.17}$$

where the subscripts clearly indicate the various phases during which energy is consumed.

The total energy used by the IM process can be calculated from the equation:

$$E_{process} = E_{inject} + E_{pack} + E_{clamp} + E_{eject} \tag{8.18}$$

where E_{inject} is the energy needed to fill the mold, E_{pack} is the energy needed to pack the mold after injection, E_{clamp} is the energy needed to hold the mold shut during injection, and E_{eject} is the energy needed to eject the product from the mold.

This pressure can be estimated from the following equation:

$$P = \frac{12\mu L}{h^2 t_{inject}} \tag{8.19}$$

where P is the pressure flow in a channel; h is the nominal thickness of the cavity; L the distance from the cavity through metal flows; μ the apparent viscosity; and t_{inject} the inject time.

Once the pressure is calculated from the above expression, we can easily find out the energy required to fill the mold:

$$E_{inject} = PV_{mold} \tag{8.20}$$

where V_{mold} is the volume of polymer injected. This can be calculated with known mass and density of the polymer.

E_{pack}, E_{clamp}, and E_{eject} depend on the type and size of machine and on the mold and part characteristics. These three energy quantities usually account for less than 25% of the process.

The average cooling power per square meter, $E_{cooling}$, is

$$E_{cooling} = \rho C_p h (T_m - T_{inject}) \tag{8.21}$$

where T_m is the melting temperature, T_{inject} is the ejection temperature, C_p the specific heat, and ρ is the density of the polymer.

The energy, E_{melt}, required to melt the amorphous polystyrene can be calculated from

$$E_{melt} = mC_p\Delta T \tag{8.22}$$

where m is the mass of material to be melt and ΔT the difference between nozzle temp. and hopper temp.

8.3 Computational system to evaluate sustainability of manufacturing processes

8.3.1 Introduction of graphic user interface system

A graphic user interface (GUI) system to evaluate sustainability for manufacturing will be presented to help industry quantify sustainability indicators. The system utilizes knowledge base of different manufacturing processes in terms of part design, material properties, process parameters, and machine profiles. The presented computer system combines coded empirical relations, user input, and knowledge from available databases for the determination of sustainability indicators.

The presented sustainability analysis algorithm is divided into the following sections that are mention below,
- Part basic information module,
- Mathematical processing module,
- Result display module.

These modules are explained in the following sections.

8.3.1.1 Part basic information module

In this module, basic information of part is entered, which are relevant to the specific manufacturing process. For example, for a molding or a casting process, these parameters would be the volume of cavity, clamping force, flow rate, density, and other relevant parameters. These input parameters are provided manually and the system takes other relevant data from the available databases.

8.3.1.2 Mathematical processing module

In this module, processing of data is done by using logical reasoning and mathematical calculations; some of which have been described above. The results thus obtained are displayed and stored in the system.

8.3.1.3 Result display module

Results determined by the processing module are displayed by the system. Further output in the form of bar charts is presented by this module. A report for the analysis of sustainability is generated by the system. This report is stored in the database and is retrieved on demand.

8.3.2 Database for system

System uses various such databases as material, machine, and runner volume percentage database. Databases are stored in the form of XLS files. The databases are interlinked to all the modules, and the results obtained from the processing unit are stored in these databases for future reference.

8.3.3 GUI of sustainability analyzer

This section shows a snapshot of the GUI for different manufacturing processes (Figure 8.1). The sustainability analyzer consists of various panels such as the selection of manufacturing process family, selection of subprocesses, and selection of unit manufacturing processes.

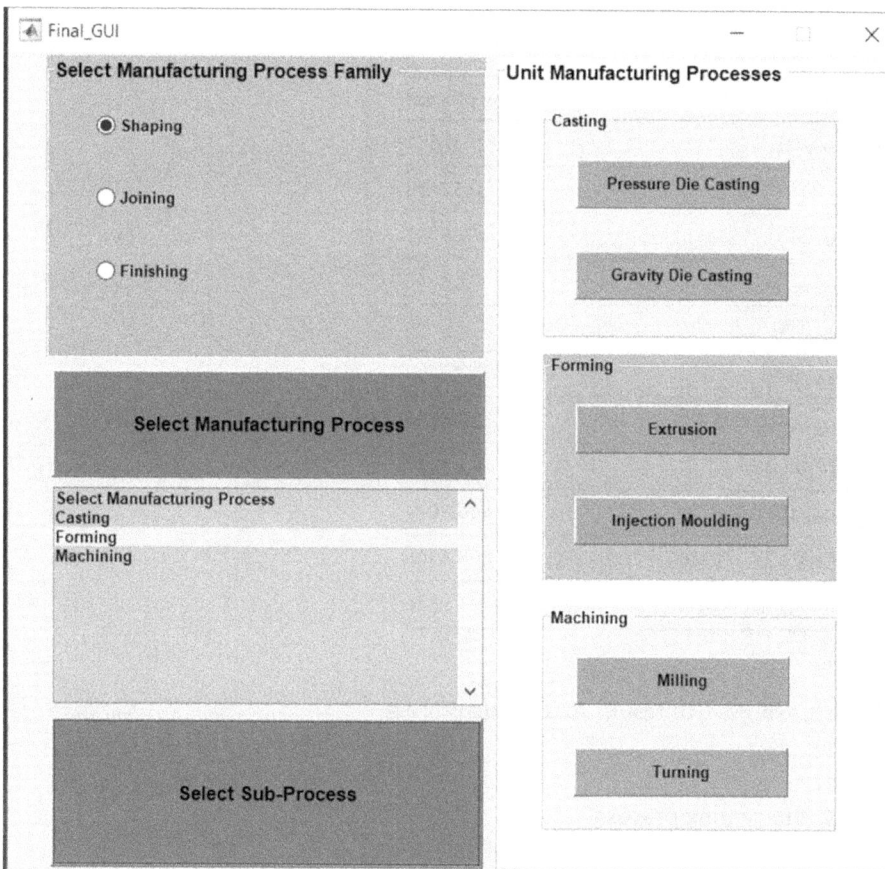

Figure 8.1: GUI of different manufacturing processes.

8.3.4 Case studies

For the validation of proposed GUI system, three case studies that are considered are machining, forming, and casting processes.

8.3.4.1 Turning process

A cylindrical-shaped aluminum alloy workpiece is used for turning process case study. All the initial parameters and length of cut are mentioned in the input module to assess sustainability. The interface provide user to vary initial parameters to achieve better sustainability results. The major objective is to analyze and estimate the energy consumed in different machining operations by using workpiece dimensions (Table 8.1).

Table 8.1: Outputs in tabular form for turning process.

Calculated parameters	Turning process
$Time_{turning}$ (s)	0.57301
$Time_{idle}$ (s)	0.57432
$Time_{basic}$ (s)	75.5743
$Power_{turning}$ (kW)	53.8908
$Power_{idle}$ (kW)	10
$Power_{basic}$ (kW)	7.5
$Energy_{turning}$ (kJ/cut)	30.8802
$Energy_{idle}$ (kJ/cut)	5.7432
$Energy_{basic}$ (kJ/cut)	566.8074
Total $energy_{turning}$ (kJ/cut)	603.4308
Total $power_{turning}$ (kW)	7.9846

Figure 8.2 shows GUI results for turning process.

8.3.4.2 Die-casting process

Sustainability indices for a pressure die casting of a zinc-alloyed tap handle is determined with the help of sustainability analyzer presented in this chapter. Zinc

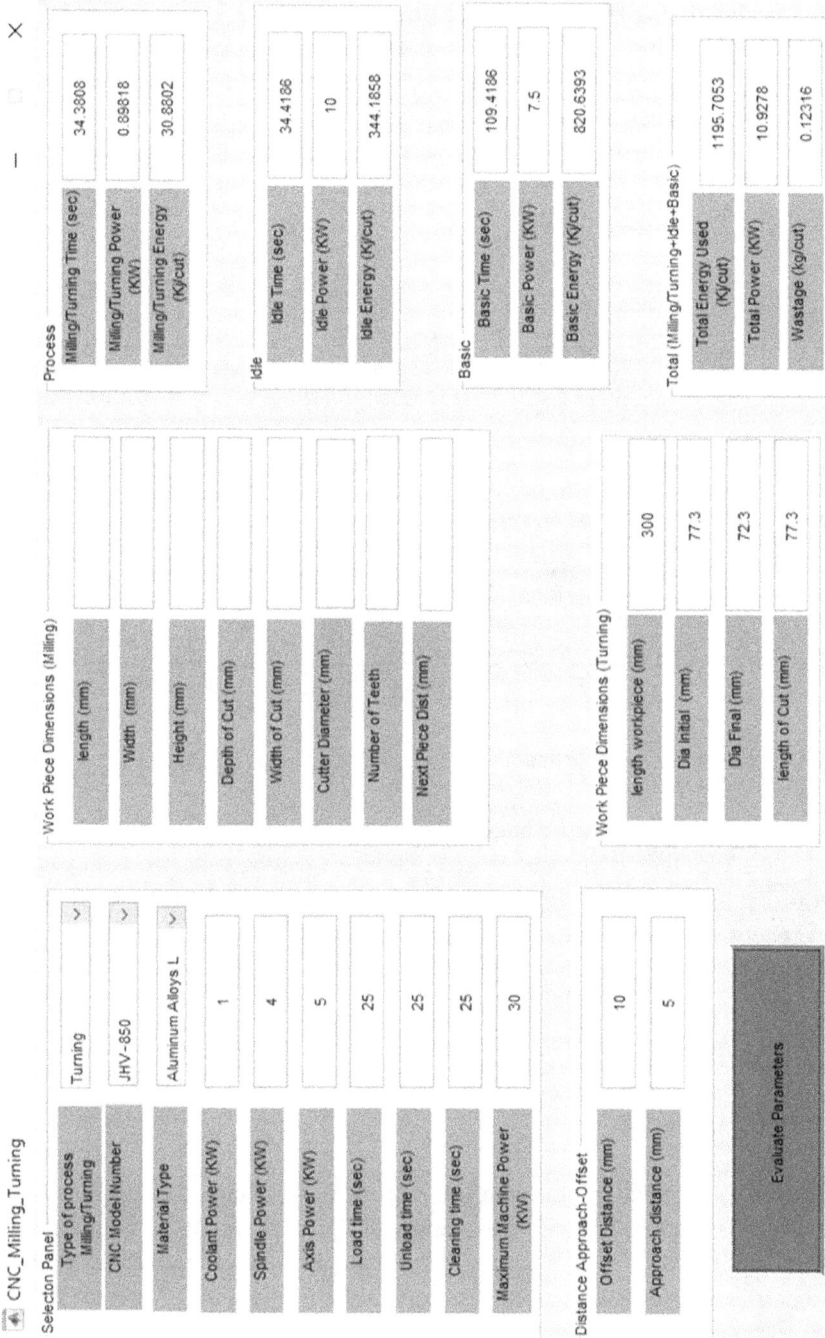

Figure 8.2: A snapshot of sustainability analyzer for turning process.

alloy is considered as a base material for this case study. The furnace serves the purpose of both melting and holding the molten alloy. The poured metal is then transferred to the casting machine, whereas after solidification casting is machined.

GUI determines such sustainability indices as carbon emissions, electrical energy required, fuel required, and solid waste by processing the input parameters provided by the user (Table 8.2). The sustainability analysis is performed per shift basis as shown in the result display panel depicted in Figure 8.3.

Table 8.2: Outputs in tabular form for pressure die-casting process.

Unit-processes	Carbon emissions $kgCO_2$	Electrical energy consumption (kWh)	Fuel consumption (lt.)	Wastage (kg)
Melting and holding furnace	38.048	10.6	530.699	22.5
Pressure die-casting machine	152.039	180.99	0	0.560
Microlevel activities	6.10	7.64	0	0
Total	196.19	199.44	530.699	23.060

8.3.4.3 Injection molding process

This case study features a plastic bucket that is manufactured by using a gravity IM process. The computer system, introduced in this work, processes the input parameters provided by the user to determine such sustainability indices as solid waste, electrical energy consumed, amount of fuel combusted, and resulting carbon emissions. Figure 8.4 shows the sustainability analyzer output panel for the IM process featured by the case study (Table 8.3).

8.4 Conclusions

A GUI for the determination of sustainability in manufacturing processes has been presented in this chapter. The proposed model calculates sustainability by processing theoretical mathematical models with available databases of the process parameters of various manufacturing methods. The sustainability indicators featured in this work are carbon emissions, energy and fuel consumption and material wastage. The input parameters are enter either manually or values for process parameters from existing databases. The system then processes the input data and produces the results in form of sustainability indicators. The system information shows that the proposed

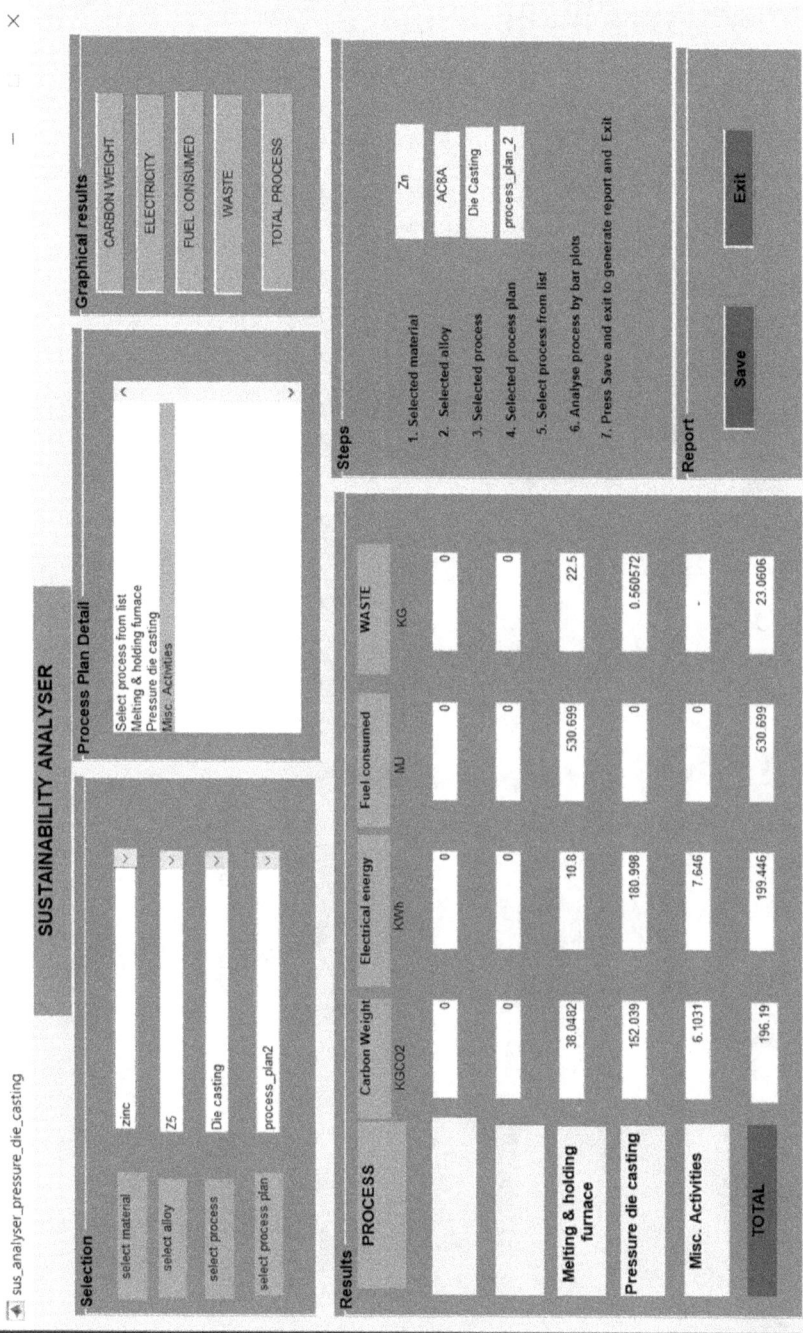

Figure 8.3: Graphic user interface model for pressure die-casting process.

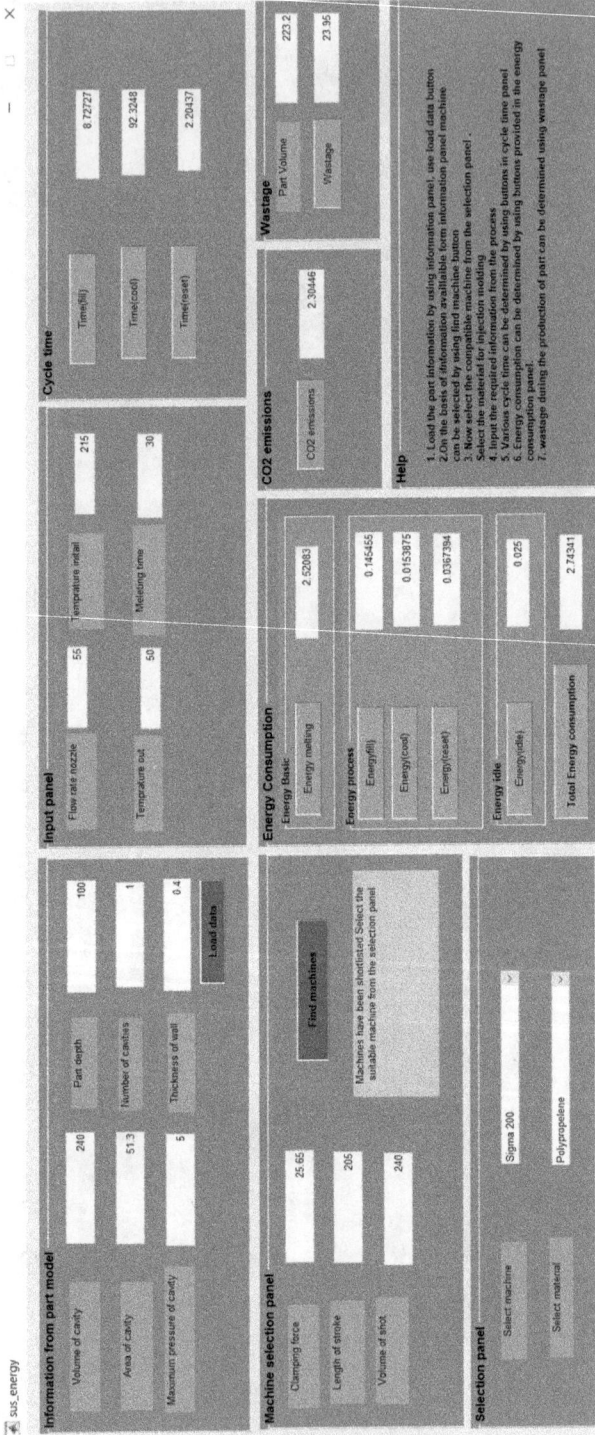

Figure 8.4: Graphic user interface model for injection molding process.

Table 8.3: Results for injection molding process.

Sustainability indicators	Sustainability analyzer results
Energy consumption (kWh)	$E_{melting} = 2.52083$
	$E_{fill} = 0.145455$
	$E_{cool} = 0.0153875$
	$E_{reset} = 0.0367394$
	$E_{idle} = 0.025$
	$E_{total} = 2.74341$
Air emissions (kgCO$_2$)	2.30446
Solid waste (kg)	23.96

system is valid and useful for the intended purpose. Moreover, the sustainability results determined by the system based on existing databases are close to these calculated on the basis of actual measurements of process parameters. The system can also be used to compare two process plans for manufacturing a part and can be used to measure progress of a company in terms of energy and material uses at various stages.

References

[1] Gutowski, T., Dahmus, J., and Thiriez, A. Electrical energy requirements for manufacturing processes. In13th CIRP International Conference on Life Cycle Engineering, 31May–2 June, 2006, 2006, Belgium, Leuven.
[2] ASTM, 2. (2017, 05 02). WK35705 New Guide for Sustainability Characterization of Manufacturing Processes. Retrieved from http://www.astm.org/WorkItems/WK35705.htm.
[3] US Department of Commerce. Sustainable Manufacturing initiative. In Proceedings of the 2nd Annual Sustainable Manufacturing Summit 2009, 2009, Chicago, USA.
[4] Singh, K., and Sultan, I. Framework for Sustainability Performance Assessment for Manufacturing Processes- A Review. IOP Conference. Series: Earth and Environmental Science, 2017, 73.
[5] Singh, P. P., Madan, J., Singh, A., and Mani, M. A computer aided system for sustainability assessment for die-casting process. Proceedings of the ASME International Manufacturing Science and Engineering Conference (MSEC 2012), 2012, Notre Dame, IN, USA.
[6] Bhanot, N., Rao, P. V., and Deshmukh, S. G. An integrated sustainability assessment framework: a case of turning process. Clean Technologies and Environmental Policy, 2016, 18, 1475–1513.
[7] Bourhis, F. L., Kerbrat, O., Dembinsk, L., Hascoet, J. Y., and Mongal, P. Predictive model for environmental assessment in additive manufacturing process. 21st CIRP Conference on Life cycle Engineering, Procedia CIRP 15, 2014, 26–31.
[8] Chen, D., Thiede, S., Schudeleit, T., and Herrmann, C. A holistic and rapid sustainability assessment tool for manufacturing SMEs. CIRP Annals – Manufacturing Technology, 2014, 63, 437–440.

[9]   Ciceri, N., Gutowski, T., and Garetti, M. A tool to estimate materials and manufacturing
      energy for a product. IEEE/International Symposium on Sustainable Systems and Technology,
      May 16–19.2010, Washington DC, USA.

[10] Culaba, A. B., and Purvis, M. R. A methodology for the life cycle and sustainability analysis of
      manufacturing processes. Journal of Cleaner Production, 7, 435–445.

[11] Dalquist, S., and Gutowski, T. Life Cycle Analysis of Conventional Manufacturing Techniques:
      die Casting. Proceedings of the 2004 ASME IMECE, 2004, Anaheim, CA.

[12] Dalquist, S., and Gutowski, T. Life Cycle Analysis of Conventional Manufacturing Techniques:
      Sand Casting. Proceedings of the ASME International Mechanical Engineering Congress and
      RD and Exposition, 2004, Anaheim. California, USA, November 13–19, 2004.

[13] Huanga, A., and Badurdeena, F. Sustainable Manufacturing Performance Evaluation:
      Integrating Product and Process Metrics for Systems Level Assessment. Procedia
      Manufacturing, 2017, 8, 563–570.

[14] Jeswiet, J., and Kara, S. Carbon emissions and CESTM in manufacturing. CIRP Annals
      Manufacturing Technology, 2008, 57(1), 17–20.

[15] Kim, B. D., Shin, J. S., Shao, G., and Brodsky, A. A decision-guidance framework for
      sustainability performance analysis of manufacturing processes. International Journal of
      Advanced Manufacturing Technology, 2015, 78, 1455–1471.

[16] Madan, J., Mani, M., Lee, J. E., & Lyons, K. W. Energy performance evaluation and
      improvement of unit- manufacturing processes: injection molding case study. Journal of
      Cleaner production, 2014, 1–14

[17] Moldavska, A., and Welo, T. Development of Manufacturing Sustainability Assessment Using
      Systems Thinking. Sustainability, 2016, 8(5), 1–26.

[18] Ocampo, L. A., and Clark, E. E. Developing a Framework for Sustainable Manufacturing
      Strategies Selection. DLSU Business and Economics Review, 2014, 23(2), 115–131.

[19] Thirez, A., and Gutowski, T. An environmental impact assessment of injection molding. IEEE
      International Symposium on Electronics and the Environmental, 2016. San Francisco, CA, USA.

[20] Vimal, K. K., and Vinodh, S. LCA Integrated ANP Framework for Selection of Sustainable
      Manufacturing Processes. Environment Model Assess, 2016, 21, 507–516.

[21] Feng, S. C., and Joung, C. B. An overview of a proposed measurement infrastructure for
      sustainable manufacturing. Proceedings of the 7th Global Conference on Sustainable
      Manufacturing, 2009.

[22] Seow, Y., and Rahimifard, S. A framework for modelling energy consumption within
      manufacturing systems. CIRP Journal of Manufacturing Science and Technology, 2011, 4(3),
      258–264.

[23] Madan, J., Mani, M., Lee, J. E., and Lyons, K. W. Energy performance evaluation and
      improvement of unit- manufacturing processes: injection molding case study. Journal of
      Cleaner production, 2014, 1–14.

[24] Lee, J. Y., Kang, H. S., and Noh, S. D. MAS2: an integrated modeling and simulation-based life
      cycle evaluation approach for sustainable manufacturing. Journal of Cleaner production,
      2014, 66, 146–163.

[25] Zhang, H., Zhu, B., Li, Y., Yaman, O., and Roy, U. Development and utilization of a Process-
      oriented Information Model for sustainable manufacturing. Journal of Manufacturing
      Systems, 2015, 37, 459–466.

[26] Carbon trust, 2011. Organizational carbon footprints [online].Available from: http://www.car
      bontrust.co.uk/cut-carbon-reduce-costs/calculate/carbon-footprinting/pages/organisation-
      carbon-footprint.aspx [Accessed April 20, 2018]

[27] Bill, A., 2005. Die Casting Engineering: a hydraulic, thermal and mechanical process, Hi Tech
      International Inc., South Haven, Michigan, USA: CRC Press.

Index

Acetone 98–99, 101
Additive manufacturing (AM) 1–6, 9–18, 20–21,
    25–38, 40, 44–56, 114–115, 199
Application 59–60, 66–71, 76, 82, 85, 89–92,
    94, 98–99, 101–102
Aspect ratio 253
ASTM 60, 64
Automotive industry 118

Basic design of drill for cored hole 145
Binder jetting process 22
Biodegradable 152
Biodegradable polymer 59, 87–88, 90

Carbon dioxide emissions 277
Carbon emission 272
Cardiovascular 246
Chemical treatments 94, 98–101
Chip evacuation 252
Chip-clogging 252
Chip-evacuation torque 259
Coatings 80, 94, 98
Computer-aided design (CAD) 2, 4–5, 8, 10,
    13–14, 17–18, 20–21, 23, 26–27, 31–32,
    37, 49, 60, 67, 72, 94, 199–201, 203–204,
    207, 229–231
Computer-aided manufacturing (CAM) 8, 23
Control of surface finish 93–101
Cored holes 113
Cross-link(ed/ing) 230–231, 235
Cutting torque 259

Deep hole drilling 251
Defects of FDM 81–84
DICOM 199–200, 204–205, 207–210, 223–224
Die-casting process 275–277
Direct energy deposition (DED) 22, 28
Direct heating 94, 96–97
Direct light processing (DLP) 18
Drill design 113
Ductile materials 251

Ecosustainable industry 59
Effectiveness 267
Electric energy consumption 275
Electrical impedance 256

Energy consumption 271
Energy density 174, 178
Entrance stability 145
Environmental aspects 84
Evolution of additive manufacturing 61–64
Excitation frequency 257

Fatigue life 171
Fluctuation 260
Force balance 116
Fossil fuel consumption 275
Fuel consumption 275–276
Fused deposition modeling (FDM) 59–102, 229
Fused material addition 97

Galvanoplasty 94, 98
Graphic user interface (GUI) 280

Half-wavelength 255
Hard tissue(s) 243
High deposition efficiency 167
Horn 253
Human bone implants 171

Immersion in liquid solvents 94, 99–100
Injection molding process 278–279
Internal porosity 178

Langevin 254
Large parts 168
Large-scale 154
Large-sized 162, 167
Laser energy density 178
Laser parameters 177–178
Laser wavelength 237, 247
Layer thickness 178–179
LENS 22
Lightweight 149
Local energy density 178
Local peaks 257
Low cost 154, 156

Machining process 277–278
Manufacturing 59–84, 97–98, 102, 271
Manufacturing processes 271
Materials in FDM 85, 93

https://doi.org/10.1515/9783110549775-009

Mechanical treatments 94–95
Medical model(s) 199–203, 215, 217–218,
    224–225
Modal analysis 256
Monitoring 168

Orthopedic implants 183

Paired *t*-test 263
Parameters in FDM 79–81
PLA 59–102
Polylactic acid 59, 84–85, 88–90
Polymer 59–60, 64, 66–67, 71–84, 87–93
Polymeric additive manufacturing 71–84
Porosity 174
Postprocessing 65, 78, 83, 93–96
Powder bed fusion (PBF) 18, 28, 33, 35, 37
Predict 263
Probability distributions 265
Propagation 232–235
Properties 66
Properties of the powders 169

Rapid fabrication 186
Rapid injection molding (RIM) 10
Refine grains 158
Refined grains 158
Refining grains 158
Renal 244, 246
Resonant frequency 254

Scanning speed 174
Solid waste 277
Sonotrode 254

Statistical tests 264
Steam softening 94, 100–101
Stereolithography (SLA) 4–5, 18, 30–31,
    37–38, 47
Stick-slip torque 259
Strength 171
Subtractive manufacturing (SM) 1–3, 6–7,
    9–10, 13–14, 16, 32, 40, 44, 47, 54, 56, 114
Sustainability analyzer 281
Sustainability assessment 271
Sustainability indicators 284
Sustainable development manufacturing 272
Swept frequency 256

Thermal conductivity 159
Thermal treatments 94–97
Thickness 178
3D laser scanner 199
3D scanner 200, 205, 218–219, 224
3D-printed 204–205
Tool life 113
Topological optimization 30, 32
Transducer 253
Tuning 257

Ultrasonic consolidation 185
Ultrasonic-assisted drilling (UAD) 252
Ultraviolet 229, 233

Virtual model 60, 65

Wastage 272
Weight reduction 151–153

De Gruyter Series in Advanced Mechanical Engineering

Volume 3
J. Paulo Davim (Ed.)
Drilling Technology, 2017
ISBN 978-3-11-047863-1, e-ISBN 978-3-11-048120-4,
e-ISBN (EPUB) 978-3-11-047871-6

Volume 2
J. Paulo Davim (Ed.)
Progress in Green Tribology, 2017
ISBN 978-3-11-037272-4, e-ISBN 978-3-11-036705-8,
e-ISBN (EPUB) 978-3-11-039252-4

Volume 1
J. Paulo Davim (Ed.)
Metal Cutting Technology, 2016
ISBN 978-3-11-044942-6, e-ISBN 978-3-11-045174-0,
e-ISBN (EPUB) 978-3-11-044947-1, Set-ISBN 978-3-11-045175-7

www.degruyter.com

www.ingramcontent.com/pod-product-compliance
Lightning Source LLC
Chambersburg PA
CBHW061804210326
41599CB00034B/6872